Why Men Won't Ask for Directions

RICHARD C. FRANCIS

Why Men
Won't Ask for

DIRECTIONS

THE SEDUCTIONS OF SOCIOBIOLOGY

PRINCETON UNIVERSITY PRESS
PRINCETON AND OXFORD

Copyright © 2004 by Princeton University Press
Published by Princeton University Press, 41 William Street,
Princeton, New Jersey 08540
In the United Kingdom: Princeton University Press, 3 Market Place,
Woodstock, Oxfordshire OX20 1SY

Library of Congress Cataloging-in-Publication Data

Francis, Richard C., 1953–
Why men won't ask for directions: the seductions of sociobiology/
Richard C. Francis.
p.   cm.
Includes bibliographical references and index.
ISBN 0-691-05757-5 (alk. paper)
1. Evolutionary psychology.   2. Evolution (Biology).
3. Sex.   4. Sex differences.   I. Title.

BF698.95.F73 2004
155.7—dc21   2003042879

British Library Cataloging-in-Publication Data is available

This book has been composed in Goudy

Printed on acid-free paper. ∞

www.pupress.princeton.edu

Printed in the United States of America

1   3   5   7   9   10   8   6   4   2

For Tamara

CONTENTS

List of Illustrations   ix

Acknowledgments   xi

CHAPTER 1
Darwinian Paranoia   1

CHAPTER 2
An Orgasm of One's Own   10

CHAPTER 3
Sex without SEX   19

CHAPTER 4
Transgendered   36

CHAPTER 5
Alternative Lifestyles   51

CHAPTER 6
Social Inhibitions   75

CHAPTER 7
Why Does the Mockingbird Mock?   102

CHAPTER 8
Brain Ecology   124

CHAPTER 9
Why Men Won't Ask for Directions   150

CHAPTER 10
A Textbook Case of Penis Envy?   175

CHAPTER 11
Darwin's Temptress   192

Notes   201

Bibliography   257

Index   311

LIST OF ILLUSTRATIONS

Figure 3.1 Hybridogenesis and gynogenesis   27

Figure 3.2 The reproductive cycle in the
parthenogenetic lizard *Cnemidophorus uniparens*   30

Figure 4.1 The size advantage model of sex change, illustrating
conditions for protogynous (A) and protandrous (B) sex change   42

Figure 5.1 Some examples of dromedary and Bactrian humps   54

Figure 5.2 The hypothalamic-pituitary-gonadal (HPG) axis   68

Figure 5.3 The migration of GnRH neurons from the olfactory
placode in the mouse   72

Figure 6.1 Oral conception in *Haplochromis burtoni*   76

Figure 6.2 Six possible scenarios for the causal relationships
among social dominance, testosterone levels and GnrH neurons   85

Figure 6.3 Experimental protocols for the conversion of
territorial males into nonterritorial males   87

Figure 6.4 The hypothalamic-pituitary-adrenal (HPA) axis   90

Figure 6.5 Interactions between the hypothalamic-pituitary-gonadal
(HPG) and the hypothalamic-pituitary-adreanal (HPA) axes   92

Figure 7.1 Representative sonographs for the zebra finch (a),
white-crowned sparrow (b), and canary (c)   103

Figure 7.2 Stages of song learning in an idealized songbird, two
age-limited learners (the zebra finch and the white-crowned
sparrow) and an open-ended learner (the canary)   114

Figure 7.3 The song nuclei of songbirds   116

Figure 8.1 Cross section of the human brain showing the
hippocampus in relation to other limbic structures (amygdala and
mammilary body) and cerebral cortex   132

Figure 10.1 Urogentital system of female spotted hyena showing
womb and birth canal   188

ACKNOWLEDGMENTS

When I was first approached about writing this book, Bill Clinton was still in his first term as President and Jack Repcheck was still at Princeton University Press. Jack was the more significant of the two in getting the project off the ground, and the one with whom I had a personal bond, so the psychological blow I experienced upon his departure was the more acute. Hence, I am grateful to Jack not only for guiding the project through its initial stages but also for his continued encouragement and advice.

Fortunately, I eventually fell into the very capable hands of Sam Elworthy, an editorial transition that, to my mind at least, worked out much better than the presidential one. Sam was especially helpful with respect to the book's structure. The fact that it took a considerable amount of time for me to see the wisdom in Sam's suggestions is testimony to his patience and perseverance.

I have also greatly benefited from the comments of David Crews and Kim Sterelny, who read the manuscript in its entirety. An anonymous reviewer also provided very detailed and constructive criticism of the entire manuscript. I am grateful as well to Andrew Bass, Ivan Chase, Richard Lewontin, Elizabeth Lloyd, Susan Oyama, and Rasmus Winther for comments on individual chapters. I am especially grateful to Peter Godfrey-Smith, with whom I have met most Tuesdays at the Stanford Gym for the last ten years to discuss, while we work out, all manner of subjects, including many related to this book. Peter has proved to be an ideal sounding board, because, though our views often differ, they don't differ so much that we end up talking past each other. (Except, that is, when it comes to the arts, an area in which his enthusiasms are sometimes misplaced.) Peter graciously read both the earliest and last versions of the manuscript and much in between. He also incorporated material from this book into courses he taught at Harvard and Stanford, which provided me some valuable feedback.

Finally, I owe my deepest thanks to Tamara Bushnik for her Penelope-like patience.

Why Men Won't Ask for Directions

1

Darwinian Paranoia

◆━━━◢◆◣━━━◆

THERE IS an old joke that goes something like this:

QUESTION: Why does it take millions of sperm to fertilize one egg?
ANSWER: They won't stop to ask for directions.

What makes this joke funny of course is its allusion to the conventional wisdom that among the foibles of the male members of our species is a reluctance to admit they are lost, especially when they are in the company of women. Moreover, the joke implies that given the same circumstances, a woman would have no such inhibitions. The fact that putative sex differences such as this are so often the subject of jokes speaks to their inherent interest to so many of us. So it is not surprising that scientific investigations of human sex differences feature prominently in the popular media. Just last week I awoke one morning to the news on my clock radio informing me that scientists had discovered the neural basis for another sex difference that has inspired many jokes: the "fact" that women remember the details of past arguments much better than men. It seems that modern brain imaging techniques revealed that the "emotional centers" of the brain "light up" during arguments to a greater extent in women than they do in men. This finding, coupled with the fact that the more emotional salience a memory has the better it is remembered, apparently explains women's elephantine memory for verbal slights.

Fun stuff. But even if you take these putative human psychological sex differences unsalted, it is not obvious that they require an evolutionary explanation. Nature abounds with robust sex differences compared with which the sex differences we are talking about here are piddling and inconsequential from an evolutionary perspective. Take songbirds, for example. In most species, males do most if not all of the singing. Moreover, we know that male songbirds sing in order to attract mates and ward off male competitors, clearly evolutionarily relevant functions. Then there are the fishes known as wrasses. In more than one wrasse species, male and female members look so different, even to the trained eyes of professional ichthyologists, that they were originally classified as separate species. Now that's a sex difference! It is one we should bear in mind in considering the claims of human sex difference enthusiasts, such as evolutionary psychologists, who, without going so far as to embrace the popular doctrine that human males and females had separate interplanetary origins, certainly tend to treat each sex as a distinct species of humanoid.

The tendency of evolutionary psychologists to overinterpret human psychological sex differences stems directly from the fact that they have embraced with unprecedented fervor a particular evolutionary perspective known as adaptationism. Adaptationists are distinctive among natural scientists in the way they query nature; they use "why-questions"—not just generic why-questions, as in "why is the sky blue," but why-questions of a particular sort. Here are a few examples of their why-questions that we will examine in this book: Why does the male mockingbird mimic the songs of other bird species? Why do female spotted hyenas have male-like phalluses? Why do women have orgasms? And yes, why won't men ask for directions?

What these why-questions have in common is that they require answers of a particular form; answers that refer to ends as opposed to means and to effects as opposed to causes. Answers of this sort are often referred to as reasons, functions, intentions, or purposes, all of which fall into a category known as "teleological." Teleological thinking is the norm in our everyday interactions with each other, but in the realm of science its status is controversial. Indeed, teleology is steadfastly eschewed in most of the natural sciences, including most of biology. But evolutionary biologists of the adaptationist persuasion rely heavily on teleological thinking and increasingly flaunt it. I will argue that despite its undeniable heuristic value, teleology is a distorting lens through which to view the evolutionary process. I will further argue that questions such as "Why won't men ask for directions?" are often the wrong questions to be asking.

Such questions seem so natural, however, that many people find it difficult if not impossible to resist asking them. They seem so natural because all of us, scientists and nonscientists alike, are teleologists at the outset. As children, we found it most natural to explain things by connecting them to the intentions, purposes, and goals of some agent analogous to ourselves. Hence, our curiosity is expressed in the form of teleological why-questions. Among the why-questions posed to me by my 10-year-old son of late are: "Why are there mosquitoes?" "Why do mockingbirds sing at night?" and "Why do I have to go to bed now?" Each of these questions is a request for a teleological explanation, an explanation that explains by identifying intentions, purposes, reasons, or goals. It is only gradually that he will learn to differentiate those questions for which the teleological perspective is appropriate, such as why he must go to bed now, from those for which the teleological perspective is misguided, such as why there are mosquitoes. Learning when the teleological perspective is warranted and when it isn't is an important part of normal cognitive development. In part, this learning entails knowing when teleological why-questions should be resisted and replaced by questions of a different sort, "how-questions," for example. How-questions, as a child learns, are requests for a quite different sort of explanation: what the cause is; or how it came to be.

This transition is not easy; it takes a certain kind of discipline. Many adults, in fact, never learn how to resist looking for teleological explanations. I have been asked more than once by adult acquaintances why mosquitoes exist. This question was motivated less by my perceived biological expertise than by the teleological perspective of my interlocutors. They presumed that, as a biologist, I could explain the benefit mosquitoes provide in the grand scheme of nature, their purpose. God, or Mother Nature, must have had a reason for creating mosquitoes; they must be doing something useful. My attempts to answer this question from outside the teleological perspective—for example, that mosquitoes exist because they are good at what they do—were not well received. My answers seemed lame, the equivalent of "these things happen."

Though we all pass through a teleological stage, most of us come to recognize that at some point the teleological perspective can become problematic; in fact, excessive teleology is an important aspect of the pathology known as paranoia. The paranoid is convinced, in a visceral way, that everything happens for a reason of the sort requested by "why" questions. The paranoid's reasoning is deemed unreasonable not just because it reflects a sense of persecution, nor because it is so ego-centered, but because of the assumption that behind every occurrence there are intentions, a purpose, a design.

You don't need to be paranoid to have a paranoiac mindset. Wherever the teleological perspective is used to devalue causal explanations, wherever it is believed that the answers to teleological why-questions trump all others, you will find evidence of the paranoiac explanatory style, and its characteristic distortions. Most religions, for example, foster the paranoiac mindset in their insistence on teleological "ultimate explanations." But scientists, on the whole, are especially careful to guard against these teleology-induced distortions; indeed, scientists tend to be the least paranoiac citizens among us. By the time a child becomes a scientist, he or she has usually abandoned the teleological perspective, at least for the purposes of doing science. The glaring exception is evolutionary biology; in that field, some adaptationists rely heavily on teleology in their own search for ultimate explanations. For them this "teleological" quest for ultimate explanations seems to follow from Darwin's principle of natural selection. As we will see, however, you can whole-heartedly embrace Darwin's principle of natural selection without embracing teleology. In fact, there are some distinct advantages to nonteleological Darwinism, not least in avoiding the paranoiac pitfalls to which all teleologists are prone. These pitfalls are especially obvious in the Grandfather of Darwinian teleology, Bishop William Paley.

From Paley to Darwin

It is at least ironic that the teleological perspective of the staunch creationist William Paley continues to exert so much influence on Darwinians. Paley was one of the most important advocates of a particular theological tradition known as Natural Theology. What distinguished Natural Theology from other theological traditions was its attempt to accommodate and assimilate science, and more broadly, reason, by way of promoting a religious agenda to an educated audience. Perhaps the central problem for Natural Theology was to reconcile God's presumed benevolence with his omnipotence. By the middle of the eighteenth century, European intellectuals were increasingly vexed by the so-called problem of evil: how it was that a God who is assumed to be both benevolent and omnipotent could allow so much that is evil to exist. One popular solution to this problem was the view that evil is only apparent and that ours is, in fact, the best of all possible worlds. One way of demonstrating such, popular among English theologians, was to provide evidence for the "fit" between animals and their environments. William Paley was perhaps the most famous of these naturalist-theologians who took to the English countryside in order to demonstrate God's beneficent design.

Paley everywhere found signs of God's beneficence; his was an ecstatic view of nature: "It is a happy world after all. The air, the earth, the water, teem with delighted existence. In a spring noon, or a summer evening, on whichever side I turn my eyes, myriads of happy beings crowd upon my view."[1]

Paley was not at all perplexed by the abundant evidence of discomfort and suffering, nor did he believe that we must defer until death any reward for our suffering. According to Paley, our suffering is more than compensated for by the heightened pleasure we experience during the interludes, a pleasure which "few enjoyments exceed."[2] Paley would have no problem with the "why are there mosquitoes?" question. For Paley, mosquitoes exist because mosquito bites are actually blessings, without which we would be denied the pleasure of the relief that comes from scratching them. Paley could also use this strategy to effectively deal with much deeper sorts of suffering, including his own. When he wrote those words, he was slowly dying a painful death from intestinal cancer.[3]

It seems odd that a man like Paley, given his agenda, could be so loudly exalted by such an outspokenly atheistic Darwinian as Richard Dawkins[4] and could inspire another neo-Darwinian, John Maynard-Smith, to proclaim that "The main task of any evolutionary theory is to explain adaptive complexity, that is, to explain the same facts that Paley used as evidence of a creator."[5] Surely, Maynard-Smith is not seeking to explain *all* of the "facts" that Paley used as evidence of a creator; nor does it seem likely that Dawkins's notoriously dark view of life is compatible, no matter how far we stretch it, with Paley's conviction that "it is a happy world after all." No, there is something else about Paley that makes him attractive to these neo-Darwinians: his teleological adaptationism, his conviction that every feature of living things reflects the imprint of a designing or design-like force, and that this is most evident in the way that living things are adapted to their environment. This form of adaptationism comes straight out of Paley's creationism.

In this approval of Paley, Maynard-Smith and Dawkins can legitimately say that they are following in Darwin's own footsteps. Though Darwin rejected Paley's creationism, he largely accepted Paley's adaptationist teleology, a fact he quite openly acknowledged: "I was not able to annul the influence of my former belief, then almost universal, that each species had been purposely created; and this led to my tacit assumption that every detail of structure, excepting rudiments, was of some special, though unrecognized, service."[6]

This premise of "some special, though unrecognized, service" is Darwin's explicit recognition of the teleological mindset he acquired by virtue of his indoctrination in Natural Theology. Darwin, however, because of his conflicts with advocates of Natural Theology over the issue of evolution (descent with modification), was much more aware of the pitfalls in the teleological perspective than are modern Darwinians such as Dawkins and Maynard-Smith. Nature's imperfections provided him some of his best evidence for descent with modification, because this was the hardest evidence for Natural Theology to assimilate. It was because of their preoccupation with demonstrating evolution that early Darwinians were concerned to disassociate themselves from the idea of design in nature.

As the fact of evolution became more apparent to one and all, save those who chose to ignore the evidence, the original conflict between Darwinians and Natural Theologians over descent with modification largely receded to the background and a greater emphasis was placed on the assumption common to both—though increasingly unacknowledged as such—about the centrality of adaptation. Many of those evolutionists who are self-consciously "Darwinian" have become increasingly prone to design thinking and much more sympathetic toward Paley than Darwin ever was. For these "adaptationists," Mother Nature has replaced God as the guarantor of suitability or adaptation, through her omnipotence as a natural selector. Nature could still be interpreted as if designed, not all at once, but over time—"designed on the installment plan," to use Dewey's droll phrase.[7]

Obviously, a lot rides on how closely natural selection approximates Paley's God in its powers of design. If the approximation is close, the teleological perspective is quite powerful; if it is not close, the teleological perspective will mislead. Throughout this book, I will argue that those neo-Darwinians who describe themselves as adaptationists tend to exaggerate the similarities between natural selection and Paley's beneficent God.

Some adaptationists are more prone to paranoia than others, of course, depending on their attitude toward teleology. For some, it is merely of heuristic value, sometimes useful, sometimes not; for others, however, the neo-Paleyans, teleology is embraced much less consciously. The great danger of teleology, even for those who use it primarily as a heuristic, is that it tends to take over one's mental landscape. Kenneth Weiss has aptly likened teleology to kudzu, that rampant vegetational scourge of the southeastern United States, which rapidly comes to dominate the landscape unless it is diligently beaten back.[8] It is these kudzu-like properties of teleology that lead to paranoiac regression. This pitfall has always been more apparent in those adaptationists who study behavior—behavioral ecologists and sociobiologists—than in those who study anatomy and physiology. Sexual behavior, in particular, has proved a wellspring for some of the most paranoiac thinking ever to pass itself off as science, the apotheosis of which is a new adaptationist discipline called "evolutionary psychology." On the premise that, given the obvious link between reproduction and evolutionary fitness, every detail of sexual behavior must reflect design, evolutionary psychologists have provided teleological explanations of all manner of sexual phenomena ranging from wet dreams to rape.

Fortunately, there are effective ways to control the kudzu-like tendencies of adaptationist teleology; one of the most effective was devised by George Williams. According to Williams, we should view adaptation as an onerous concept, one that carries with it the burden of proof, not the presumption of truth. In order to be confident that a trait is an adaptation, we need strong evidence of natural selection for that trait, evidence that the trait exists because bearers of that trait experience a competitive advantage over those who lack that trait. If we lack compelling evidence to this effect, Williams advises us to assume that this trait is not an adaptation but a mere "effect"—or to use the term I prefer, a *byproduct*—of some other processes and hence requires a different sort of explanation, a how-explanation.[9]

Most adaptationists at least pay lip service to Williams's distinction between adaptations and byproducts, but many fewer take it to heart, evolutionary psychologists least of all. Instead, they are guided by another distinction, this one guaranteed to promote a kudzu-like proliferation of adaptationist teleology. This distinction, which we owe to Ernst Mayr, is very much in the spirit of Paley.

CAUSES AND AGENDAS IN MODERN BIOLOGY

Mayr proposed that there are actually two radically distinct explanatory agendas within biology; one for evolutionary biologists, who ask "Why?" and seek teleological answers to these why-questions,[10] and another agenda altogether for the physiologists, morphologists, geneticists, and developmental biologists who ask "how?" and use material and causal conditions as their explanatory resources. We can call the biology of teleological explanations "why-biology" and the rest of biology with its material/causal considerations, "how-biology." Mayr referred to the material/causal explanations of how-biology as "proximate explanations" and to the teleological explanations of why-biology as "ultimate explanations."[11] The terms *proximate* and *ultimate* reveal Mayr's teleological bias and his allegiance to Paley. In his estimation, asking "why" gets you to the end of the explanatory line; proximate explanations are just mileposts along the way. Mayr further proposed that in fulfilling the teleological why-agenda we can jump to the end of the line and bypass the proximate material/causal how-biology considerations altogether. Put another way, why-biology's teleological explanations trump all how-biology's material/causal explanations.

At best, on this view, how-biology can help us flesh out the teleological explanations of why-biology. There is no question of adaptations versus by-products here; rather, how-biology can only complement the teleological accounts of adaptations provided by why-biology. This would be true if we lived in a perfectly Paleyan universe; in that case, teleological explanations would indeed trump any how-biology considerations. We do not live in such a universe, however, or anything approximating it. And in the non-Paleyan universe in which we find ourselves, there are three possible outcomes when why-biology meets how-biology. First, sometimes the material/causal explanations from how-biology do indeed complement and support the why-biology explanations. In such cases, something like Mayr's proximate/ultimate distinction is appropriate.[12] In other cases, however, the how-biology considerations fundamentally alter the explanation provided by why-biology. And finally, in still other cases, how-biology explanations should be viewed as outright replacements for teleological explanations; this outcome is implicit in Williams's distinction between adaptations and byproducts. In these latter two cases, Mayr's proximate/ultimate distinction is not at all germane; rather it constitutes a barrier to a better understanding of the evolutionary process.

It is only by bringing how-biology considerations to the foreground that we can remove the dead hand of Paley from evolutionary biology, and hence Darwinian paranoia. This process will require a reintegration of Mayr's two biologys, an increasing trend within evolutionary biology in any case. For example, the most dynamic and rapidly expanding area of evolutionary biology as a whole, evolutionary developmental biology (known as *evo-devo*) sits squarely at the intersection of Mayr's two biologies.[13] And so, too, does my own area of special interest, evolutionary neurobiology (or *evo-neuro*, by way of analogous shorthand). For both evo-devo and evo-neuro, the how-biology is paramount, in contrast to evolutionary psychology (*evo-psycho?*), which lightly skips over these considerations in favor of pure teleology.

The value of the view of evolution informed by how-biology goes well beyond this negative program of correcting teleological excesses within Darwinism. In fact, the real payoff of this approach comes from connecting this how-biology to evolutionary history, or phylogeny, as represented in the tree of life. A picture of evolution quite different from that of the neo-Paleyans then emerges, one that places more emphasis on the contingent historical or genealogical connections within this tree and less emphasis on the ecological conditions in which each species on that tree finds itself.

All living things are connected through a branching genealogy (or phylogeny); as a result, all living things share traits with even their most distant ancestors. This genealogical-based resemblance increases, of course, as we move from more distant to more immediate ancestors, no matter what environment a given species inhabits or what sort of natural selection it is subject to. On this view, natural selection does not at all resemble a divine engineer, but rather a tinkerer who must work with the materials at hand and who is fundamentally limited with respect to how deeply his tinkering can penetrate.[14]

Because genealogy limits the capacity for natural selection to adaptively modify an organism, it is often viewed as a constraint on adaptation.[15] Biologists who emphasize the role of genealogy in evolution point to the predictive power of taxonomy, as reflected in the hierarchical branching of the "tree of life," rather than to ecological factors in explaining the traits of living things. They tend to emphasize similarities within branches rather than selection-induced dissimilarities.

Recognizing a role for genealogy in evolution is only a first step in developing an alternative to unrelieved adaptationism. The next step is to identify the particular biological properties that are conserved within particular branches, or lineages, that constitute the tree of life, and this is where the how-biology comes in. In multicellular organisms such as ourselves, these properties are often subsumed under the category of "development." Any evolutionary alteration of a given species requires an alteration in an existing highly integrated developmental process inherited from the species' ancestors; this requirement greatly limits the set of viable adaptive responses to any environment because any deep modification is overwhelmingly likely to mess things up. This fact has given rise to the notion of "developmental constraints" and development-based alternatives to adaptationist explanations. But it is important to remember that developmental constraints are how-biology imperatives born of genealogical connections.

In this book, I focus on only one branch of the tree of life—that of the vertebrate animals—because, frankly, I know more about this branch than any others. I will further confine myself to sexual phenomena, broadly construed. The how-biology of sex is not confined to the gonads and their secretions; it also includes the neural and the hormonal systems, both of which function to integrate and coordinate the body's activities and behavior. At a finer scale, it includes biochemicals called hormones and other biochemicals called neurotransmitters, by means of which the endocrine and neural systems affect and are affected by events elsewhere in the body; it also includes other biochemicals called genes, which serve in the production of hormones, neurotransmitters, and other biochemicals.

We will find that once we incorporate even the most basic considerations about how brains and gonads work, how they interact, and how genes and hormones act and interact, we will have a different view of the evolution of sex differences and the other subjects of this book. But let us begin, not with a sex difference, but with what, from an adaptationist perspective, seems like a problematic similarity in human males and females.

2

An Orgasm of One's Own

THE EROTIC sculptures adorning the temples at Khajuro in central India have long shocked and titillated observers from Judeo-Christian and Islamic cultures. But these illustrations of passages from the Hindu sacred text the Kama Sutra were meant to instruct devout Hindus in what is for them an important aspect of the religious experience, one that became the focus in some Tantric traditions. Eroticism, however, has never been integral to the religious experience of devotees of Levantine religions. In fact, eroticism has generally been viewed as antithetical to an appreciation of the sacred. St. Paul was quite explicit on this subject, and his dim view of erotic pursuits has permeated Christian doctrine, as in, for example, the exaltation of celibacy and proscription of birth control by the Catholic Church.

St. Paul's admonitions reverberated well beyond official church policies; they broadly influenced, to varying degrees, cultural attitudes throughout Christendom, culminating most famously in Victorian England. Nothing could be further from the spirit of Khajuro than Victorian attitudes; so it is somewhat ironic that the Victorians were the first Europeans to view this art—a culture shock of epic proportions. Contrast the fabled Victorian mother's advice to her newlywed daughter to "lie still and think of England" with the exuberantly erotic description of the consummation event of a well-bred Hindi woman in the Kama Sutra:

> But when Nala inserts his member into the sanctuary of Eros, Damayanti discovers sensual pleasure unknown before. She realizes that it is a source of wonderful pleasure and they remain for a long time clasped in each others arms.[1]

Particularly galling to the Victorians must have been the vivid sculptures at Khajuro, and equally explicit descriptions in the Kama Sutra, of homoerotic couplings among women.[2]

The problem with this art, from the Victorian perspective, was that, given their utilitarian view of sex, eroticism could not be an end in itself; hence, the sexuality displayed at Khajuro was considered gratuitous—unseemly at best, perverse at worst. Even Richard Burton, certainly among the most liberated Victorians, felt compelled to omit much of the account of lesbian sex from his English translation of the Kama Sutra.

Just as the Victorians found problematic the gratuitous sex in the Kama Sutra, so too do some latter-day Darwinians. For Darwinians, of course, all behavior must ultimately serve a procreational end, but sexual behavior most of all. So any sexual behavior that seems gratuitous, with respect to the Darwinian imperative to reproduce, is problematic. The gratuitous-seeming sexual behavior that I focus on here is the female orgasm. It will be instructive to examine the ways adaptationists have attempted to make the female orgasm safe for their latter-day Victorianism.

The Victorian Mind-Set in Biology

The least sophisticated tactic adopted by evolutionary Victorians confronted with gratuitous-seeming sexual behavior is to deny that the behavior is actually sexual. If you deny that the gratuitous-seeming sexual behavior is actually sexual behavior, you eliminate the need to explain why, given your Victorian principles, it continues to exist. Though this solution to the problem may seem draconian, even preposterous, it has been tried. It has proved especially attractive as a way to deal with lesbianism.

The philosopher Elisabeth Lloyd called attention to a nice example of the denial tactic in a study of our closest living relatives, the bonobos.[3] Bonobos (*Pan panmiscus*), sometimes referred to as pygmy chimpanzees, rival humans when it comes to gratuitous sex. They have an impressive repertoire of sexual behavior, including one technique referred to as "the swing" in the Kama Sutra, and more prosaically, as genito-genital (or GG) rubbing by ethologists. It is noteworthy that among bonobos, as among humans, the swing is practiced by females with other females. During these episodes, the females hold each other and swing their hips from side to side, while maintaining contact between the vulva and erect clitoris.[4] The swing certainly looks like sexual behavior, since it involves genital stimulation. Yet, because it is not "reproductive," the ethologist who first described it concluded that, genital stimulation notwithstanding, it could not be sexual behavior.[5] No possibility of conception—ipso facto, no sexual behavior.

But this way of delineating sexual behavior, based on teleological background assumptions, seems excessively narrow, and even most adaptationists would actually acknowledge that it is. Rather than adopt the denial tactic when confronted with gratuitous-seeming sexual behavior, they choose to demonstrate that the gratuitous-seeming sexual behavior is actually not gratuitous at all, because, despite appearances to the contrary, it does serve, more or less directly, reproductive ends. This has been the approach favored by those whose aim it is to make the female orgasm safe for teleological Darwinism. Their task, as they see it, is to identify the Darwinian function of the female orgasm.

The function of orgasms in males is obvious. Except for those highly skilled in the techniques of certain tantric practices, some of which are detailed in the Kama Sutra, it is part and parcel of ejaculation; and ejaculation is, from a male's evolutionary perspective, the raison d'être of sexual intercourse. It is not at all obvious, however, that women should require orgasms in order to reproduce effectively. In fact, there is no evidence that nonorgasmic women are any less adept at producing babies than are multiorgasmic women. Why, then, do female orgasms exist? What is their evolutionary function? Or, perhaps, this is the wrong question to ask.

Let us examine some attempts to identify the evolutionary function of female orgasms. These conjectures have been quite diverse, comprising a continuum from male-centered (androcentric) to female-centered (gynocentric). Not surprisingly, the male-centered adaptationist view is the more traditional. In his popular book *The Naked Ape* (1967), Desmond Morris proposed that the orgasm has an adaptive function in maintaining a female's interest in continued sexual activity, which is in turn essential in ensuring that her mate remains bonded to her. A sexually satisfied mate is less likely to have a wandering eye.

But the Morris hypothesis was composed before the advent of sociobiology; in the postsociobiological adaptationist framework, it is assumed that human males will have wandering eyes no matter how sexually responsive their mates may be. In the current vernacular, males are always amenable to "extra-pair copulations," or EPCs, as the debunkers of apparent monogamy affectionately refer to them. Females, in this view, have a completely different agenda. A female's reproductive success is directly related to how much paternal care she can cajole from her mate. Hence, in the more recent "pairbond" accounts of orgasms, it is alleged that female orgasms function as a way of ensuring paternal investment in her offspring. John Alcock, for example, argued that orgasms evolved to indicate to the female the likelihood that the male with whom she is copulating will be a good father.[6] A male who is sensitive enough to attend to his mate's sexual enjoyment will be a trustworthy partner.

Not surprisingly, feminist functionalists have a somewhat different perspective on female orgasms. Sarah Hrdy, for example, disdains the Victorian tendency to underestimate female sexual agency. She also takes a broader view of female orgasms as characteristic of primates in general, not just of humans.[7] According to Hrdy, female primates are as willing to engage in extra-pair copulations as males; and, given the fact that "it takes two to tango," this certainly makes sense. But females do not benefit from their EPCs in the same way as males. Whereas a male is just out to disseminate his sperm as widely as possible, a sexually assertive female uses these trysts as an opportunity to confuse the male about paternity and thereby manipulate him into at least tolerating her offspring.[8] This motivation is important because males are prone to kill infants that are not their own.[9] Orgasms, therefore, are designed to keep a female's motivation high for the considerable expenditure of time and energy required to keep males confused, paternitywise. The responsive clitoris is natural selection's way of compensating her for the greater physical power of males.

A problem common to these very different adaptationist explanations of the female orgasm is well stated by Elisabeth Lloyd: "Not to put too fine a point on it, if orgasm is an adaptation which is a reward for engaging in frequent intercourse, it does not work very well."[10] About 10 percent of women never experience orgasm, and 30 percent *never* do so during intercourse. Masturbation is a far more reliable route to orgasm than intercourse. Though slow to assimilate this fact, functionalists have recently come to recognize that the relationship between orgasm and intercourse is complicated at best. The more recent female-centered functionalist accounts involve much more complicated adaptive scenarios.

Poleax or Upsuck?

Morris had more than one adaptive explanation for female orgasms in *The Naked Ape*. In addition to his pair-bonding idea, which explains its longterm function, Morris proposed that orgasms have a short-term function as well: to keep the female horizontal. Our celebrated bipedalism seems to pose a problem when it comes to insemination. If, as in most animals, a female is tempted to return to her daily activities soon after intercourse, Newton's inexorable physical force will cause the sperm to leak out. According to Morris, orgasm prevents her from moving, because it induces a state of lethargy—ergo no leakage. This is now referred to as the poleax hypothesis, because on this view a postorgasmic woman is as if felled by that implement.

The current rival to the poleax hypothesis is the upsuck hypothesis (I did not invent these terms), according to which the orgasm-induced muscle contractions draw the sperm into the womb.[11] This idea has a long history; its basic outline was sketched out by Aristotle:

> the mouth of the uterus is not closed when the discharge takes place which is usually accompanied by pleasure in women as well as men, and when this is so there is a readier way from the semen of the male to be drawn into the uterus. (*Generation of Animals* 739a20–21, 31–35)

A virtue of the upsuck view, according to its adherents, is that it applies more broadly in mammals, most of which are quadripedal and do not have a leakage problem; hence, they do not need to be poleaxed. One variant of the upsuck hypothesis was formulated in order to account for the fact that masturbation is a far superior orgasm inducer than coitus. According to Robin Baker and Mark Bellis, the female orgasm is an essential component of an elaborate evolutionary strategy that adapted females use to modulate the probability of becoming impregnated.[12] The authors first note that much of the seminal fluid deposited by the male during intercourse gets ejected, along with female tissue and secretions, as "flowback," whether or not the female has been poleaxed. Orgasms that occur at or near the time of ejaculation result in a higher level of sperm retention than is the case if there is no climax, or climax outside of this period. Hence, by climaxing when and with whom she finds optimal, a female can influence the amount of sperm retained and, hence, the probability of becoming impregnated. Unfortunately for husbands, these sperm-retaining orgasms are more likely to occur during adulterous sex than in the marriage bed.[13]

Maximizing sperm retention is but one of the orgasmic "strategies" available to adapted females, according to Baker and Bellis. In their account, autoerotic orgasms are every bit as adaptive as those obtained during intercourse. Masturbation is an essential tool in any adapted female's arsenal. The adaptiveness of masturbation is related to the phenomenon known as a sperm block. Baker and Bellis cite evidence that any sperm deposited in the vagina during copulation hinder the sperm deposited in a subsequent copulation; they act as a sperm block. But the efficiency of the block declines with time. An "inter-copulatory orgasm"—i.e., masturbation—arrests the decline in efficacy of a sperm block. Hence, by masturbating while still retaining sperm from a previous copulation, a female can reduce the likelihood of becoming pregnant as a result of the next copulation.

In order to see why this might be a good thing to do, you must consider how the female benefits from trysts outside the marriage. And it has nothing to do with a lack of attention from her mate. Rather, the adapted female engages in adultery only with males who are genetically superior to her mate. (Assume, for now, that there is some straightforward way of assessing relative genetic value.) So she would rather become impregnated by her lover than her mate. But, because the latter must be kept around in order to help rear the child, she must continue to engage him in intercourse as well, lest he become suspicious. Her mate's sperm must be kept safely at bay while she intersperses adulterous sex with pair-bond sex. She manages to do so in three steps. First, she has a huge orgasm with her EPC partner; she then masturbates before his sperm block starts to wane significantly, thus practicing selective birth control on her mate. And, though Baker and Bellis do not mention this obvious extension of their logic, perhaps as the final flourish for this elaborate adaptive hoax, she fakes an orgasm with her husband.[14] As a pair-bonded male, I can only say, Yikes!

Perhaps we should view this deception as poetic justice given the so-called sexual double standard (boys are expected to be wayward, but women must remain chaste), the explanation for which is considered a triumph of sociobiology. What androcentric sociobiology giveth, gynocentric sociobiology taketh back, and then some. But I am more impressed with the similarities than the differences in androcentric and gynocentric sociobiology. Something has gone wrong in both the androcentric and gynocentric stories of adaptive orgasms. Everything seems to be explained effortlessly, but with only the flimsiest empirical evidence. For example, what about the many women who never have orgasms, and the much larger number who never do so during intercourse? According to Baker and Bellis, this too is an evolutionary strategy. What is lost by way of flexibility is recouped "at least partially, through greater 'crypsis'" with respect to the partners' knowledge of fertilization.[15] The best way for a female to keep her mate in the dark paternity-wise is to avoid orgasms altogether. What the frigid female lacks in upsuck is compensated for by the mystery that attends her intercourse. (Baker and Bellis do not explain why the frigidity "strategy" was so much more popular during the Victorian era than ever before or since).

And what about orgasms in virgins? Surely that is not adaptive. But, according to Baker and Bellis, virgin orgasms are every bit as adaptive as EPC orgasms; they just serve a different function. A virgin who masturbates is practicing good hygiene, because the orgasms expel unwanted bacteria. Call it antibiotic behavior.

ARISTOTLE ON WET DREAMS

There is yet another aspect of sexual behavior that cries out for adaptive explanation and that was, once more, noted by Aristotle: "What are called wet dreams occur by night with women as with men."[16] The reader may have guessed by now that nocturnal emissions are not difficult to handle on the Bellis and Baker hypothesis. They are just another type of intercopulatory orgasm, all the more effective because the most cryptic of them all. But as Aristotle also noted, "the same thing happens to young men also who do not yet emit semen, and to those who do emit semen but whose semen is infertile."[17] Baker and Bellis, primarily concerned as they are with female orgasms, do not tackle this one; but this last of the orgasmic mysteries— ironically, one that concerns males—was not a mystery for long.

In a 1995 paper, Randy Thornhill, Steven Gangestad, and Randall Comer filled the sole remaining lacuna in Darwinian explanations for orgasmic phenomena in humans: those nonejaculatory orgasms that occur in young virgin males during sleep.[18] According to Thornhill and his colleagues, these dry wet dreams are Mother Nature's way of preparing young males for the deceptions of their future mates. By experiencing what is essentially a female orgasm, males can better empathize with the female orgasmic experience. This empathy not only makes them better orgasm inducers, it enhances their ability to detect fake orgasms (the adaptive significance of which skill should be obvious by now). The battle of the sexes can be very subtle indeed.

This kind of thinking will keep private detectives working overtime. Moreover, they will need to extend their services: in addition to the standard procedures for demonstrating infidelity, they will need to begin monitoring those masturbations, heretofore a sign of a healthy sexual life but now part of a sinister (albeit unconscious) strategy. (Here's a question: If this is an unconscious strategy implanted by natural selection, would it be right for a husband to get angry about it? Or should he just throw up his hands and mutter "girls will be girls"?) The most enlightened husbands, the ones who understand the evolutionary dynamics here, will be a private detective's best clients. But before you put money down for a retainer, you should consider an alternative perspective on female orgasms.

KINSEY VERSUS THE VICTORIANS

Alfred Kinsey was as unlikely a person as one can imagine to fundamentally alter the way we think about sexual behavior. He spent his early scientific career detailing the anatomical minutiae of minute insects called gall wasps in order to elucidate their taxonomic relationships. Perhaps this is where he acquired his eye for detail, which was to serve him so well when he became a sexologist. His two major surveys and analyses of human sexual behavior, the first on males and the second on females, were certainly testimony to his the-devil-is-in-the-details approach.[19] For the most part, these considerable tomes are as dry as his publications on gall wasps. Given the resistance to any open discussion of sexuality at the time, this tone worked to his advantage, but only so far as his work on males was concerned. Though you would think that only the purest Puritans could accuse Kinsey of attempting to titillate, the publication of the research on female sexuality prompted such outrage that his funding was withdrawn.

It was not only Victorian sexual attitudes to which Kinsey found himself opposed. He also had to counter some pernicious ideas promulgated by the arch anti-Victorian Sigmund Freud, who was not one to let contrary evidence dampen his enthusiasm for his carefully constructed theories. Freud was particularly unimpressed with his patients' accounts of their own experience whenever they contradicted what he knew apriori. That dogmatism is the only way to make sense of his view that the vagina could be the locus of the female's deepest, most satisfying orgasms. He certainly could not have learned that from his female patients. But Freud had this notion about a transfer of the orgasm from the clitoris to the vagina, a transfer that occurred when a woman reached a certain stage of psychic health, and he wasn't about to let a few patients convince him otherwise.

It was Kinsey who decisively refuted this Freudian notion, which did much to improve the quality of life for Western women. Kinsey demonstrated that virtually all orgasms involve the clitoris and that Freud's orgasmic transfer could not occur without major surgical reconstruction. But Kinsey did not stop there; he then went on to argue that, given the location of the clitoris, the female orgasm is unrelated to reproduction. This view greatly displeased not only the Catholic Church but evolutionary adaptationists as well, and for essentially the same reason: God (or Mother Nature) would not countenance such gratuitous pleasure.

According to Kinsey, the clitoris is the primary focus of pleasurable sexual sensations in women. But it was not designed to be so; it just, fortuitously, is. Adaptationists find the latter claim particularly unpalatable; doubly so for some feminist adaptationists. How did Kinsey come to the conclusion that the female orgasm, for all of its significance, does not reflect design? First, and most obviously, the clitoris is typically not stimulated during intercourse. Second, given his training in biology, Kinsey recognized that the male penis and the female clitoris are homologous. As such, the clitoris will have the same circulation and neural circuits as the penis, barring some selective imperative for alteration. Hence, we should expect the clitoris to respond to tactile stimulation in much the same way as the penis. And it does: during female orgasm the clitoris even reflexively contracts at exactly the same frequency as the penis does during the male orgasm.[20]

Elisabeth Lloyd, who brought attention to the problems inherent in functionalist descriptions of bonobo sexual shenanigans, also did much to revive Kinsey's explanation of female orgasms as an alternative to the functionalist speculations.[21] Donald Symons had earlier developed this argument in his book *The Evolution of Human Sexuality*.[22] He cited as analogous the evolution of male nipples in mammals, including humans. Nipples serve an obvious function in females; but why do males need them? As Stephen Jay Gould detailed in an essay in *Natural History* magazine entitled "Freudian Slip,"[23] the question had long perplexed English biologists laboring within the functionalist framework of Natural Theology, including Erasmus Darwin, grandfather of Charles. Gould, following Symons and Lloyd, argued that male nipples have never had a function. Male mammals have nipples simply because (1) female mammals require them, and (2) given the obvious homology of male and female nipples, males are going to have nipples unless there are very strong negative fitness consequences. Gould further argued, again following Symons and Lloyd, that the capacity for orgasms in females is, in turn, a byproduct of selection for ejaculation in males—the Kinsey hypothesis.

Gould's column evoked a spirited response from Alcock, who characterized his own view as that of an "ardent adaptationist."[24] He suggested that, in comparing female orgasms with male nipples, Gould was slighting or

devaluing orgasms. But, as Gould correctly responded, the claim that orgasms are not adaptations can be perceived as denigrating only by those
whose values have been shaped by their ardent adaptationism. A feminist
functionalist can find the Kinsey explanation demeaning only if she is a
functionalist first. Painting, playing the piano, and solving algebra problems
are just a few human activities that are of no less value for not being adaptations. Alcock, Thornhill, and Baker and Bellis epitomize the mind-set of
hard-core adaptationists, the true believers, who insist that there is a functional why-explanation for every organismic trait or condition. But, in this
case, the work of Kinsey, Symons, and Lloyd seems to have rendered the
search for an adaptive explanation of female orgasms quite unnecessary.
Once we have removed our adaptationist blinders, the most parsimonious
explanation for female orgasms is that they are an evolutionary byproduct, a
"freebie." There is no teleological ultimate explanation for these orgasms,
only the proximate how-biology. This is then a case in which the how-
biology explanation replaces outright the teleological why-biology explanation. But there is much resistance to this view, for reasons identified long
ago by William Shakespeare, one of the most astute observers of human
psychology, in A *Midsummer Night's Dream*.

> Such tricks hath strong imagination
> That if it would apprehend some joy,
> It comprehends some bringer of that joy.[25]

Ardent adaptationists such as Alcock and Thornhill fail to appreciate
Shakespeare's insight. Rather, their intuition is epitomized in the following
quote from two sociobiology-infected psychotherapists: "Can something that
is now so important be an accidental side effect of some other selective
force?"[26] Though they obviously consider self-evident the answer to this rhe-
torical question, I hope the reader will conclude otherwise. You don't have
to look hard to find "accidental side effects of other selective forces." Your
ability to read this book is one of them.

3

Sex without SEX

⊷ ⊷ ❖ ⊷ ⊷

We are in the northernmost portion of the Chihuahuan desert—the Big Bend Region of southwestern Texas—looking down at the Rio Grande about 2,000 feet below. It is a morning in late May, the middle of the rainy season. Last night we pitched our tents on an exposed cliff, for the view. The unimpeded panorama of a vast stretch of the desert floor extending well into Mexico was spectacular, especially when the lightning show began many miles to the southeast. But the lightning, and then the thunder, were moving in our direction and soon were upon us, forcing us inside the tent. Inside, trying to imagine what the tent looked like from the outside, from a thunderhead's-eye-view, our choice of campsite seemed less opportune. Tonight, we'll try a more sheltered spot.

With the rains, though, the desert plants have stirred. Everything is blooming. Plant sex is in the air—enough to make you sneeze. Little mammilaria and large prickly pear are aflower, so too the chollas, sotols, nolinas, yuccas, and agaves. The rainbow and claret cup cacti are particularly beautiful. Many annuals are also flowering, in a brief but exuberant display. But we have not come here for the flowers; it is the animals we came to see, particularly the local reptiles. Deserts are the best places on the earth to observe snakes and lizards, and the Chihuahuan desert has some species that cannot be found anywhere else.[1] We are particularly interested in one group of lizards, called whiptails (genus *Cnemidophorus*, pronounced *nee mih DAH for us*), because we have heard of some strange goings-on, sexwise, among these creatures.

TANTRIC DOUGHNUTS

We have come at the right time of year; when plant sex is in the air, lizard sex soon follows. Within a very short period, we spot a male *Cnemidophorus inornatus* on the prowl for a mate, and settle in to observe things. He approaches a female, flicking his forked tongue, tasting her in the air. He is lucky; she is receptive, remaining still as he approaches, until his head lies above hers. We are somewhat taken aback when he grips her neck in his jaws. She looks uncomfortable and starts to squirm, but the male presses her down to the ground, scratching her with his claws in the process. After she is completely immobilized, the male begins to twist his body, in an obvious attempt to copulate. But he can twist only so far, and it does not seem far enough. Suddenly, he shifts his jaw-grip from the neck to the pelvic region, adopting an extremely contorted "doughnut" posture, by means of which he achieves a productive climax. He maintains this posture, a tribute to his yoga-like flexibility, for five or ten minutes, before releasing the female. Then they go their separate ways, without any formal leave-taking.

As we observe other whiptail species while moving north and west of the Big Bend, we find that the courtship of C. *inornatus* is typical of whiptails in general. Nonetheless, we are surprised to observe, when we arrive in the area where Texas borders both Mexico and New Mexico, this same court-ship ritual in the desert grassland whiptail (C. *uniparens*). We are surprised because C. *uniparens* is a species in which mothers produce daughters that are clones of themselves, without any contribution from male sperm. Hence, both doughnutee and doughnuter must be females. But why would an asex-ual species perform the same sexual behavior as a sexual species? The answer to this question, it turns out, provides much insight into a problem that has long vexed adaptationists—the prevalence of sexual reproduction in nature. Whereas the adaptationist explanations for female orgasms in humans were unprincipled and ad hoc, the same cannot be said for adaptationist explana-tions of sexual versus asexual reproduction. The adaptationist explanations we discuss in this chapter are much more compelling than those we dis-cussed in the previous two, and they deserve to be taken much more seri-ously. Nonetheless, we will see that the adaptationist explanations for the prevalence of sexual reproduction are at best incomplete. Even John May-nard-Smith, one of the most respected adaptationists in evolutionary biol-ogy, concedes that "one is left with the impression that some essential fea-ture of the situation is being overlooked."[2] Whiptail lizards, it turns out, nicely illustrate that one such overlooked "essential feature" is the histori-cally contingent how-biology of reproduction. Ironically, it is the asexual whiptails that provide the key to explaining why asexuality is so rare in vertebrate animals. As exceptions that prove the rule, they provide clues to the hurdles that must be overcome if asexuality is to evolve into a lineage where sexual reproduction is entrenched.

For most of us, commonsense suggests that sexual reproduction is such an obviously good thing as to require no explanation. As is so often the case, however, this common sense attitude precludes a deeper understanding of nature. Sex does require an explanation. The problem: If you could repro-duce without a mate, you could pass on much more of your genetic self—twice as much, in fact. Given this marked advantage of asexuality, the wide-spread occurrence of sexual reproduction is problematic. Sex should be much less common than it is. In certain groups, such as vertebrates, sexual reproduction is almost universal. Many neo-Darwinians consider this the central problem in evolutionary biology.

Sex versus SEX

Before proceeding further, we need to unpack the various connotations of the term *sex*. In the vernacular *sex* refers to behavior that culminates in copulation, which reaches its apotheosis in bonobos and their close relatives *Homo sapiens*. The technical, biological meaning of sex refers to the process whereby new combinations of genes are generated, known as mixis.[3] Mixis involves two distinct but related processes: outcrossing and recombination. Outcrossing is the process whereby a new genome is generated consisting of two sets of chromosomes, one from each parent. When the resulting daughter is in turn ready to reproduce, through a process called meiosis, she makes eggs that consist of only one set of chromosomes, which are a random mixture of those inherited from her parents.[4] The same is true of male sperm. So, when egg and sperm unite, the two-set (diploid) condition is restored. Hence, each parent's genetic contribution to its offspring is 50 percent. This is halved to 25 percent (on average) in the grandchildren, and so on. In this way, novel gene combinations are constantly being generated.

During one phase of meiosis, the homologous chromosomes align in pairs, one from each parent. At this time, a gene (allele) on the maternal chromosome may switch places with its counterpart on the paternal chromosome. This is referred to as recombination or crossing over. If the maternal and paternal alleles are different, a new genome results.[5] Recombination generates novel genotypes only when there is outcrossing.

In order to avoid confusion, I will use sex to denote the vernacular connotations and SEX the more technical, biological connotations of the term. Moreover, throughout this chapter, I use the terms *sexual* and *asexual* in the technical, not the vernacular, sense. Asexuality means no mixis. One form of asexuality is parthenogenesis, and individuals that reproduce in this way are called parthenogens. Asexually reproducing species or lineages consist entirely of females and are often referred to as unisexual. Sexually reproducing species require two mating types, or genders, and are often referred to as bisexual, which again has a completely different vernacular connotation.

Why SEX is a Problem for Adaptationists

For the purposes of this discussion, we must distinguish between two separate issues, the evolutionary origin of SEX and the maintenance of SEX. I will focus only on the latter: why is sexual reproduction so common, given its apparent disadvantages relative to asexual reproduction?

Many adaptationists take the view that asexual organisms provide a glimpse into something like a golden age, a time of innocence before sex reared its ugly head and your children were no longer you. In this scenario, asexuality is not only the original condition, but also one with massive inertia. Hence, the widespread occurrence of SEX indicates that it must confer benefits that outweigh its obvious disadvantages relative to asexuality. But once the initial inertia of asexuality was overcome, there was a more or less fair competition on an even playing field between sexually and asexually reproducing species that resulted in the current distribution of these modes of reproduction throughout the tree of life. On this conception, once sexual reproduction had evolved, there was little inertia in the transition from sexual to asexual reproduction or vice versa, so we can ignore the contingent historical factors that resulted in the particular form of the tree of life. Any twig on the tree of life has the same prior probability of being sexual or asexual as any other twig on the tree of life, no matter how distant. I will argue that this ahistorical approach is untenable, because once we take only the most rudimentary consideration of the how-biology of sexual and asexual reproduction, it will be obvious that there can be substantial inertia in the transition from one to another. I will illustrate this imprint of history in one branch of the tree of life, the vertebrates.

SEX in the Black Box

There are two basic sorts of ahistorical explanations of SEX. One sort of explanation emphasizes the deleterious consequences of asexuality, the other emphasizes the benefits of sexuality.[6] The disadvantages of asexuality were first identified by the geneticist Henry J. Muller, who noted that, in asexual populations, once a deleterious mutation evolves, there is no way of getting rid of it short of the extinction of the clonal lineage in which it originated.[7] As such, deleterious mutations inexorably accumulate through time in a given clone, a phenomenon known as Muller's ratchet. Because of this ratchet-like accumulation of deleterious mutations, asexual populations become increasingly less viable through time and are at a competitive disadvantage against sexual populations, which do not suffer from the ratchet effect. This is the explanation for SEX favored by many geneticists.

Muller's ratchet emphasizes the disadvantages of asexuality and hence, only indirectly, the advantages of sexuality. Other explanations of SEX focus more directly on the advantages of sexuality. This category includes a variety of hypotheses, but here I want to examine those most favored by adaptationists, which focus on ecological considerations.[8] Most of these explanations emphasize the fact that sexual reproduction produces variation, which is assumed to be a good thing. Variation, after all, is the engine of natural selection. The nineteenth-century biologist August Weismann was perhaps the first to link sexual reproduction to the virtues of variation.[9] Among recent evolutionary biologists, George Williams is the most Weismannian in spirit. He has proposed a number of models in which it is assumed that the more the genetic deck is shuffled, the better.[10] Williams differs from most other adaptationists in his claim that sexuality actually hinders adaptation rather than facilitates it. According to Williams, SEX prevents populations from becoming too closely adapted to a particular environment, and this makes their extinction less likely when things around them change.[11] On this account, SEX, though not adaptive per se, is meta-adaptive.

Though most other evolutionary functionalists also emphasize the importance of variation, they are convinced that sexuality must be adaptive in a more straightforward way. As always, they have looked to environmental factors to explain its existence. The basic premise is that to adapt effectively to variable environments, you need genetic variability. The many variational adaptationist hypotheses can be distinguished by the nature of the environmental variability that is believed to drive the evolution and maintenance of sex. These fall into two basic categories: (1) those that emphasize spatial variation in the environment and (2) those that emphasize temporal variation in the environment.

Let's first consider spatial variation. Michael Ghiselin, who is fond of economic metaphors, proposed that in a saturated economy, in which the environment is at or near its carrying capacity, it pays to diversify; but in an unsaturated economy, or seller's market, it is better—because it is more efficient—to produce just one kind of product. Hence, he proposed that sexuality is an adaptation to saturated environments; asexuality, on the other hand, is more effective in unsaturated environments that are well below their carrying capacity.[12] Graham Bell's version of this idea, which he refers to as the "tangled bank," puts more exclusive emphasis on spatial variation than does Ghiselin.[13] He proposes that when environments vary over small spatial scales, sexual reproduction is advantageous in providing wide varieties of offspring that are able to exploit the various niches. This diversification also serves to minimize competition between relatives.

Environments can also vary in time. When things change, especially when things change quickly, asexuality can become a liability. On the standard view, clones evolve more slowly than sexual populations because of their lack of recombination; as a result, asexual populations have more difficulty coping with environmental change.[14] This is especially true if we consider the biotic environment—all the other life forms, whether predator, prey, competitor, or parasite—that may affect viability, even as they themselves are evolving. This situation has been likened to that of the Red

Queen in *Through the Looking Glass*, who had to keep running as fast as she could just to stay in place.[15]

One particular version of the Red Queen emphasizes the role of parasites in the evolution of sex. The logic is as follows: Parasites, which have significant negative impacts on fitness, evolve much more rapidly than their hosts because of their shorter generation times.[16] They target the most common phenotype in a population, thereby conferring an advantage to the uncommon phenotype,[17] until, that is, the uncommon becomes the most common, at which point it now becomes the primary target. Because of this coevolutionary cycling of host and parasite, the next generation of hosts will inherit an environment that is different from that of the previous generation. The net result is selection for offspring variability and hence sexual reproduction.

Both Red Queen and "tangled bank" explanations have an intuitive appeal. To adapt, you need variation, and SEX is a great variation provider. The problem is that this ability to adapt more effectively, and hence to contribute disproportionately to succeeding generations, must provide an advantage that exceeds the reduced genetic contribution to each offspring. And this compensation must occur in the short term, not just in the long run.[18] Unfortunately, given these considerations, SEX is much too common in nature, according to the variational models. Moreover, long-lived, low-fecundity animals, such as vertebrates, that should benefit the most by asexuality are almost all sexual. This is a big problem for the ecology-variational models, one acknowledged by Williams.[19] He therefore suggested an entirely different explanation for the prevalence of SEX in vertebrates and many other animal taxa—a historical explanation. Williams proposed that the reason SEX is so common in many taxa is that they got stuck with it at some point in their evolutionary history, and once so encumbered they could not easily get out from under it. Williams himself is an adaptationist and therefore is much less inclined to invoke historical factors than someone like Gould, so this concession indicates the magnitude of the problem that SEX poses for adaptationists.[20] True believers, like Bell, accused Williams of defeatism.[21] Notwithstanding the manifest inadequacy of existing adaptationist explanations, Bell proudly proclaimed his faith in them: "I will continue to believe that history is bunk."[22]

But Williams was surely right. Here is an important fact to consider: All multicellular taxa, whether animals, plants, or fungi, are primitively sexual.[23] That is, the baseline condition for all three of these taxonomic kingdoms is sexual—not asexual—reproduction.[24] Moreover, extant vertebrate animals are twigs on a branch of the tree of life, where sexual reproduction has been the rule for hundreds of millions of years. This fact suggests that there is considerable inertia in the transition from sexual to asexual reproduction in vertebrates. Acknowledging this imprint of history is just the beginning. The real rewards from adopting a historical, or phylogenetic, perspective is in linking it to the how-biology. In determining, that is, which aspects of the historically contingent how-biology preclude the evolution of asexuality in vertebrates. It is to this how-biology that I now turn.

Constraints on Asexuality

Of the hurdles to overcome in the evolution of asexuality, the cellular process known as meiosis may be the most fundamental. Meiosis, the foundation for sexual reproduction in all multicellular organisms, first evolved in bacteria, in a non-reproductive context.[25] In multicellular organisms, including all vertebrates, this meiotic machinery has come to constitute the very foundation for development,[26] so it cannot be compromised. The few vertebrate species that have managed to evolve asexual reproduction have surmounted the meiotic barrier without significantly altering meiosis itself. More surprisingly, asexual vertebrates have retained, unaltered, the sexual development of their sexually reproducing ancestors. In fact, all extant asexual vertebrates exhibit fairly typical female development and are anatomically and physiologically indistinguishable from sexually reproducing females of closely related species. But perhaps most surprisingly, in many cases these asexual females need to be fertilized by males, yet another indication of the importance of evolutionary history.

Consider, for example, some fascinating fishes from the family Poeciliidae. Poeciliids (pronounced *pee SIH lee ids*) range from the southern United States to northern South America. Their center of abundance, with respect to both number of species and number of individuals, is Central America, where they inhabit water ranging from pristine mountain streams to the most revolting gutters of Belize City.[27] Most members of the family exhibit internal fertilization and bear live young, rare qualities among fishes, and traits that make them extremely useful for genetic studies. More important, for our purposes, some poeciliids have abandoned SEX.

Amazons

The legendary ichthyologist Carl Hubbs discovered the first asexual vertebrate in 1932, near the border of Texas and Mexico.[28] It was a member of the genus *Poecilia*, closely related to a species favored by tropical fish hobbyists, known as the sailfin molly (*Poecilia latipinna*).[29] The newly discovered molly (*P. formosa*) consisted entirely of females and was dubbed the Amazon molly, after the legendary race of formidable female warriors who poured out of the Caucuses to torment the ancient Greeks. This Amazon molly is an evolutionary product of mistaken identity, the mating of a male *P. latipinna*

with a female of the closely related *P. mexicana*. Unlike most hybrids produced in this way, the first Amazons were not only viable but also could
produce offspring when mated to males of either parent species. The offspring, however, were always female and looked exactly like their mom, no
matter who their dad happened to be. Dad's sperm were getting in, but his
genes were not getting through.[30] This state of affairs is obviously not a good
thing for the duped males, from an evolutionary perspective. Nor, however,
does it seem optimal from the perspective of the Amazons: they can exist
only where they have access to heterospecific male sperm, a considerable
inconvenience at best.

The mythical human Amazons have always played the role of villains.
Aeschylus, who was one of the most notable Amazon detractors, referred to
them as "men haters."[31] His opinion was no doubt inspired by reports—
rumors, actually—of Greek men being detained against their will as sex
slaves. But our knowledge of the Amazon molly should inspire more sympa-
thy for their legendary human counterparts. Perhaps like the mollies, these
estimable women were just insufficiently liberated. They too may have
found themselves evolutionarily addicted to sperm. No sperm, no daughters,
no more Amazons. This should not lessen our compassion for the sex slaves,
however, because they were being badly used from an evolutionary point of
view—parasitized, in fact. Their daughters could hardly be called their own.
One can only hope that for both the Greek sex slaves and the unwitting
male mollies, sex, even without SEX, is its own reward.

A closely related group of fishes of the genus *Poeciliopsis* (pronounced *pee
sih lee OP sis*) occur in several river drainages in western Mexico. Male
mollies of this genus regularly become confused as to which females constitute suitable mates. (Evidently, the females of these species look as much
alike to courting males as they do to us.) As a result, hybrids are generated
on a regular basis. Some of these hybridizations have generated all-female
clones.[32] These clones also require sperm for their continued existence. In
some cases, however, they let the male genes in—for a while. The daughters
produced by the mating of a unisexual *P. monacha lucida* with a male *P.
lucida* do get a genetic contribution from dad, but when it comes time for
these daughters to make eggs of their own, their father's genes are excluded.
The paternal genome of the male with which she mates is, in turn, excluded
before the next reproductive cycle, and so on. In this way, the maternal
genome is inherited clonally over successive generations, but the male genomes are not transmitted to successive generations (figure 3.1).[33] Though
this is a subtler form of sexual parasitism than that practiced by the Amazon
molly, the males are again being used.

The mating system of these mollies is polygynous. Successful males are
those who manage to inseminate lots of females, often in the context of a
group scramble. Under these conditions, it does not pay to be too discrimi-

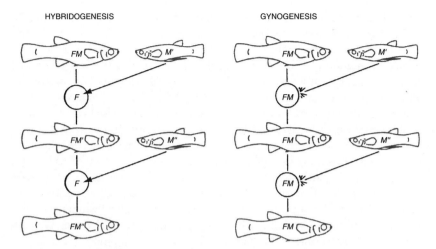

Figure 3.1 Hybridogenesis and gynogenesis.

F and M refer to female and male genomes, respectively. In hybridogenesis, the male genome is transmitted to the zygote but is omitted from the haploid gametes, so only the female genome is transmitted clonally. In gynogenesis, the male makes no genetic contribution to the zygote
(Adapted from Vrijenhoek 1993, p. 389)

nating: a deliberative male will be a lifelong virgin. It is therefore not surprising that unisexual parasites can sustain themselves by exploiting this male trait, which evolved in a different context. If the males encounter these females relatively rarely, the waste of energy and sperm is no big deal. In fact, however, *P. monacha* males sometimes find themselves in situations where unisexual females are quite common. And this **is** a problem, one that considerably complicates their sex lives. It causes them to become more discriminating, a trait that, in turn, creates selection for clones that more closely resemble the females of the bisexual species, and so on—an interspecific battle of the sexes.[34]

The primary moral to take from these mollies is that they are exceptions that prove the rule: history matters. The continued sperm dependence and hybrid origin of these asexual fishes are indicative of the hoops that a vertebrate must jump through to become asexual. It is not surprising that few have managed to make these leaps.

Sex without SEX without Sperm

A few vertebrates have managed to become asexual that do not require the sperm of bisexual progenitors, including our doughnuting whiptails. Of the approximately 45 species of whiptail lizards, about one-third are unisexual.[35] As do the mollies, all unisexual whiptail "species" owe their existence to mistaken identity, the mating of a male from one sexual species with a female from another. The parthenogenic whiptail *Cnemidophorus uniparens* resulted from the union of a female little-striped whiptail (*C. inornatus*) with a male rusty rumped whiptail (*C. burti*).[36] The hybrid produced by this event had three sets of chromosomes instead of two (a triploid genome) and could reproduce without any males whatsoever, a form of asexuality known as parthenogenesis. Parthenogenesis is a considerable advance over the sort of asexuality practiced by mollies, because it is more autonomous.[37] But, though whiptail parthenogens don't need to exploit the sperm of sexual species, it appears that they still need to be courted and seduced. Their continued survival as a clone seems to require that unisexual whiptails experience being doughnuted. But why?!! Why would God or Mother Nature put them in that position? Perhaps it is because their reproductive systems need to be primed in the same manner as that of their sexual progenitors.

COURTSHIP AND REPRODUCTION

Animal courtship rituals are often fascinating for their bizarreness, which is why they are a staple of television nature programming. Behavioral ecologists typically focus exclusively on the flamboyant male displays and the role they play in attracting mates or repelling rivals, but that is only part of the story. In many species, these rituals also serve to coordinate the internal reproductive physiology of the partners so as to maximize the chances of fertilization. Boobies (family Sulidae) are diving birds of the marine tropics, closely related to the gannets of higher latitudes. Blue-footed boobies (*Sula nebouxii*) are popular subjects of wildlife videos because of their comical mating rituals, which continue long after these monogamous birds have paired. The male does a carefully choreographed, slow-motion dance to show his bright blue feet to his mate, while she contributes vocal commentary and carefully timed contortions of her head and neck. This odd ritual serves to synchronize their reproductive rhythms; without it many unfertilized eggs would be laid.[38]

For bisexual whiptail lizards, doughnuting behavior probably serves the same role as booby courtship; it induces the female to ovulate while the most sperm are available.[39] Unisexual whiptails, of course, do not need to be fertilized; they do, however, require an analog of male sexual behavior in order to ovulate and initiate embryonic development. This requirement seems inefficient at best; it is not the way you would design an asexual species from the ground up. But actual evolved organisms are not designed; they are jerrybuilt through selective tinkering at the surface of existing, highly integrated developmental processes. It is not surprising, therefore, that unisexual whiptails and mollies evidence the imprint of their evolutionary history as members of a sexually reproducing lineage.

In light of their evolutionary history, the receptive behavior of parthenogenic whiptails is not so mysterious. But why would any individuals play the male role? This is a mating system that should reward only the doughnutee not the doughnuter. The time spent in the male mode would seem to be wasted, because it does not increase a parthenogen's fecundity. In fact, each whiptail typically spends part of its reproductive cycle acting like a female and part of the cycle acting like a male.[40] But this strategy still seems problematic from an evolutionary perspective. Time spent behaving like a male is time (and energy) wasted, if there are no sperm to be delivered. If the whiptail on the receiving end is a member of your clone, however, her daughter is the equivalent of your daughter—and of you, for that matter. So contributing to the clone by laying eggs and contributing by causing your sister to lay eggs amount to the same thing.

An adaptationist would be content with this sort of explanation. In fact, she would probably refer to it as the "ultimate" explanation of this strange behavior, the "why" of it. But as we shall see, it is the "how" of whiptail sexual behavior that does all of the work in answering the "why" of it as well. The reproductive physiology of whiptails is such that unisexual females will doughnut no matter who is on the receiving end.

The How and Why of Sexual Behavior in Asexual Whiptails

The fact that unisexual whiptails are evolutionarily addicted to sexual behavior by virtue of their ancestry indicates a problem with teleological accounts of SEX. In a perfectly teleological universe, or even a universe dominated by teleological imperatives, that should not happen. And the fact of its happening helps answer two related questions: Why is asexuality so rare in vertebrates? and How have those few asexual vertebrate species managed to become so?

But the teleological perspective would still seem important in providing an explanation for the occurrence of asexuality among otherwise sexual vertebrates. Why are those particular species of whiptails asexual and not some other lizard species? What is it about their ecology that sets them apart? Moreover, the fact that unisexual whiptails can manage, with one sex, what bisexual whiptails can accomplish only with two sexes suggests that the invisible hand of natural selection has been hard at work here. In order to assess how much effort was required of natural selection, we need to take a look at the reproductive physiology that undergirds sexual behavior in typical bisexual whiptails then look for the alterations that occurred in the

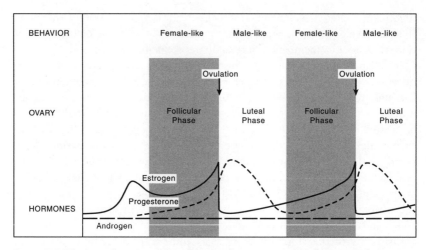

Figure 3.2 The reproductive cycle in the parthenogenetic lizard *Cnemidophorus uniparens*, illustrating the relationship between ovarian state, steroid hormone levels, and sexual behavior
(Adapted from Crews 1987, p. 112)

unisexual species. Let us first compare unisexuals with their female counterparts from bisexual species.

In a typical whiptail female from a bisexual species, as in vertebrates generally, there are two distinct phases to the sexual cycle (figure 3.2). During the first, or *follicular*, phase, when the eggs mature and become yolked, estrogen levels are high and progesterone levels are low.[41] During the second, or *luteal* phase, when the eggs enter the oviduct, progesterone levels are high and estrogen levels are low. Female receptivity is high when estrogen levels are high but diminishes when estrogen levels decline and progesterone levels increase.

Sex steroid hormones such as estrogens influence sexual behavior through their effects on specific regions of the brain, particularly the hypothalamus. Certain evolutionarily conserved neuronal populations in the hypothalamus concentrate these hormones. Not surprisingly, those neuronal populations that concentrate estrogens tend to be larger in females, whereas those that concentrate androgens (such as testosterone) tend to be larger in males. The ventromedial hypothalamus (VMH), which functions primarily in female (receptive) sexual behavior, is larger in females during the breeding season.[42] The preoptic area (POA) and anterior hypothalamus (AH), on the other hand, are larger in males, as a result of the stimulatory effects of testosterone.[43] These two areas are thought to be of particular significance for male sexual behavior, especially copulation.

Female parthenogens have retained this biphasic hormonal pattern of their sexual progenitors: high estrogen levels during the follicular phase and high progesterone levels during the luteal phase. The parthenogens, like the females of bisexual species, are sexually receptive when estrogen levels are high during the follicular phase, but they are not receptive during the luteal phase when estrogen levels are low. The sexually dimorphic brain areas of female parthenogens also resemble those of sexually reproducing female whiptails in every respect. They, too, have a smaller POA than males and a larger VMH. Moreover, the size of the two brain structures is comparable in female parthenogens and females of sexual species.[44] The utter femaleness of unisexual lizards would seem to present a problem when it comes to male copulatory behavior. But worse, these parthenogens seem handicapped even with respect to female copulatory behavior because they have only about 20 percent of the estrogen levels of females from sexually reproducing species.[45] With not even a smidgen of testosterone and paltry estrogen levels, even quasi-conception in these lizards seems miraculous. Perhaps natural selection has designed an alternative mechanism for copulatory behavior in this species to accommodate their asexuality.

In order to better understand why this is not the case, we need to consider the actions of sex steroid hormones in a bit more detail. We also need to consider the role of steroid hormone receptors. These receptors are proteins that become activated once they bind the hormone; the activated receptor then moves into the nucleus, where it binds to a specific DNA sequence, by means of which it regulates the expression of a particular gene (or genes).[46] Among the genes that are regulated in this way are those "coding" for the receptors themselves. The receptor type that binds testosterone and other androgens is called androgen receptor (AR); those that bind estrogens and progestins are referred to as ERs (estrogen receptors) and PRs (progesterone receptors), respectively.

Let's first consider how parthenogens manage to engage in receptive copulatory behavior. Estrogen receptors can be found in many brain regions, but, as you might expect, they are particularly abundant in the VMH.[47] Actually, they are abundant in the VMH only if estrogen is present. Estrogen induces the expression of both ER and PR.[48] Increased ER levels, in turn, cause a decrease in estrogen levels, creating a negative feedback loop.[49] It turns out that estrogen receptor expression in females of the parthenogenic C. uniparens is much more sensitive to estrogen than in sexually reproducing C. inornatus females.[50] Is this the key adaptation that allows them to mimic the female receptive behavior of females from bisexual species? No! The greater sensitivity of ER levels to estrogen in unisexuals is merely a passive effect of the increased genetic dosage for ER (and all other proteins) in unisexuals that results from having three sets of chromosomes rather than two. Their low baseline levels of estrogen relative to typical whiptail females is also an unselected consequence of their extra set of chromosomes.[51] Thus, we don't need to invoke selection to explain female sexual behavior in parthenogens.

Natural selection, however, would seem to be required to explain the expression of male sexual behavior in unisexual whiptails. In considering the alterations required for the evolution of asexuality, it is useful to distinguish between two aspects of the transition from female to male behavior. First, receptive behavior must be turned off; second, doughnut-seeking behavior must be turned on. Surprisingly, progesterone seems to play a role in both. Progesterone is widely known as one of the "female" hormones. In rats, it acts in concert with estrogens to promote receptive sexual behavior, or lordosis.[52] In addition, progesterone is known to interfere with the action of testosterone and other androgens; for that reason it has been used to control the libido of male sex offenders.[53] These feminizing and demasculinizing actions of progesterone are only part of the story, however; as it turns out, this hormone can have several other effects, which depend critically on context.

Consider first, progesterone's role in inhibiting female receptivity at the onset of the luteal phase. Recent research on rats and guinea pigs indicates that progesterone initially stimulates lordosis and only later inhibits it.[54] This dual effect is related to changes in the endocrine context, particularly changes in the number of hormone receptors and estrogen levels. In both sexual and asexual female whiptails, receptive behavior decreases when progesterone is artificially elevated.[55] The similarity of these progesterone effects to those found in rodents indicates that this mechanism of regulating female sexual behavior is evolutionarily conserved. Little or no modification of the existing mechanisms of steroid mediated sexual behavior was required to turn off receptivity in female parthenogens. They need only retain the ancestral condition.

Still left unexplained, however, is the remarkable ability of parthenogens to perform the male role in doughnuting. In bisexual whiptails, testosterone is known to stimulate doughnut making, and not only in males. When testosterone is administered to female whiptails of bisexual species, they too become doughnut fiends.[56] But this mechanism is of no use to unisexual whiptails, because they entirely lack testosterone. Here at last, we seem to have identified a role for natural selection: finding a testosterone substitute in order to make male copulatory behavior possible. But, yet again, we are confronted with an adaptationist mirage. It turns out that progesterone can do the job quite nicely, and not only in parthenogens. When parthenogenic *C. uniparens* females receive progesterone treatments, they go into the doughnut mode.[57] But progesterone also stimulates doughnuting in males of the sexual species *C. inornatus*. Moreover, progesterone is as effective as testosterone in eliciting doughnuts in castrated males. So, far from being a simple maleness inhibitor, progesterone can actually be androgenic. It turns out that this is not just a quirk of lizards. Progesterone also restores male-typical behavior in castrated male rats,[58] perhaps because it can bind to androgen receptors.[59] But androgen receptors in asexual whiptails have no higher affinity for progesterone than those in sexual whiptails.[60] Again, there is no evidence for an adaptive modification of the basic vertebrate neuroendocrine mechanisms regulating reproduction in order to make parthenogenesis possible.

Parthenogenic whiptails, and perhaps all parthenogenic vertebrates, are the result of an evolutionarily instantaneous transition from a sexual to an asexual state, not the slow sculpting of natural selection to accommodate particular ecological contingencies. They have retained all of the neural and hormonal mechanisms regulating reproduction found in the females of their sexual ancestors, down to the last detail. The apparent differences in sexual and asexual whiptail reproductive physiology merely reflect the plasticity and context sensitivity of generic vertebrate neuroendocrine processes. There is no need for a functional explanation. Why do asexual whiptails doughnut? Because, it turns out, they must do so in order to reproduce, as inefficient as it obviously is. Asexual whiptails must doughnut because their sexual ancestors did—because of their genealogical connection to their sexual ancestors, which has bequeathed to them the causal machinery for sexual reproduction. Asexual whiptails must work with this existing causal machinery in order to reproduce at all, even asexually.

The fact that whiptails and other vertebrate parthenogens evidence so little difference from their sexual relatives also suggests an answer to the larger why-question: Why are there so few asexual vertebrates when the adaptationist models tell us there should be many? The short answer is: evolutionary history. Given this evolutionary history, the transition from the baseline condition of sexual reproduction to asexual reproduction is not an easy one to make. In fact, it has been made only as a result of improbable hybridizations. Evidently, however, the transition by way of the "hopeful monster" route is more probable than the pathway in which reproduction is reengineered for asexuality by natural selection, even under the most favorable ecological conditions.

SEXUAL HANG-UPS AND THE EVOLUTION OF ASEXUALITY

Adaptationists have traditionally looked to the ecological context in explaining the "why" of sexuality in vertebrates, its "ultimate cause." But if "ultimate" explanations are truly evolutionary explanations, and not just adaptationist explanations with an attitude, phylogeny is as important as ecology, and this phylogeny is reflected in the details of the how-biology. Ardent adaptationists, such as Graham Bell, are seeking ecological explanations of SEX that transcend history. To find them, he must black-box all of the anatomical, physiological, and developmental particulars of organisms ranging from amoebae to humans, their how-biology. He must, in fact, treat organisms as ciphers such that terms like *Cnemidophorus inornatus* and *Paramecium aurelia* are only bookkeeping labels for entities that are essentially equivalent for the purposes of his analyses, without any of the rich connotations that most biologists typically derive from them. Because he abstracts over the biology of organisms, all of the explanatory load must be carried by ecological factors. Structuralists take the opposite approach, preferring to bracket ecological factors in their theories. But, for both an adaptationist like Bell and a structuralist like Brian Goodwin, a predilection for working at a high level of abstraction is tied up with their view that the science of biology should approximate physics in its lawabidingness.

It seems bizarre on the face of it to even contemplate the ahistorical evolutionary theory of anything—evolution without phylogeny. In fact, what Bell is after is not an evolutionary theory at all; it is an ecological theory, albeit one that relies heavily on natural selection. But natural selection, in and of itself, does not an evolutionary theory make. Only when it is connected to a temporalized view of the relationships of living things can the principle of natural selection contribute to an evolutionary explanation. It follows directly from Darwin's idea of descent with modification that natural selection cannot fundamentally re-engineer organisms. Instead, existing ecological imperatives cause the modification of previously evolved structures and processes. A priori, the deeper these modifications, the less likely they are to result in a viable organism, because of inevitable influences on nontarget processes. These side effects are overwhelmingly likely to mess things up. Given these considerations, it is not at all surprising that asexual whiptails continue to engage in sexual behavior or that asexuality is so rare in a taxon like the vertebrates, in which so much of development is predicated on sexual reproduction.

Maynard-Smith, a somewhat less ardent adaptationist than Bell, refers to the presence of previously evolved SEX-dependent structures and processes in asexual species—such as sperm dependence and copulation dependence— as "sexual hang-ups."[61] He seeks an adaptationist explanation for the evolution of sex in those lineages that lack such hang-ups. The problem is that, once the hung-up lineages are eliminated, there may not be a whole lot left to work with. Although it is certainly true that some taxa are more hung-up than others, true virginity is hard to find, and it is utterly absent in animals. A history of sexual reproduction has left its mark on all multicellular organisms, and this history cannot be ignored in any explanation of the pattern of sexual and asexual reproduction that warrants being taken seriously. How-biology considerations suggest that the paradox of sex is partly an artifact of adaptationist teleology. More important, these how-biology considerations dramatically alter our understanding of the distribution of SEX on the tree of life. It is not that the why-biology explanations are completely wrong; they are just inadequate. So this is not a case of replacement like the one we discussed in chapter 2, but rather one involving a fundamental alteration in our understanding of the evolutionary process born of how-biology considerations.

4

Transgendered

CORAL REEFS are the marine equivalents of tropical rain forests. But in contrast to rain forests, where much of the action is hidden from view, coral reefs are transparent. With only a snorkel and a mask, you can directly experience the awesome abundance and diversity of life on the reef, an environment so different from those in which we spend most of our lives that by comparison, the Gobi Desert, Antarctica, the Congo, and New York City are just variations on a single theme. The most spectacular coral reefs occur in an arc extending westward from Indonesia to the Philippines and south to New Guinea. The reefs off the north coast of New Guinea are some of the richest of the rich, perhaps containing a greater variety of living things than any other habitat on the earth. Even for a well-trained marine biologist, this variety exceeds comprehension.

In the sea, as on land, many animals are active only at night, so in order to fully appreciate these reefs, you need to explore them after sunset as well as during the daytime. We are all familiar with some of the dusk and dawn transitions that occur on land. Bats issue forth from their caves, as the swifts vortex inward toward their nests. Songbirds become quiet after the sun sets, just as many mammals begin to stir. But these terrestrial transitions are quite gradual compared with those on the reef, where, in as little as 10 to 15 minutes, the daytime contingent disappears and the creatures of the night emerge. As the anemones retract, the basket stars unfurl; as the groupers and snappers retreat, the moray eels and octopuses begin to prowl; white-tipped reef sharks emerge from their caves, and the black-tipped reef sharks retire to theirs. At dusk, some wrasses bury themselves in the sand, while parrotfishes make a giant mucous cocoon within which to rest. Crevices that harbor squirrelfishes and soldierfishes during the day are taken over by surgeonfishes and butterflyfishes as the light wanes. The myriad damselfishes are seemingly absorbed by the coral.

Not all damselfishes seek shelter in the coral crannies, however. Some retreat instead into coral relatives known as anemones.[1] Though it looks to the uninitiated as if the fish has become prey to the tentacled creature, it is really preparing for sleep. For a daytime fish the night can be especially perilous. The anemone's tentacles provide the anemonefish protection from the dangerous creatures of the night. In the fading light of late afternoon, an anemonefish ceases to forage, and its movements become slower and more relaxed, as it settles within the retracting tentacles to await the next dawn.

Those tentacles provide protection by day as well. While the sun is up, anemonefishes hover just above their host, sallying forth only to grab tiny zooplankton that pass by or to chase away other reef fish that come nosing around. Every minute or so, an anemonefish dives into the anemone and bathes in its tentacles, rubbing against them as if its whole body itched. This is the way an anemonefish reacquaints the anemone with its chemical essence, thus preventing its host from discharging its toxic stings as it normally would upon contact by any fish. Presumably, the anemone requires frequent reminders because it has only the most rudimentary nervous system and nothing remotely resembling a brain. The payoff for this effort on the part of the anemonefish becomes evident when a potential predator draws near. The anemonefish dives into the protective tentacles, and the predator, if it has had any previous experience with anemones, will not press the issue.

All anemones possess injectable stinging elements of varying toxicity called nematocysts, which deter fish from grazing on their tentacles.[2] The mucus of anemonefishes seems to protect them from these stingers, probably by preventing the anemone from releasing the nematocysts. It is not that anemonefishes are born with immunity to anemone toxins; nor can they just dive into an anemone and expect to be treated differently than any other fish. Rather, they must slowly introduce themselves through a conditioning process that takes several hours, during which the fish does get stung but with diminishing frequency until it fails to elicit any defensive response from the anemone.[3] Even then, constant physical contact is required if the anemone is to continue to react passively to the fish. The minuscule life expectancy of an anemone-less anemonefishes proves that this effort pays off. So bound is an anemonefish to its anemone that it may spend virtually its entire lifetime (up to 15 years) within a meter or two of the same mass of tentacles.

The relationship between these fish and their anemones is a special one, a form of biological intimacy known as symbiosis. In this chapter, we explore the consequences of this symbiosis with regard to sexual development in anemonefishes. The anemonefish's dependence on anemones is an important ecological factor in what, from the human perspective, is a bizarre pattern of sexual development indeed: they change sex from male to female once they reach a certain size. But, as we shall also see, this sex change is not nearly as noteworthy when considered in the context of their fishness. Though sex change is not at all the norm among vertebrates, and therefore cries out for explanation, it is not at all uncommon among fishes, and not just fishes with the unconventional boarding arrangements of anemonefishes. Ecological factors, it turns out, can go only so far in explaining why anemonefishes change sex. A more comprehensive evolutionary account of this sex change will incorporate considerations of their genealogy and how-biology.

Symbiosis and Its Consequences

Why do anemonefishes change sex, and why do they change from male to female? Part of the answer to both of those questions is provided by an excellent adaptationist theory, known as the size-advantage model. We will see that, despite its excellence, the size-advantage model has its limitations and must be augmented by considerations of the how-biology of anemonefish sexual development.

In order to appreciate the virtues of the size-advantage model, we need more information about the natural history and life history of anemonefishes. Our particular study area is near Madang, Papua New Guinea, home to nine species of anemonefishes, more than are found anywhere else on the earth. Among those present here are four of the most photogenic species: the clown anemonefish (*Amphiprion percula*), the pink anemonefish (*A. peridairon*), the spine-cheek anemonefish (*Premna bimaculauta*); and my personal favorite, the red and black anemonefish (*Amphiprion melanopus*). The clown anemonefish, which I will henceforth refer to as the clownfish, is representative of the entire group.

As do most marine fishes, clownfish spend their early days as part of the plankton and drift for variable periods in ocean currents before settling out on a particular reef patch. During their planktonic phase, clownfish are transparent and little resemble the adults, even in shape; only an expert would be able to identify them as clownfish. They do not metamorphose until they find an anemone. Unfortunately for these young clownfish, it is overwhelmingly likely that the reef is already well settled. Vacancy rates for suitable hosts can hover right around nil; in comparison, Manhattan is an apartment hunter's paradise. Many more clownfish descend upon a given reef patch each year than it can possibly sustain, creating intense competition for suitable hosts. Adult clownfish must spend much of their time and energy defending their anemones against any potential usurpers.

Clownfish do not live alone, however. Typically, an anemone is occupied by one adult female, one adult male, and several juveniles. The female is the largest fish, and the male the second largest. The smaller juveniles are not the offspring of the adult residents; they are the progeny of pairs from other reefs that have drifted here with the current. And they are certainly not treated like offspring. Rather, they are subject to constant harassment by the adult pair. But a clownfish without an anemone is a dead clownfish, so the juveniles persevere, and one of them may get lucky and become the adult resident once a vacancy arises. Their chances are much improved by the fact that a single juvenile can become the mate of either the male or female residents.

Once a breeding pair of clownfish has become established on their host anemone, their prospects are relatively bright. They are not, however, invulnerable. The stinging tentacles of a clownfish's host anemone deters many potential predators, but every time the clownfish darts out to snag a tidbit floating by, there is some danger that it may itself get snagged. If this misfortune befalls the female, the male quickly assumes her role, and, where once there were testes, ovaries now appear.[4] The largest of the juveniles hanging about then matures into a male and becomes the partner of the former male. If it is the male that dies, the largest juvenile will again mature as a male, this time as partner to the remaining female occupant. Hence, a death of either adult can be compensated for without forcing either the widow or the widower to face the perils of leaving the anemone to find a new mate.

Although these considerations help explain the advantage of being sexually labile, they do not explain why the female should be bigger than the male or why the progression should be from male to female. The particular pattern of sex change exhibited by anemonefishes is called protandry (which is Latin for "male first": *pro*, as in "prologue" = first; *andry*, as in "androgen" = male). Among sex-changing species, this progression is much less common than female-to-male sex change, or protogyny (which is Latin for "female first": *proto* = first, *gyn* as in "gynecology" = female), an exemplar of which is the cleaner wrasse.

CLEANERS AND CLIENTS

If you dive anywhere in the tropical Indo-Pacific, a vast region extending from the Red Sea down to Madagascar and eastward to the Hawaiian Islands, you are likely to witness the following scene. Several large reef fish, some of them predators, are arrayed in a remarkably orderly queue. In this case, an impressive giant grouper is in front. His mouth is open, his gills are flared, and his color is uncharacteristically dark. He appears to be in a trance. After discretely approaching the grouper, you notice a small fish, much longer than it is wide, swim right into the grouper's mouth. But, before you have time to ponder the adaptive significance of animal suicide, the little fish emerges from one of the grouper's gills. As you continue to watch the skinny little fish, it busily courses along the exterior of the grouper, occasionally nipping at the grouper's scales. After a while, the grouper seems to awaken from his trance, his color brightens, and he begins to breathe normally. And then he is gone. But his spot is quickly occupied by a snapper who seemed to be patiently awaiting its turn, and the skinny little fish begins again this curious routine. Now that you have approached even closer, you see that this little wrasse is really quite striking. A black stripe extends through from its snout through its eye, then widens considerably on its way to the tail. Toward the tail, the black stripe is flanked, above and below, by a swatch of electric blue. You are in the presence of the famous "cleaner wrasse" (*Labroides dimidiatus*). The larger fish are lined up at his cleaning station waiting their turn.

The client fish actually seem to enjoy the cleaner wrasse's ministrations, but there is some disagreement as to whether the benefit to the client fish extends beyond the tactile stimulation. Originally, it was believed that, by removing external parasites, the cleaner fish were providing a significant health care service. Early field experiments seemed to confirm this hypothesis. When cleaner wrasse were removed from a reef, the health of the other reef fish deteriorated substantially.[5] Some studies, however, have failed to demonstrate that the client fish benefit much from being cleaned.[6] It may well be the case that the client fish are anxious to be cleaned simply because being cleaned feels good. They certainly do seem to enjoy it. In that case, the relationship between cleaner and cleaned, though mutualistic—that is, beneficial to both parties—in a quality-of-life sense, is not mutualistic in an evolutionary sense. The cleaner wrasse is exploiting its ability to create pleasurable tactile sensations in larger fish to develop a food source, but this tactile stimulation does not increase the fitness of its clients.

The cleaner wrasse social system consists of one male and several females. The male dominates all females, which themselves form a linear hierarchy based largely on size. If the male should one day enter a grouper's mouth and proceed to the stomach rather than the gills, the largest female in his harem will, within a few hours, assume his role, even courting the other females.[7] And within weeks she will be a he, not only courting her erstwhile fellow females but inseminating them as well. If the male was only temporarily removed—by, say, a graduate student—a battle will ensue upon his return, between the he and the she-he. If the interloper has not completed sex change, and the original male succeeds in reestablishing his dominance, sex change in the interloper can be arrested. But once the sex change has proceeded beyond a certain point, there is no turning back, and the battle is joined in earnest. The loser will have to establish another cleaning station to which he can attract both clients and females.

The female → male sex change of cleaner wrasse is more typical than the male → female sex change of anemonefish. The vast majority of sex-changing fishes mature first as females and become males only after they have reached a size at which they can physically dominate most others of their kind. In understanding why these fishes change sex and why the majority of sex changers proceed from female to male, it will help to consider not only the role of current natural selection but also the genealogy and how-biology of sex change.

The Functional Logic of Sex Change

A very good adaptationist explanation of sex change, dubbed the size-advantage model, was devised by Michael Ghiselin and elaborated by Robert Warner, among others.[8] The size-advantage model is both simple and straightforward (two considerable virtues). The model not only predicts what sort of sex change will occur but also at what point in the life of an individual it will occur, on the basis of sex differences in the way fecundity is related to size (figure 4.1). Fecundity generally increases with increasing size in both sexes, though for different reasons. For female fishes, the larger the individual, the more eggs she can lay. On the basis of energetic considerations alone, large females can produce more eggs than smaller females. Sperm is much cheaper to produce than eggs, however, and there is often little correlation between male size and the amount of sperm a male can produce.[9] But large males often do produce more offspring than smaller males because they monopolize matings by virtue of their social dominance.

Though fecundity often increases with size in both sexes, the rate of increase may differ markedly in males and females. This sex difference is easiest to see if we construct some hypothetical graphs, with size as the X-axis and fecundity as the Y-axis (figure 4.1). The sex differences in the rate of increase in fecundity with increasing size are expressed graphically as differences in the slopes of the lines describing the size–fecundity relationship in males and females. Notice that in figure 4.1A, the line representing female fecundity is above the line representing male fecundity, for the smaller sizes, but the reverse is true for larger individuals. Where these two lines cross can be considered the fulcrum, or pivot point. Below this size, it is better to be a female; above this size, it is better to be a male. But an individual that begins its reproductive life as a female and then undergoes protogynous (female → male) sex change when it reaches the size at which the two lines cross enjoys the best of both worlds. It will have a higher lifetime reproductive success than any individuals that do not change sex and even individuals that change sex at a different size.

Figure 4.1A depicts the conditions under which protogynous (female → male) sex change will be selected for, as in the cleaner wrasse. Protandrous (male → female) sex change, of the sort found in clownfish, is advantageous under quite different conditions (figure 4.1B). Again, female fecundity increases linearly with size. But this time, male fecundity does not increase with size or increases only very slowly. This situation occurs when males do not compete with each other for access to mates, as in many schooling

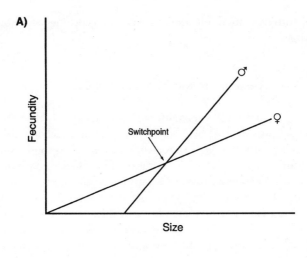

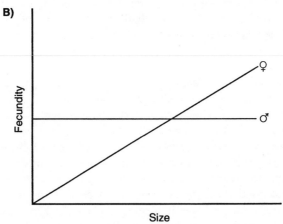

Figure 4.1 The size-advantage model of sex change, illustrating conditions for protogynous (A) and protandrous (B) sex change

pelagic fishes, where mating is a free-for-all. Large males of pelagic species cannot monopolize females in the way that larger males of territorial species can. Nor can clownfish males monopolize females, because of the constraints on their mobility that result from their dependence on widely scattered sea anemones. For the monogamous clownfish, the reproductive success of the pair depends entirely on the fecundity of the female. The larger the female, the more eggs she lays; hence, the larger the female, the better for both her and her mate. An increase in male size, however, little benefits either him or his mate. Hence, it is better for both the male and the female clownish, if the female is the larger of the two, a condition which is guaranteed by protandrous (male → female) sex change.

The size-advantage model is very effective in predicting both the direction of sex change and the size at which sex change occurs in cleaner wrasses, clownfish, and many other teleost fishes. But, as a general account of the occurrence of sex change in nature, it is much less successful. For example, based on considerations from the size-advantage model alone, we would expect to find protogynous (female → male) sex change in many vertebrates with polygynous mating systems resembling that of the cleaner wrasse, in which only a few larger males are able to monopolize matings.[10] According to the size-advantage model, polygynous birds such as prairie chickens should be female → male (protogynous) sex changers, as should polygynous mammals such as gorillas and elephant seals. But in fact, sex change, protogynous or protandrous, is confined to teleost fishes. So we confront a situation resembling that discussed in chapter 3. Just as the functionalist models lead us to expect more asexuality among vertebrates than in fact exists, so too the functionalist size-advantage model leads us to expect sex change to be more widespread among vertebrates than it is. And, once again, it is the historically contingent how-biology to the rescue.

THE HOW-BIOLOGY OF SEX CHANGE

Rather than look for a hidden functionalist rationale for the fact that sex change is common in teleost fishes but absent in mammals, we might look for basic differences in the how-biology of sexual development in these two taxa. One property of teleost sexual development stands out in this regard: its extreme plasticity.[11] All teleost fishes, whether they change sex or not, are much more sexually labile than any mammal. Put another way, their sexual development is much less canalized, to use a term favored by developmental biologists, than that of mammals. One upshot of this difference is that teleost sexual development is subject to environmental influences in ways that mammalian sexual development is not. Consider, for example, the role of sex chromosomes in the sexual development of teleosts.

Sex Chromosomes and Gender

In the vast majority of mammals, including humans, sex is determined by the presence or absence of the Y-chromosome. If you have the Y-chromosome, you are a male; if you don't have it, you are a female, no matter what the environmental conditions. As you might expect, relatively few teleost fishes exhibit this sort of sex determination. The majority lack well-developed sex chromosomes, the kind that you can distinguish from other chromosomes (autosomes) simply by viewing them through a standard light microscope.[12] Yet, a number of fish species *do* have sex chromosomes,[13] and they are particularly instructive with respect to some basic differences in the sexual development of teleosts and mammals.

Many poeciliids (the family to which the Amazons belong) have sex chromosomes, as do members of related families, including the medaka (*Oryzias latipes*), a species favored by geneticists.[14] Medaka have the XY mode of sexual inheritance found in mammals, but the developmental mechanism through which their sex is determined is not at all mammal-like.[15] The difference is evidenced by the fact that the XY contribution to sex determination in medaka, in stark contrast to mammals, can be completely nullified by hormonal treatments. When young fish are treated with estrogens they all become females no matter what their genetic constitution. Androgen treatments have the opposite effect.[16]

This sort of hormonal manipulation of sex has become commonplace in aquaculture, for which monosex populations are often preferred. In the culture of salmon, for example, females are more desirable than males. Fewer males result in less fighting, and hence more efficient meat production.[17] Estrogen treatments can result in up to 100 percent females, even in salmon species that have sex chromosomes.[18] African cichlids called tilapia (genus *Oreochromis*) are also popular in aquaculture. In the culture of Tilapia, males are preferred because females essentially stop growing when they reach reproductive age. Androgens have been used successfully to generate all-male stock, even in the face of countervailing sex chromosomes.[19]

Hormonal treatments cannot induce gonadal sex reversal in mammals. If they could, human transsexuals would not require surgical interventions.[20] Mammals lack the sexual bipotentiality of teleosts because of their canalized sexual development. The difference between teleost fishes with genetic sex determination and mammals, all of which have genetic sex determination, is that the genes that influence sex determination in mammals act earlier in sexual development and more irreversibly than those in teleost fishes. As a result, the process of sex determination in mammals is much less protracted and less subject to environmental perturbations than that of teleost fishes. We can perhaps best appreciate this fundamental difference if we take a close look at sexual development in a cloned fish.

Sex and the Single Clone

Zebrafish (*Zebra danio*) provide a dramatic illustration of teleost sexual lability. Native to eastern India, this species was among the first species imported for the tropical fish trade. Though its appearance is unremarkable and experienced aquarists would never put one on display, it remains popular if only because it is so hardy that it can survive even the most cursory care. Zebrafish are not at all fussy when it comes to breeding either. In fact, one reason zebrafish have become *the* vertebrate model for developmental genetic studies is that even molecular biologists can breed them. A less prosaic reason is that they can be easily cloned.[21] Since the 1980s, a variety of clones have been produced in which there are specific mutations of genes involved in neural development. These clones provide a powerful tool for developmental genetic studies of the embryonic nervous system.

There is a truly remarkable fact about zebrafish development, which has somehow failed to elicit much interest from the developmental biologists who make use of this animal: within any clone you will find both sexes. Think about that for a minute—one clone, two sexes; two genetically identical individuals, one female and the other male. Clones of a female mouse will always be female. The same is true of a sheep, such as Dolly, or any other mammal. But zebrafish clones invariably consist of both sexes. Some of these clones are even predominantly male. This fact is of enormous consequence even for those developmental biologists who fail to see its broader significance. Because genetically identical male and female zebrafish can perpetuate the clone through their garden-variety matings, it is possible to maintain these clones indefinitely without all the technical fuss required to maintain clones of mammals like Dolly.

The fact that both males and females can be produced from a single genome would seem to indicate that sex (gender) is not determined genetically in zebrafish. But that conclusion is an oversimplification. In fact, genes do play an important role in zebrafish sex determination. It is clear, however, that, within clones, it is some environmental factor that is the difference that makes a difference. There are several of obvious candidates. For instance, temperature is known to influence sex determination in many reptiles and some fishes.[22] In most turtle species, females predominate at high incubation temperatures, whereas more males are produced at lower temperatures.[23] The reverse is true for some lizards and crocodilians, including the American alligator (*Alligator mississippiensis*).[24] There is also a third pattern, in which females are produced at extreme (both high and low) temperatures.[25] In the laboratory, temperature manipulations have been used to override genetic sex determination in some frogs and fishes.[26] In other fishes, pH is known to effect sex determination.[27]

But both males and females can be found even within zebrafish clones that have been reared in the same tank and that, presumably, have experienced the same temperature and pH. We must therefore look beyond such obvious features of the physical environment to some environmental influence that varies within a single group of fish reared together. One such factor might be the social environment. Even genetically identical individuals reared together can experience a different social environment, for reasons I will discuss later. Unfortunately, except for sex-changing species, there has been very little research on the influence of the social environment on sex determination in fishes. I stumbled upon such an effect quite by accident during the course of my doctoral research on paradisefish (*Macropodus opercularis*), a freshwater fish native to India and adjacent parts of Southeast Asia.

I had set out to study the interaction of genotype and experience in determining social dominance in this species. In exploring genetic influences on social dominance, I artificially selected for individuals that had a higher or lower propensity to "win" paired (dyadic) encounters.[28] By the third generation, I noticed a change in my "high-dominance" line, though not in the trait under selection. The high-dominance line had come to consist almost entirely of males. Conversely, the "low-dominance" line was mostly female. Somehow, by selecting *for* social dominance I had dramatically altered the sex ratios.[29]

These results were unanticipated, not least because I had been led to believe that this paradisefish had sex chromosomes,[30] in which case sex ratios should not have been affected by selection.[31] So I went back to the library to learn more about paradisefish. There I discovered, from a fairly obscure article published in a German periodical in 1939, that paradisefish must have a non-chromosomal form of sex determination.[32] Moreover, until a certain age, all paradisefish are females; then, before maturation, some individuals become males, influenced no doubt by a combination of genetic and environmental factors. Through my selection *for* social dominance, I had clearly genetically altered the probability that my paradisefish would be deflected to a male trajectory.[33]

But, just as important, I found that I could alter the sex ratios within each line simply by manipulating the population density in the rearing tanks; the greater the density, the greater the proportion of females. This outcome suggested an important role for social interactions in determining which individuals would be deflected from the female trajectory and when, during the course of their sexual development.[34]

Later, during postdoctoral research with George Barlow at the University of California, Berkeley, we discovered a similar case of socially regulated prematurational sex change in the Midas cichlid (*Cichlasoma citrinellum*), a native of Lake Masaya in Nicaragua.[35] The occurrence of socially regulated prematurational sex change in two teleost fishes as distantly related as paradisefish and Midas cichlids suggests that socially mediated prematurational sex change is widespread in this taxon. It turns out that zebrafish also exhibit this pattern of sexual development, which helps explain how we can get both sexes from a single clone. All zebrafish begin life as females, but some are diverted to a male trajectory while they are still juveniles.[36] Whether a particular individual is diverted to the male trajectory depends on a variety of genetic and environmental—including social—factors. The social environment experienced by juveniles within a brood varies, particularly as a result of nonheritable size differences. This unshared aspect of their environment explains why, among genetically identical individuals reared under the same physical environmental conditions, some are diverted from a female to a male trajectory and some are not.[37]

More important, for our purposes, the prematurational sex change of zebrafish, paradisefish, and Midas cichlids points to a general property of sexual development in teleost fishes. In some important respect, all teleost fishes exhibit protogynous sexual development. All teleost fishes, whether they have sex chromosomes or not, probably begin to differentiate as females, in the manner of zebrafish and paradisefish.[38] If they do, maleness is a secondary condition initiated by later acting internal and external factors, which deflect development from the primary female trajectory.[39]

But what about our clownfish and other protandrous (male → female) sex changers? They would certainly seem to be exceptions to this rule. Actually, however, protandrous species provide some of the most compelling evidence for the fundamental protogyny of teleost sexual development. Experiments have shown that, although clownfish typically reach sexual maturity as males, their initial juvenile development, as in other fishes, is as a female.[40] Moreover, under the right conditions, clownfish can skip the male stage altogether and mature directly as females.[41]

GONAD SEX AND BRAIN SEX

The protogynous sexual development of teleost fishes does not, in and of itself, explain the fact that cleaner wrasse, but not elephant seals, undergo adaptive female → male sex change. In principle, protogynous sexual development could be as highly canalized as male or female development in humans, in which case it could not be put to adaptive use in cleaner wrasse or clownfish. It is the fact that the sexual development is uncanalized, or labile, that permits cleaner wrasse to undergo female → male sex change under the appropriate social conditions and that permits clownfish to undergo male → female sex change at all. So what, ultimately, is the source of this adaptive sexual lability?

The protracted nature of gonadal differentiation in teleosts is certainly a factor. This is part and parcel of their lack of sexual canalization. The later in development that gonadal fate is fixed, the more opportunity there is for environmental influence, adaptive or otherwise. But it seems unlikely that teleost fishes would be able to make adaptive use of these environmental influences if they acted directly on the gonads themselves. My own thesis is that, in teleost fishes, in contrast to mammals, all sex-determining factors, whether genetic or environmental, affect the gonad only by way of the brain. That is, brain events determine the fate of the gonads.[42] This must be true of any sex-changing species in which the sex change is initiated by social interactions. The only way social events can affect the gonads is through the brain. According to my thesis, though, this is true not only of sex-changing species but also of all or a large subset of teleost fishes. Under certain ecological conditions, this brain-mediated gonadal sex determination can be exploited for adaptive sex change.[43]

It is conceivable that sex change is bootstrapped within the gonads themselves. Presumably, the same sort of hormonal treatments used to create monosex populations in aquaculture could induce sex change as well. But, in fact, hormonal manipulations of this sort are not effective in inducing adult sex change in species that do not undergo sex change in nature. Moreover, neither anti-estrogens nor androgens are particularly effective in inducing the female → male sex change in species that do undergo this transition in nature.[44] The ineffectiveness of these hormonal manipulations is one indication that sex change is initiated outside of the gonads.[45]

Further evidence for the primacy of the brain in teleost sexual development comes from some recent research on a cousin of the cleaner wrasse. Under the right social conditions, bluehead wrasse (*Thalassoma bifasciatum*) will undergo a behavioral sex change (female → male) even when their gonads have been removed.[46] Clearly then, gonadal feedback is not required for the behavioral transition from receptive behavior to active courtship. The nature of the brain events that initiate either primary or secondary sexual differentiation in teleosts remains to be established. Much of this research has focused on gonadotropin-releasing hormone (GnRH) neurons in the hypothalamus because they constitute the final common pathway for neural influences on the gonads.[47]

For our purposes here, my particular thesis as to the how biology-of teleost sexual lability matters less than the fact of this lability. But this exploration of the differences in teleost and mammalian sexual development does concretely illustrate how we can go about using considerations from how-biology to construct testable alternatives to functionalist explanations based on considerations of costs and benefits alone.

WHY DON'T ELEPHANT SEALS CHANGE SEX?

I want to emphasize again that I consider the size-advantage model of sex change to be adaptationism at its principled best. It provides a very good adaptationist explanation for the fact that clownfish and cleaner wrasse are sex changers and for the fact that clownfish are protandrous (male → female) sex changers, whereas cleaner wrasses are protogynous (female → male) sex changers. But it is certainly not up to the task of explaining the occurrence of sex change among vertebrates in general. Moreover, much of the explanation for sex change in clownfish and wrasses lies in the general lability of teleost sexual development, without which there could be no socially mediated sex change, no matter the ecological circumstances. A number of bird and mammal species share with cleaner wrasses the salient socioecology and should, according to the logic of the size-advantage model, exhibit protogynous sex change. That they don't, indicates that something important is being ignored in adaptationist explanations of the evolution of sex change, and that something is to be found in the how-biology.

In chapter 3 we posed the question, Why don't more vertebrates reproduce asexually, given the advantages of asexuality? Here, we need to ask another "why don't" question: Why don't elephant seals and other polygynous mammals undergo protogynous sex change like cleaner wrasse? It's not for lack of adaptive rationale. Consider that the vast majority of male elephant seals never breed, not once. Only a few live long enough to attain a size sufficient to defend portions of the most desirable beaches along central California's rocky coast.[48] These few have it extremely good fitness-wise—they may impregnate hundreds of females during the course of their lives—but for most of the rest, there is next to nothing.[49]

Whereas the female elephant seals mature at around 3 years of age, the males become potential breeders at about 6 to 8 years.[50] Even then, they spend the next couple of breeding seasons in great agitation, snapping or bellowing at their fellow "bachelors" or, when the urge is overwhelming, attempting to waylay females in the surf as they return to the breeding grounds. Either way, they are left unrequited, and frustrated. Much of the cacophony of an elephant seal colony consists of the bellowing of beachmasters as they abuse the bachelor males attempting to sneak their huge corpuses into the sanctum sanctorum of the harem grounds. Whereas a 6-year-old female may be suckling her third pup, the 6-year-old male is still waiting for his first copulation, and none too patiently. After all, he will probably die first. Among the obstacles to his survival to a postvirgin state are those pesky great white sharks circling offshore beneath a few foolhardy surfers.[51]

Contrast elephant seals with cleaner wrasse. A cleaner wrasse that is too small to control a harem is called a female. And she reproduces as such until she is old and large enough to take over the harem. She may get eaten first, but she won't die a virgin. Now wouldn't elephant seals be better designed if they, like the cleaner wrasse, were protogynous sex changers? Of course they would, as would sage grouse, elk, gorillas, pronghorns, and any one of a number of polygynous birds and mammals. It is not because of anything related to their ecological niche that male elephant seals are born male and must remain so.

In order to explain why elephant seals cannot change sex, we need only apply the same logic by which we explained why they and all other verte-brates reproduce sexually rather than asexually. Just as elephant seals and all other vertebrates reproduce sexually—despite an adaptive rationale to the contrary—because of their verbrateness, elephant seals and all other mam-mals cannot change sex because of their mammalness.[52] That is, by virtue of their evolutionary history as mammals, elephant seals were bequeathed a highly canalized process of sexual development involving the early deter-mination of gonadal sex; this canalization of sexual development is incom-patible with sex change, whether it be adaptive or not. Elephant seals would have to be fundamentally redesigned in order to become sex changers. If it is problematic to consider natural selection as the invisible "designer of life," it is much more problematic to ascribe to natural selection the power of redesign.

How should we conceive of the relationship between the how-biology and why-biology explanations for the occurrence of sex change among ver-tebrates? Obviously, this is not a case in which the how-biology–based sex-ual lability explanation replaces the why-biology–based size-advantage model. Nor does this how-biology explanation merely constitute a proxi-mate explanation in relation to the ultimate explanation provided by the size-advantage model. Rather, the how-biology in this case should be viewed both as a competing alternative explanation for the distribution of sex change among vertebrates, and one that complements the why-biology ex-planation with respect to our understanding of the evolution of sex change.

5

Alternative Lifestyles

Sausalito, California, is a small, affluent community situated on the San Francisco Bay, just north of the Golden Gate Bridge. It has an abundance of what travel agents refer to as "charm," which is why it is a popular tourist destination. One of Sausalito's principal attractions is its houseboat-lined shoreline. *Houseboat* typically conjures images of a ramshackle bohemian lifestyle. Not so in Sausalito. The houseboats there run into the million-dollar range and beyond. Multistoried and well appointed, some resemble miniature Venetian palazzos.

In May 1985, the houseboat residents began complaining of unearthly droning sounds emanating from the water around them, loud enough to preclude sleep. These noises, which also caused crystal chandeliers to vibrate ominously at the local debutante balls, usually reached a crescendo around dusk. When the wealthy complain, the media listen, and the plight of the houseboat owners became something of a cause celebre in the San Francisco Bay area, with daily updates in the newspapers and on television. Various theories were advanced as to the cause of the noise, ranging from a secret military device to an underwater generator. Experts of many kinds were summoned to the scene, including some who specialized in underwater acoustics. Finally, long after the average television viewer had lost interest in the plight of the houseboat occupants, the source of the sounds was identified: a fish, known as the plainfish midshipman or toadfish (*Porichthys notatus*). The creatures responsible do not make a visual impression commensurate with the noise they produce. They are smallish bottom dwellers, who, as their common name implies, are not particularly attractive to the human eye.

The irritating noise is a mating call emitted by ardent male toadfish in order to attract females. But not all males are blameworthy. The males that make these calls, the noisy courters, defend territories where females can find a suitable spot to lay their eggs. They then dutifully guard the eggs and fry until they become free-swimming in a couple of weeks. Some males, however, that are much smaller, do not vocalize. These little silent males, often referred to as sneakers, do not bother defending a territory; doing so would be futile given their small size. Instead, they lurk at the peripheries of the territories defended by the noisy courters and attempt to shower their sperm over the courting couple at the moment of climax, at which point, understandably, the courting male's vigilance is somewhat compromised. He recovers surprisingly quickly nonetheless, and the sneaker male has a very limited window of opportunity within which to shower his sperm over the couple, before being chased away by his larger rival. Under the best of circumstances, this is an inefficient way for sperm to fertilize egg. By way of compensation, the little sneakers have huge testes, much larger in relation to their body size than those of the territorial males, and are therefore able to deliver massive volumes of sperm.

The noisy courters and little sneakers exemplify what behavioral ecologists refer to as alternative reproductive tactics, or ARTs,[1] which have garnered a lot of attention. Of particular interest are the little sneakers. Based on the principle of sexual selection, we expect males to vie with each other for dominance status and to posture to attract the attention of females. Why then is not every male a noisy courter? What is the evolutionary explanation for the unconventional little sneakers?

The adaptationist explanation for the unconventional males is that they are exploiting a vulnerability in the tactics of the conventional males. When a relative few males are able to monopolize matings through the conventional tactics of intimidation and displays of virility, there will be a pool of males, perhaps the majority, who can only hang around at the periphery, biding their time. Many will spend their entire lives in this unrequited state. Unconventional males avoid this fate by going about finding mating opportunities in a completely different way. Rather than compete directly with the big boys, they seek other means entirely to bequeath their sperm to the females.

This explanation does help us make sense of the Sausalito toadfish and many other cases of alternative reproductive tactics as well, but it suffers from the same defect as the adaptationist explanations of both SEX and sex change: Many animals that seem to meet the adaptationist conditions for alternative tactics fail to evolve alternative reproductive tactics, just as many animals that "should" change sex fail to do so. As is true of SEX and sex change, the how-biology helps us understand this failure. But before we address the how and why of alternative reproductive tactics, we need to be more explicit about how we distinguish alternative mating tactics from other sorts of variation in reproductive behavior. I will focus on the male side of the reproductive equation.

SOME GUIDELINES IN ART APPRECIATION

The term, *alternative reproductive tactic* has been applied quite indiscriminately and has come to embrace some rather dubious cases, so we need to be careful about what we include under this rubric. Two criteria are crucial in determining whether a species exhibits true alternative reproductive tactics, in an evolutionarily relevant sense. First, it must be demonstrated that the putative alternatives are in fact distinct, not merely the poles of a continuum of sexual behavior. Call this the discreteness criterion. Second, there must be evidence that those who practice the unconventional behavior are able to compete effectively against the conventional males. Call this the fitness criterion.

Let's first consider the discreteness criterion. Alternative tactics are interesting to evolutionary biologists primarily because they are discontinuous traits. Most phenotypic variation within populations is continuous and can be represented by means of the classic Bell curve, or, to use the camel by way of analogy, the "dromedary hump." This is true of height in humans, for example. Alternative reproductive tactics constitute a very special kind of variation that is radically discontinuous and best represented by a bimodal curve, or "Bactrian humps" (figure 5.1). Any adequate explanation of alternative reproductive tactics must acknowledge the fact that the one-hump (dromedary) populations, consisting of pure conventional behavior, constitute the baseline condition. Two-hump (Bactrian) populations evolve through the emergence of unconventional behavior in some males. Moreover, the two-hump (Bactrian camel) is much less common than the one-hump (dromedary) and therefore is noteworthy. In the vast majority of species, male sexual behavior varies continuously, and usually extreme behaviors at each end of the hump are rarer than those near its center. In some species, such as humans, this variation can be quite large. Only when this variation can be partitioned into two distinct humps should we suspect alternative mating tactics.[2] Bactrianism does not constitute the limit of humpedness. Dr. Suess–like three-humpers—call them "Swiss camels" or "swamels"—also exist.

The reproductive behavior of toadfish males certainly meets this discreteness criterion; there are clearly two humps. One reason we can be confident that the sneakers are practicing a distinct tactic is that they differ from the conventional males not only behaviorally but anatomically and physiologically as well. The noisy courters are much larger than the silent sneakers, and they exhibit a distinctive suite of characteristics not found in the sneakers, ranging from the microscopic structure of their sonic muscles to their neural circuitry. The humming noises are produced by sonic muscles attached to the swim bladder. When these muscles are contracted, they cause percussion on the swim bladder, which acts as a drum. These muscles are much larger in the noisy courters than in the silent sneakers.[3] Noisy courters have four times as many sonic muscle fibers as sneakers, and the diameter of each averages five times larger. There are other differences in these muscles that are evident only with the aid of the high magnification provided by electron microscopes. For example, mitochondria, the organelles that are the cells' engines, are far more numerous in muscle cells of the

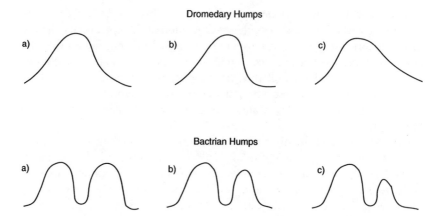

Figure 5.1 Some examples of dromedary and Bactrian humps

noisy courters.[4] There are also a number of neural differences in the two male types, especially with respect to the vocal-motor circuits.[5]

Although toadfish satisfy the discreteness criterion, other putative cases of alternative reproductive tactics do not. Consider treefrogs, for example. Most male treefrogs emit mating calls in order to attract females within range, but some do not, parasitizing instead the calls of other males. These parasites are sometimes referred to as waylayers. Callers and waylayers are otherwise physically and even behaviorally identical. Indeed, individual treefrogs can switch back and forth from calling and waylaying during the course of a single evening. But, more important, there is an important sense in which all treefrogs are behaving the same, whether they are calling or waylaying. All treefrog matings, including those of callers, involve waylaying. Both callers and waylayers seem to accost their mate in attempting to achieve amplexus. The most notable difference between callers and waylayers is that callers call and waylayers do not. But do a behavior and the absence of that behavior constitute alternative tactics? If they do, we could claim alternative tactics for any behavior we choose to define. We need to be careful not to include conceptual artifacts, such as a behavior and its absence (calling and noncalling) under the two-hump rubric. If we do not exclude conceptual artifacts, we are likely to find that the two-hump condition is nearly universal, a sure sign that something has gone wrong. The inflation danger is especially acute in those species, such as treefrogs, in which individuals switch back and forth between courting and not courting.[6] In such cases, it is especially important that the putative ART meet our second criterion, the fitness criterion.

Males of many species can be assigned to one of two categories: those that breed and those that don't. These are not alternative reproductive tactics, because not reproducing is not a reproductive tactic. This fact remains true even when the nonbreeders employ any one of a number of ruses and stratagems to gain access to females. Consider the bachelor male elephant seals discussed in chapter 4. They attempt to waylay females, especially in the surf, while they are out of sight of the beachmasters. They also try to sneak into the harems of beachmasters. But these unconventional mating behaviors should be considered reproductive tactics only to the extent that they succeed. In lieu of such evidence, we should consider the waylaying behavior of the subordinate male elephant seal males a mere byproduct of an unrequited libido rather than a product of natural selection for a waylaying tactic.[7]

But adaptationists have tended to describe unconventional reproductive behavior, such as waylaying in young elephant seals, as an alternative reproductive tactic, without worrying much about whether it succeeds. Such behavior, which is also often described as "making the best of a bad job," does sometimes qualify as an ART, but more often it does not. It qualifies as an ART only if (1) it is not part of a continuum of reproductive behavior and (2) those who engage in the activity have experienced selection to do so because they derive some fitness benefit.

If we are not careful in judging which cases of making the best of a bad job qualify as ART, the ART category tends to become very squishy and overly inclusive. We can be much more confident in applying this term to cases, such as that of the toadfish, in which the reproductive success of the unconventional males extends well beyond making the best of a bad job—those cases in which unconventional males hold their own in competition with conventional males.

With the discreteness and fitness criteria in mind, let us look a bit more closely at some other putative cases of alternative tactics.

Courting and Sneaking

One of the first well-documented cases of ART was discovered in the bluegill sunfish (*Lepomis macrochirus*), beloved of young fishers throughout much of North America. The bluegill mating system resembles that of the toadfish.[8] Males stake out territories in clumps and spend much time and energy fending off challengers and wooing females. Once the eggs have been laid, the male bluegill assumes all parental responsibility; his primary task is to protect the eggs from predators, most notably other bluegills. But, as in toadfish, some bluegill males are sneakers; they too are relatively small but have large gonads. Like their toadfish counterparts, they dispense with the territorial defense and parental duties entirely. Also, like the unconventional toadfish males, they lurk around the territories defended by the parental males, maintaining a low profile until the opportunity arises for them to dispense their sperm while the territorial male is engaged in the climactic phase of courtship. The bluegill mating system has an additional twist, however. The unconventional bluegill males exhibit two distinct behavioral patterns.[9] Immediately after maturation, when they are smallest, they fertilize eggs by sneaking. They continue to grow, however, and when they reach the size of mature females, they become female mimics and use this ruse to get near spawning territorial males and their female partners.

There are some other interesting variations on this basic theme. Most Pacific salmon (e.g., coho, *Onchorhyncus kisutch*) males mature after 3 to 5 years at sea and return to their native stream with a distinctive hook-shaped jaw called a kype, which they employ when they fight other males for access to females. Some males, however, mature precociously (after only one year at sea). They return to spawn at a much smaller size and sans kype.[10] Rather than fight for access to females, these "jacks" lurk under rocks and logs, awaiting their chance to dash into the middle of a spawning event, spew some sperm, then hastily depart. Some male Atlantic salmon (*Salmo salar*) do not even bother to go out to sea. Rather, these small males mature before the seaward migration and compete for fertilizations with the returning male migrants by sneaking and spewing sperm.[11]

Both bluegill and salmon meet the discreteness criterion: the salmon as a Bactrian (two-humper) and the bluegill as a swamel (three-humper). There are also both direct and indirect evidence that the unconventional males of these species compete successfully with conventional males for their share of the fitness pie.[12]

Alternative Tactics and Sex Change

In chapter 4, I briefly discussed the bluehead wrasse (*Thalassoma bifasciatum*), a female → male (protogynous) sex changer related to the cleaner wrasse. Sexual development in bluehead wrasse is actually more complicated than in cleaner wrasse, because blueheads exhibit two male types, referred to as initial phase and terminal phase males. Initial phase males mature as males without undergoing sex change. Terminal phase males mature as females and become males only as a result of sex change. Not surprisingly, the two male types behave quite differently when it comes to mating. Terminal phase males are much larger and more colorful than initial phase males; they aggressively defend territories and court females. The smaller initial phase males, in contrast, don't bother to defend a territory or court females; they attempt to fertilize females in mass spawnings or by sneaking.[13] Terminal phase bluehead wrasse correspond to the toadfish noisy courters, and the initial phase male wrasses correspond to the unconventional little sneakers.

The Caribbean stoplight parrotfish (*Sparisoma viride*) presents yet another wrinkle.[14] Like the bluehead wrasse, it is a female → male sex-changing species with two male types. In this case, however, all individuals mature first as females. The majority of females transform into males when they are relatively large. These are terminal phase, or courting, males. The sex change is accompanied by a dramatic color change from a cryptic female pattern to one with bright blue, green, and purple hues and the characteristic yellow spot on the gill cover for which the species is named. Some females, however, transform into males at a relatively small size, while retaining their female coloration. These female mimics associate with the true females in order to sneak spawnings. These female mimics later transform into conventional terminal phase males when they get larger.

Clearly, the males of both the bluehead wrasse and the stoplight parrotfish can be readily partitioned into two distinct male types; both species exhibit the Bactrian (two-hump) condition, not the single dromedary hump. The male types are physically, not just behaviorally, distinct, and the unconventional males are successful in fertilizing females, not just making the best of a bad job. Hence, we can be confident that the unconventional males represent an alternative reproductive tactic.

Calling versus Waylaying

Among amphibians, bullfrogs (*Rana catesbiana*), are better ART candidates than treefrogs. These creatures are familiar to most inhabitants of the temperate areas of North America, if only through their deep, bellowing mating calls. For many people, these sounds, more than any other, evoke the pleasant feeling of a warm summer night. For the bullfrogs, however, they evoke sex. The calls are strictly the provenance of the males and carry information about the size of the caller. The deeper the pitch of the call, the larger the male, and the more attractive the male to the females. Courtship in these animals is pretty much confined to these calls. Once a female has been lured into the vicinity of the calling male, he tries to mount her, clasp her with his front legs, and position his cloaca next to hers. This position is known as amplexus. Male bullfrogs are really rather indiscriminate as to whom or what they will mount—basically, any other bullfrog within a certain size range. If, as is often the case, it happens to be another male bullfrog, the mountee simply emits an unfeminine grunt, the mounter then dismounts, and both continue their quest for amplexus with more suitable partners. The large calling males will fight on occasion—a sort of wrestling match in which the goal seems to be to put the opponent on its back—but they reserve most of their energy for calling and, especially, for amplexus, which can last for hours and during which the male may have to fend off other males with his large hind legs.

Some smaller males skip all but the amplexus. Females would detect their small size as soon as they opened their mouth, so they keep quiet. They do not even try to stake out a calling territory, because if they did so, they would soon be pinned (probably after having first suffered the indignity of being mounted) by the larger males. Instead, these "satellite" males lurk at the fringes of the calling arenas and attempt to waylay females attracted to the calls of the larger territorial males.[15]

The smaller bullfrogs behave much like toadfish sneakers, but there are some important differences. Because of the way frogs mate, it is not possible to fertilize eggs by showering sperm in the vicinity of the mating couple. Hence, a true sneaking tactic is impossible. Moreover, in contrast to the sneaking toadfishes, it is mainly the younger individuals that adopt the unconventional tactic; the older (and larger) individuals tend to be the callers.[16]

The case for bullfrog alternative reproductive tactics, although better than that of tree frogs, is not as compelling as that of the fish species described. The putative tactics may well represent an age-dependent continuum of reproductive behavior and a single dromedary hump, rather than true bactrianism. Moreover, it is not clear that the waylaying of the small males counts as more than making the best of a bad situation, in which the best is not very good.

He-Males and She-Males

We now come to the reptiles, where one of the most curious ART candidates occurs in the garter snake, *Thamnophis sirtalis*. This species occurs throughout much of the western United States and Canada, wherever surface water is available. One subspecies, the red-sided garter snake (*T. sirtalis parietalis*) ranges from Manitoba to Texas.[17] Red-sided garter snakes in Manitoba (the northern limit of their range) emerge in spring from hibernation ready to mate. In fact, virtually all females are mated within 3 weeks of their awakening, many within an hour.[18]

Courtship in this species is fairly rudimentary—a chin rub and then it gets serious. Males scramble madly to entwine themselves with females and maneuver to get in a position to inseminate. Getting in position can be somewhat complicated for the male, because he might have to contend with dozens of other libidinous males for the privilege. The female is often completely obscured by the writhing mass of males. The invisibility of the female does not deter other males from joining the fray because, even if they cannot see her, they can smell her.[19] Not all of the female smells are emitted by females, however. A small number of males also give off female odors. These female mimics, called she-males, are thought to obtain an advantage over their competitors by distracting them from their target.[20]

Do the she-males represent a distinct unconventional reproductive tactic? The fact that they give off an unusual (for a male) odor certainly suggests that they might indeed constitute a separate hump from that of the so-

called he-males. Whether this stratagem is successful remains to be established. There would seem to be a distinct disadvantage to any male finding
itself the center of attention of so many other ardent males. If you are a
male garter snake entwined by other males, how are you supposed to mate?
This sort of "transvestitism" would seem to have an obvious cost.[21]

A more straightforward reptilian ART is exhibited by the tree lizard
(*Urosaurus ornatus*), a species in which there are two distinct male types
that differ both physically and behaviorally.[22] The territorial males are aggressive and possess an orange dewlap with a central blue spot. Other males
are nomadic and less aggressive, and they lack the central blue spot. These
nomadic males seem to practice a sneaking tactic.[23]

Things are a bit more complicated in the side-blotched lizard (*Uta stansburiana*), which evidences the swamel (three-hump) condition. Some males
have an orange throat and defend large territories; some males have a dark
blue throat and defend smaller territories; and still other males have a yellow stripe on their throat and do not defend a territory at all—they are
sneakers.[24] Each morph is competitively superior to one morph but inferior
to the other, an arrangement that results in a rock-paper-scissors dynamic.
Blue-throated males successfully guard their females against the depredations of the yellow-throated sneakers, but their mate-guarding is ineffective
against the ultradominant orange-throated males, who overpower the blue-
throats. But the orange-throats themselves are vulnerable to the yellow-
throated sneakers. Interestingly, many of the yellow-throated offspring are
sired posthumously, suggesting that their sperm is competitively superior to
that of the other morphs.

Both tree lizards and side-blotched lizards meet the discreteness criterion
(as Bactrians and swamels, respectively). Evidence as to the fitness criterion
is particularly compelling in the side-blotched lizards, in which there are
predictable fluctuations in hump sizes.[25]

Leaping Lekkers

There are very few examples of ART among male birds, but a good case
exists in the European ruff (*Philomachus pugnax*), an atypical member of the
sandpiper family (Scolopacidae).[26] A glance through a field guide to the
birds of North America or Europe will reveal that sexual dimorphism in
most members of this family ranges from nil to moderate. Ruffs, however,
are extremely dimorphic. The males—and males alone—possess a pronounced crest and a ring of variably colored, long feathers around their
neck, from which the species gets its common name.

Male ruffs gather at traditional sites known as leks to strut and pose for
females in rituals that resemble human body-building contests. Their long
neck feathers are accentuated during their bent-legged leaps into the air,

when, on the way down, they flare out like a collar. The females make their choice on the basis of the impression the males make during this spectacle. Other males take a different approach, the satellite tactic. The satellite males have lighter and much less flamboyant plumage than the leaping lekkers. These satellite males intercept females on their way to the center of the lek, where the highest quality males reside.

Ruffs fare well as practitioners of ART according to our two criteria. The leaping lekkers and the satellites occupy two distinct humps, not one. There is good evidence, in fact, that alleles at a single genetic locus determine whether a male will be a leaping lekker or a satellite.[27] Moreover, there is good evidence that the fitness criterion is also met. The size of the two humps is kept relatively constant because the greater the proportion of the population practicing either the satellite or the lekking tactic, the lower the fitness of that tactic, a phenomenon known as frequency-dependent selection.[28]

Of Mice and Men

Well-documented alternative reproductive tactics of the sort found in toadfish are rare in mammals, though some putative cases have been proposed, especially in rodents.[29] These proposals are generally not based on field data; rather, they have emerged from adaptationist interpretations of a widely reported phenomenon in laboratory animals: the in utero effects of the hormones produced by a fetus of one sex on littermates of the opposite sex.[30] There are a number of reports that fetuses of both sexes are masculinized or feminized according to their position in the uterus. A male fetus that finds itself positioned between two females can become feminized to some degree, and a female fetus positioned between two males can become masculinized.[31] The most pronounced effects are behavioral. Females so masculinized are more aggressive and less sexually receptive than typical females.[32] Males who have imbibed their sisters' hormones, on the other hand, are said to be less aggressive. Differences in other aspects of sexual development have also been noted.[33] These differences have led some observers to speculate that the feminized males and masculinized females constitute alternative reproductive tactics.

Let us now apply the discreteness and fitness criteria to these putative cases of alternative tactics in rodents. To date, there is little evidence that any rodents are two-humpers The fact that some males are feminized through exposure to their female littermates, relative to males not so exposed, should not lead us to conclude we are dealing with a two-hump phenomenon rather than a one-humper. We need, in addition, evidence of a behavioral discontinuity. Actually, given this proposed mechanism, we should have a three-hump, or swamel, condition, because most males will be positioned between one male and one female. At this point, there is no reason to assume that male sexual behavior, even in rodents so affected, is not continuous.

Now consider the fitness criterion. We must demonstrate that the uncon-ventional males are successful at what they do. Assume for the moment that these rodent populations are three-humpers (swamels), consisting of femi-nized males, typical males, and hypermasculinized males. How would we demonstrate that the three humps are the result of selection for three-humpedness? We would first need to establish that the sizes of the humps are stable or fluctuate in a predictable manner as in the rock-paper-scissors side-blotched lizards. But we must, in addition, demonstrate that the humps are stable because they are being maintained by selection. The three-humped swamel condition in rodents could very well result simply from the fact that, given a 1:1 sex ratio, and a circular arrangement of fetuses in the uterus, 25 percent of the males will develop between two females, 25 percent between two males, and 50 percent between a male and a female.[34] Hence, we should expect to get 50 percent "typical" males, 25 percent "feminized" males, and 25 percent "hypermasculinized" males, no matter their respective reproduc-tive prospects. At this point, there is no compelling evidence that the in utero influences of sex steroids result in swamels, much less stable swamels, much less stable swamels maintained through selection. Hence, there is no evidence of true alternative reproductive tactic among these rodents.

A much better example of a mammalian ART, perhaps the best, occurs in the orangutan (*Pongo pygmaeus*). Typical adult males are twice the size of females, develop wide cheek pads called flanges, as well as a large throat sac, and emit loud vocalizations called long calls by means of which they adver-tise their virility to females in the vicinity. Other sexually mature males, however, are much smaller, about the size of females, and lack the flanges and throat sacs.[35] These smaller unflanged males do not emit the loud calls; indeed they don't attempt to court females at all but rather physically force unwilling females to copulate, a behavior that looks very much like rape.[36] This behavior occurs both in the wild and under captive conditions.

The unflanged males seem to exhibit arrested development of a sort— arrested with respect to secondary sexual characteristics (body size, flanges, and throat sacs) but not arrested with respect to puberty and other aspects of sexual development. The cause of their arrested development appears to be the presence of larger flanged males.[37] Unflanged males may remain in this arrested state for up to 20 years after reaching sexual maturity but ulti-mately develop into large flanged males. Most important for our purposes, the unflanged males successfully reproduce while in this arrested state, so they meet both our discreteness and fitness criteria.

The Adaptationist Framework

Once we use the discreteness and fitness criteria to separate the wheat from the chaff, we notice a distinctive taxonomic pattern: alternative tactics are common in (teleost) fishes, less common in lizards and snakes, and rare in birds and mammals. We need to keep this taxonomic pattern in mind in evaluating the adaptationist approach to explaining such tactics.

Game Theory

In chapter 4, I discussed a truly excellent adaptationist theory of sex change, the size-advantage model. There is nothing remotely as compelling as the size-advantage model to frame adaptationist thinking about alternative reproductive tactics. The overarching theoretical framework though, is game theory, which was first formalized by von Neumann and Morgenstern in order to explain human economic behavior.[38] The central assumption of classical game theory is that the "players" will behave rationally by some criterion of self-interest. Given this criterion and a specified set of tactics available to the players, the best "solution" to a game, that is, the best tactic, can be determined in a straightforward manner.

John Maynard-Smith was particularly instrumental in extending the game theory framework to evolutionary biology.[39] He did so by replacing the rationality criterion with population stability or equilibrium and by replacing the self-interest criterion with Darwinian fitness. He referred to the "solution" for these evolutionary games as the evolutionarily stable strategy, or ESS. The central advance of evolutionary game theory, over and above the optimality framework also favored by adaptationists, is that the ESS will depend on what other members of the population are doing.

Evolutionary game theory was originally applied to the analysis of animal conflicts.[40] But Maynard-Smith and other game theorists quickly came to apply the framework to a much broader array of phenomena, including alternative reproductive tactics. Indeed, the term *alternative tactic* is itself a byproduct of the sort of strategic thinking that game theory engenders; so this phenomenon would seem ideal for analysis within this framework.

Let us consider again our toadfish. Recall that toadfish either court or sneak. A toadfish that, from birth, is capable only of courting exhibits what is known as a "pure" strategy. The same goes for a toadfish that only sneaks. A toadfish that is potentially capable of both exhibits a "mixed" strategy. We would expect garden-variety sexual selection to promote the conventional courting behavior. The fact that sneaking exists, however, indicates that the pure courting tactic is not the ESS. It is only under conditions such as this, when no pure tactic is an ESS, that alternative tactics evolve.

Given a baseline condition of pure conventional behavior, there are two distinct routes to alternative reproductive tactics in males. One route involves a mutation that results in a pure sneaking tactic. If the pure sneaking tactic were always better than the pure courting tactic, sneaking would be the ESS and we would still not have alternative tactics. Similarly, if sneaking were always inferior to the pure calling tactic, we would not get alternative tactics. But a stable mixture of pure courters and pure sneakers would constitute alternative mating tactics.

The second route to alternative tactics from the baseline condition of pure courters requires a different sort of mutation. This time the mutation results in a toadfish that is capable of both courting and sneaking, a mixed tactic. Alternative tactics will result from this mutation if, and only if, this mixed tactic is the ESS in a contest against pure courters. In determining whether a mixed strategy is the ESS in such evolutionary games, we would, of course, need to know something about how the mixed strategist allocates his tactics. We would not expect a mixed strategist that courts or sneaks at random to be as effective as one that, say, courts if it is relatively large and sneaks if it is relatively small. The latter sort of mixed strategy is said to be "conditional." Most examples of alternative tactics evolved when mixed strategists with conditional "decision rules" outcompeted pure (conventional) strategists.

Conditional mixed tactics, which I will henceforth refer to simply as conditional tactics, are not confined to species in which individuals can switch back and forth from conventional to unconventional tactics. Indeed, the best examples of conditional ARTs are found in species, such as the toadfish, in which individuals exhibit a conventional or unconventional tactic throughout all or a good portion of their lifetimes. In such cases, the decision-rule is applied early in development, before sexual maturity, and the "choice" is irreversible. This seems to be the case in toadfish.[41] So, even though a toadfish will be either a courter or a sneaker throughout its adult life, courting and sneaking are not pure tactics, because at some point in the development of the toadfish, it could have gone either way.

There remain, however, some deep ambiguities in the application of game theoretic categories such as pure and mixed strategies in an evolutionary context. For example, the basic evolutionary game theory models, because they treat the ESS as a population attribute, cannot distinguish a population consisting entirely of individuals practicing a mixed strategy that results in, say, courting 70 percent of the time and sneaking 30 percent of the time, from a population consisting of a mixture of 70 percent pure courters and 30 percent pure sneakers. Which is to say, these models cannot distinguish between the two distinct pathways to alternative tactics that I described earlier. Adaptationists have sought to rectify the problem by making strategies either a property of individual organisms or a property of hypothetical genes, but they have done so inconsistently. This inconsistency has generated much confusion in game theoretic treatments of alternative mating tactics. Recently, Mart Gross, who is something of a pioneer in this area, has sought to clarify things, though unsuccessfully in my view.[42]

Gross did, however, make explicit the selection regime under which conditional tactics, such as those displayed by the toadfish, will evolve. He also provided useful guidelines, based on a fitness criterion, for deciding whether unconventional reproductive behavior represents a true alternative tactic. Before I discuss the fitness criterion for conditional tactics such as are found in the toadfish, it will help to first describe the fitness criterion for those alternative tactics that consist of two pure tactics.

The ruff will serve as our exemplar here. Recall that whether an individual becomes a leaping lekker or a satellite depends on a genetic polymorphism maintained by frequency-dependent selection. As such, lekking and waylaying represent pure, not mixed (conditional), tactics. At their equilibrium frequencies (hump sizes), satellites and lekkers have equal average fitness.

Things get trickier when we consider (mixed) conditional tactics. Alternative conditional tactics may or may not be maintained by frequency-dependent selection. But all true alternative conditional tactics, such as those of toadfish, do exhibit what Gross calls "status-dependent selection." That is, the "decision" as to whether to become a courter or a sneaker is based on some aspect of an individual's status at some point in its development. *Status* could refer to any of a number of traits for which an individual could be assigned a rank relative to other members of the population. We therefore need to identify the conditional status cue. Say the cue is relative size as a juvenile, which is often the case. We then need to identify the switch-point for the relative size status cue, above which the individual should be a conventional male (i.e., court) and below which it should be an unconventional male (i.e., sneak). The average fitnesses of the alternative tactics, such as courting and sneaking, *are not* equal, but the fitnesses of these two tactics at the switch-point *are* equal. This situation is formally identical to the size-advantage model for sex change discussed in chapter 4. But for sex change, it is easy to identify the switch-point and when, during the individual's development, the decision is made. For conditional mating tactics, however, it is often much more difficult to identify the switch-point, because the decision is often made long before the behavior becomes manifest. Usually, the switch-point can only be inferred indirectly if at all.

In lieu of evidence of an equal fitness switch-point, we must rely on more indirect and circumstantial evidence in inferring conditional tactics. The least convincing sort of circumstantial evidence is the mere existence of an unconventional behavior such as waylaying in bullfrogs or elephant seals. As we have seen, waylaying could well be the unselected byproduct of an unrequited libido in smaller males. We should be especially suspicious of putative alternative tactics such as waylaying in bullfrogs and elephant seals, which are confined to younger and weaker males, unless we have compelling evidence that practitioners of the unconventional tactic are successful in inseminating females. Unfortunately, however, some adaptationists, including those who study elephant seals, have been willing to credit species with alternative mating tactics given only the slightest evidence of successful unconventional behavior, from sneaking to rape.[43] How do we avoid making this category too inclusive?

Pure versus Conditional Tactics

Without direct evidence of an equal fitness switchpoint, it is all too easy to inflate the category of conditional alternative reproductive tactics through paranoiac storytelling. One way to avoid this fate is to use pure tactics as something of a standard in evaluating putative conditional tactics, in light of our two ART criteria. Consider first the discreteness criterion. For alternative pure tactics such as those of the ruff, all individuals, not just the tactics themselves, can be assigned to one of two (or more) humps, and they cannot change humps. For putative conditional tactics such as those of treefrogs, on the other hand, each individual easily moves from one hump to the other; it is only the tactics themselves that are considered discrete, in the sense that they can be assigned exclusively to one of two humps. Sometimes it is perfectly legitimate to infer conditional tactics in treefrog-like cases with much switching back and forth, but only when there is good evidence of an equal fitness switch-point. That inference requires that we identify the status cue that the frogs themselves use in assessing their rank in the population at any given point in time and then demonstrate that they switch behavior when their rank drops below the switch-point. Without such evidence, we tend to get conceptual artifacts masquerading as conditional tactics.

Such is not the case for toadfish, however. Even without direct evidence of an equal fitness switch-point we can be confident that toadfish males exhibit conditional tactics, because these tactics much more closely approximate the pure tactics of ruffs with respect to the discreteness criterion. Each individual, not just the tactics, can be assigned to one of two humps. Conditional tactics such as those of toadfish are much less likely to be the artifacts of an adaptationist behavioral taxonomy than those of treefrogs, because the alternative tactics of male toadfish involve much greater behavioral and physical differences than those of treefrogs.

Now consider the fitness criterion. Here again, our confidence that putative conditional tactics reflect selection increases as these tactics approach the standard set by alternative pure tactics. Waylaying ruffs have higher fitness when they are relatively rare and increasingly lower fitness as they become more common within a population. This frequency-dependent selection eventually results in a stable equilibrium consisting of a certain proportion of lekkers and a certain proportion of satellites; this proportion corresponds to the respective hump sizes. To date, there have been few direct investigations of the fitness of conventional and unconventional males practicing conditional tactics. We should, however, look for evidence of frequency dependence in order to avoid inflating this category.

The conditional tactics of toadfish stand up well to scrutiny according to both the fitness and discreteness criteria. There can be no question that there is a phenomenon here that requires explanation, and one that game theory may help illuminate. But when we focus on toadfish-like conditional tactics, a limitation in this why-biology approach becomes apparent. It is to this limitation that I now turn.

The How-Biology of Alternative Tactics

Game theory helped illuminate the adaptive rationale for alternative reproductive tactics, the selective milieu under which we would expect them to evolve. But it does not seem to help us understand the distribution of conditional tactics on the tree of life. Why, for example are ARTs so common in fishes but so rare in mammals? We are in a situation very much resembling that described for sex change in chapter 4. And again, we need to turn to the historically contingent how-biology in order to understand these taxonomic differences. I suggest that we begin with the most generic considerations then move to increasingly specific factors.

The first thing to note is that in most fishes (and amphibians) fertilization takes place outside of the body, whereas in reptiles, birds, and amphibians, fertilization is internal. This fact alone undoubtedly goes a long way toward explaining the dearth of alternative tactics in mammals: internal fertilization precludes unconventional tactics such as sneaking that are so common in fishes.[43] This cannot be the whole story, however, because some fishes with internal fertilization, such as platys and swordtails (family Poeciliidae), have nonetheless evolved alternative tactics.[44]

Another generic difference between fishes and mammals is their pattern of growth. Mammalian growth is much more constrained and predetermined than that of fishes, which are often said to exhibit "indeterminate growth"; that is, they continue to grow throughout their lifetime. Indeterminate growth results in much greater size variation, among sexually mature males, in fishes than in mammals. Such size variation is often the engine for the evolution of alternative male tactics. Again, however, this cannot be the whole story. Platys and swordtails not only fertilize internally but exhibit determinate growth as well.

Although both external fertilization and indeterminate growth are probably key factors in understanding why ARTs are so common in fishes, there must be other factors at work here. I suggest that one such factor may again be their labile sexual development. Let's see how this lability might be related to the evolution of conditional tactics.

Sexual Developmental and the Evolution of Conditional Tactics

Conditional alternative tactics of the sort found in toadfish require that each individual potentially have access not only to two distinct behavioral states but also to two distinct anatomical and physiological states. Each individual must be born with the potential to go either way, tactic-wise, depending on its condition relative to other males. For any alternative tactics approaching toadfish-like sophistication, the "choice" must be made sometime before maturity. It is a developmental choice, a choice between two disparate developmental trajectories. So, in seeking to understand the evolution of conditional tactics of this sort, we need to examine the relevant developmental mechanisms. The fact that we are dealing with sexual behavior suggests that we attend in particular to mechanisms of sexual development in explaining the paucity of conditional alternative tactics in mammals.

We begin with the most general properties that might distinguish sexual development of mammals from that of teleost fishes. I discussed one such general property in chapter 4: the relative lability of sexual development in teleost fishes. Mammals have the most canalized sexual development among vertebrates. Both the advantages and the disadvantages of highly canalized sexual development stem from the fact that it is not easily perturbed by either genetic or environmental alterations. This stability is advantageous because, for example, variable environmental conditions are not likely to screw things up. (For example, environmental pollutants that feminize male fishes do not affect male mammals in this way.) It is disadvantageous because it makes adaptive modifications unlikely as well. Animals with canalized sexual development sacrifice adaptability for stability. This sacrifice was made early on during the course of mammalian evolution, and there is no going back.

We now need to identify the component processes in mammalian sexual development that might resist the sort of adaptive modifications required for the evolution of alternative tactics. A good place to begin is the hypothalamic-pituitary-gonadal (HPG) axis: gonadotropin-releasing hormone (GnRH), which is manufactured in the hypothalamus, causes the release of gonadotropins from the pituitary, which in turn, stimulate gonadal growth and the release of steroid hormones such as testosterone and estrogen, which then regulate the further release of GnRH (figure 5.2). To understand why toadfish-like conditional tactics are absent in mammals, it will be

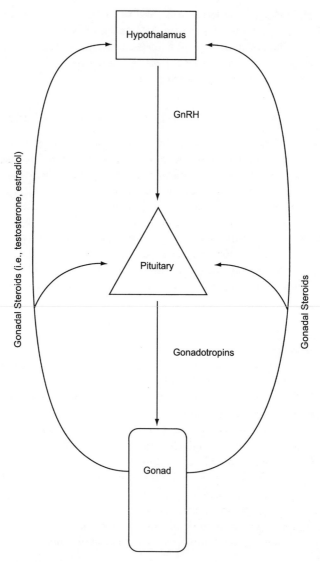

Figure 5.2 The hypothalamic-pituitary-gonadal (HPG) axis

informative to look for taxonomic differences in the development of this axis.

Michael Moore, whose research on tree lizards was described earlier, was among the first to explore the developmental basis of alternative tactics. He focused on the gonadal component of the HPG axis, specifically, the role of gonadal steroid hormones.[45] But here I want to focus on the hypothalamic component of the HPG axis, because it is farther "upstream."

Sexual Maturation and Alternative Reproductive Tactics

Most of the cases for which we can be confident that unconventional male behavior is actually adaptive and not just the expression of unrequited libido involve accelerated sexual maturation—the transition from a juvenile state to adulthood or reproductive competence. Maturation in all vertebrates is triggered by a marked increase in levels of GnRH in the hypothalamus.[46] The crucial role of GnRH expression in maturation is most apparent when GnRH expression is thwarted. The human malady known as Kallmann's syndrome is caused by abnormally low levels of this hormone. As a consequence, there are low levels of gonadotropins, and the gonads never develop beyond a very inchoate state (hypogonadism). Hence, sexual maturity is never reached, and there is no development of secondary sexual characteristics such as facial hair and a deepened voice in males.[47] A more widely studied anomaly in GnRH expression occurs in the "hypogonadal mouse," a model system for the study of neuroendocrine processes. Like those human males with Kallmann's syndrome, these mice have extremely small and undifferentiated gonads. In this case, however, the problem can be traced to a mutation in the GnRH gene itself.[48] Normal function can be restored to these mice through a form of gene therapy, in which normal GnRH genes are introduced early in development.[49] There are other ways to increase GnRH levels in hypogonadal mice without resorting to gene therapy. The most popular is to graft hypothalamic tissue from normal mice into the hypothalamus of hypogonadal mice.[50] When grafting is done properly, GnRH neurons from the transplanted tissue will send processes to the pituitary and cause the release of gonadotropins, which will, in turn, stimulate gonadal development.

The above research, based on anomalies in sexual development, demonstrates the crucial role of GnRH in sexual maturation. Now consider a case of natural variation in GnRH expression and how that variation is related to sexual maturation.

GnRH Expression and Alternative Male Types

Members of the freshwater fish family Poeciliidae have long been used in genetic studies.[51] One species, in particular, the platyfish (Xiphophorus maculatus), has been subjected to intense genetic scrutiny. Platyfish have long been bred in the tropical fish industry for their coloration. The same factors that make them amenable to artificial selection—they are easily bred and

have a short generation time—have made them ideal candidates for classical Mendelian genetics analyses.

Male platyfish mature at highly variable rates, some as early as 5 weeks,
others as late as 2 years.[52] Once a male has matured, his growth rate is
dramatically reduced, so the early-maturing males are smaller than the later-
maturing males. The later the male matures, the larger it becomes. In this
respect, they resemble the toadfish, salmon, and many other fishes. Their
mating system, however, is completely different. Whereas most toadfish
males attempt to lure females to their carefully prepared and actively defended conjugal sites, reproduction in platyfish is a mad scramble, with several males harassing the larger female, no matter, seemingly, her degree of
receptivity. In this scramble, they deploy their penis analog, called a
gonopodium, a remarkably maneuverable projection that is a modification of
a typical pelvic fin. The larger fish display earnestly, with their body contorted in various ways in order to best emphasize their colorful fins. Given a
choice, the females prefer these large males. But the smaller males, which
may have matured as much as 22 months earlier, have had more time to
compete for mates, compensating, to some degree, for their size handicap.

The age at which a male platyfish matures depends largely on alleles at a
single genetic locus.[53] This locus was named the P locus because it was
believed to influence the maturation of certain cells in the pituitary that
make gonadotropins, the hormones that directly regulate gonadal function.
More recent evidence indicates that the P locus alleles act a bit farther
"upstream," influencing the timing of GnRH expression in the hypothalamus. It is these hypothalamic GnRH neurons that project to the pituitary
and thereby regulate the output of gonadotropins, which, in turn regulate gonadal steroid hormones.[54] The early-maturing males mature earlier
than the late-maturing males because GnRH is expressed earlier in their
development.

In fishes, there is typically an increase in the number or size, or both, of
hypothalamic GnRH neurons with increased body size. Larger (and usually
older) individuals have more or larger GnRH neurons than smaller individuals. Toadfish generally conform to this pattern, except that silent sneaker
males have more GnRH neurons than they should, given their small body
size. That is, the rate of addition of GnRH neurons with body size is higher
in the silent sneakers than in the noisy courters.[55] As such, a small silent
sneaker will have the same number of GnRH neurons as a much larger
noisy courter.[56] This fact would explain how the unconventional sneakers
can mature earlier and at a smaller size than their conventional counterparts.

But what is the mechanism underlying the accelerated rate with which
hypothalamic GnRH neurons are added in the sneakers relative to the courters? There are two ways to recruit GnRH neurons: (1) by activating the

GnRH gene in neurons already in place or (2) through the migration of GnRH-expressing neurons from elsewhere in the brain. Evidence indicates that we should focus on migration.[57] Study after study attests to the dynamic nature of neuroanatomy at the cellular level. Within the embryonic vertebrate brain there are two types of germinal zones from which new neurons originate. Most neurons are born in areas adjacent to the ventricles. Some neurons, however, originate in placodes, which are embryonic structures outside the brain. Neurons of placodal origin tend to move long distances from their place of birth, and neuroscientists are only beginning to understand how this movement is accomplished. Those GnRH neurons found in the hypothalamus originate in the olfactory placode, which lies just outside of the extreme anterior portion of the forebrain.[58] This is a truly epic migration by neuronal standards, involving complex cell-to-cell communications and coordination. A failure in this process has dire consequences for sexual development in all vertebrates. Kallmann's syndrome itself may result from just such a failure, as indicated by one seemingly anomalous symptom: afflicted individuals often lack a sense of smell. This association begins to make sense if we consider that the GnRH neurons begin their migration along the olfactory nerve. Hence, a defect in that structure could also throw a monkey wrench into the mechanism for GnRH neuronal migration.[59]

The basic mechanisms involved in this neuronal migration are probably conserved in all vertebrates, but taxonomic differences in the timing and duration of this migration may be a key developmental reason that fishes are more likely than mammals to exhibit alternative tactics. In mammals (e.g., laboratory mice), where this process has been especially well characterized, GnRH neurons appear early in development (by day 9) and immediately begin their trek along the ventral forebrain to the hypothalamus. By day 14 they have reached the terminal nerve in the telencephalon, where some remain (figure 5.3); but most press on to their final destination in the hypothalamus, which they reach by day 22.[60] In teleost fishes, the process seems to be much more protracted.

The early onset and compressed nature of this migration in mammals are in keeping with their canalized sexual development. Because this process is completed so early in development, often in the womb, the mammalian HPG axis is essentially already prepared for maturation long before maturation will occur. It awaits only the appropriate signal to commence and quickly complete the maturation process. Within-species differences in the timing of mammalian maturation are simply a function of these signals, not the result of individual differences in the development of the HPG axis itself. This early migration severely limits opportunities to evolve adaptive alternative maturation trajectories and, hence, toadfish-like alternative tactics. For example, because the migration of GnRH neurons is completed well before maturation, it cannot be further accelerated to enable the evolu-

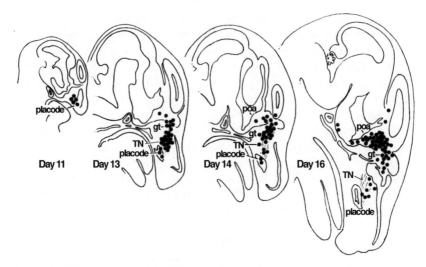

Figure 5.3 The migration of GnRH neurons from the olfactory placode in the
mouse: day 11 to day 16
(TN = terminal nerve; gt = gonadotropin; poa = preoptic area)
(Adapted from Schwanzel-Fukuda and Pfaff 1989, p. 162)

tion of a developmental trajectory such as that found in unconventional
toadfish males. Even in orangutans, the best case for conditional tactics
among mammals, the unconventional males fail to mature early. Orangutans
do demonstrate that early maturation is not essential for the evolution of
ARTs, but the very rarity of orangutan-like ARTs among other mammals,
and indeed other vertebrates, suggests that theirs is a more difficult developmental pathway to that end.

In contrast, the protracted migration of GnRH neurons characteristic of
teleost fishes provides ample opportunity for natural selection to adaptively
alter its rate or to generate individuals for which two distinct migration
rates—and hence developmental trajectories—are (conditionally) available
in their development.

This discussion of neuronal migration of GnRH neurons is intended only
to illustrate how how-biology considerations might figure in explaining the
evolution of alternative reproductive tactics. At this point, it is not possible
to effectively judge the importance of the differences in the pattern of migration of GnRH neurons as an explanation for the distribution of alternative reproductive tactics in the tree of life. But, without this sort of how-
biology research, the taxonomic differences in the frequency of ARTs will
remain a mystery.

Alternative Tactics and Historical Contingency

How should we conceive of the relationship between the why-biology explanation for alternative reproductive tactics provided by game theory and the how-biology/historical account? With respect to the particular case of the toadfish, the how-biology complements the why-biology explanation of their ARTs. At a more general level, however, when we consider the occurrence of ARTs among vertebrates as a whole, we have a situation very much resembling that described in chapter 4 concerning sex change. At this more general level, the complementary relationship breaks down. It is not the case that the how-biology explanation, conceived as proximate, merely augments the why-biology explanation, conceived as ultimate. Nor is this a case of replacement. Rather, we again find ourselves somewhere between these two poles. The how-biology explanation, when combined with a historical or genealogical analysis, competes with that of game theory as an explanation for the occurrence of ARTs in vertebrates. In any case, our understanding of the evolution of alternative reproductive tactics is enhanced by a consideration of the how-biology.

One of the attractions of the purely game theoretic approach in evolutionary biology is its apparent generality. It can be applied uniformly to all organisms, from ferns to primates. The generality of a schematic why-biology framework such as game theory is, however, largely a function of its abstractness, not its actual explanatory power. And, as we have seen, the generality of this approach is only apparent; it does not apply equally to fishes and mammals, much less to ferns and primates. It does not apply equally across taxa because of historically contingent differences in the how-biology of these animals.

For game theory enthusiasts, the account provided here will seem disappointingly particularistic. They seek general rules to which all organisms must conform and which transcend their how-biology. But this is a fool's errand. There is no transcending this how-biology, this evolutionary history. In this chapter, I have provided a very preliminary account of what the relevant material processes might be with respect to the evolution of alternative tactics. I have noted a nested suite of how-biology considerations worth attending to, culminating in the migration of GnRH neurons.

An ardent adaptationist will not be impressed. There is, after all, more than one way to skin a cat. For instance, there may be other ways of adaptively modifying sexual development in vertebrates that do not involve GnRH; if not GnRH then something else. The proper reply to this claim is that we certainly do need to learn more about sexual development before placing too much emphasis on the role of variant patterns of GnRH expression in the evolution of alternative reproductive tactics. But considerations of this sort are more likely than ecology to explain the failure of those unrequited male elephant seals, bellowing their frustration on the beaches of Ana Nuevo, to demonstrate effective alternative tactics for inseminating females, just as such considerations are more likely to explain their failure to be females, or to reproduce asexually, for that matter. There may indeed be more than one way to skin a cat; and, to use a less grisly metaphor, there is also more than one way to build a barn; but once the scaffolding is up, the options are limited. And once the hay is inside, it is too late to redesign the roof.

6

Social Inhibitions

EAST AFRICA's rift lakes are so called because they are located along a seam
in the continent running north to south, which is the result of the rifting,
or separating, of two parts of Africa, the same tectonic forces that dismem-
bered Gondwanaland. The largest of the rift lakes is Lake Tanganyika, second only to Russia's Lake Baikal in volume. It is also extremely deep. Below
about 200 feet it is anoxic and lifeless, but the shallower water teems with
life, especially near the rocky shores. The warm water is quite clear; visi-
bility can extend beyond 100 feet. The clarity of the water, combined with
the profusion of fish life, provides something like the experience of a coral
reef environment for snorkelers and scuba divers.

The majority of fishes a snorkeler will encounter belong to a single family
and are known as cichlids (pronounced *SICK lids*). Cichlid fishes are celebrated among aquarists for their beauty and their fascinating social behavior.
Among evolutionary biologists they are celebrated as the paragon of what is
often referred to as an "adaptive radiation."[1] Within a geological eyeblink,
cichlids from several of East Africa's large rift lakes have speciated to a
degree unparalleled among vertebrates. For example, all of the approximately 3,000 species inhabiting Lake Victoria evolved within the last 12,000
years, which is almost fast enough for "New Earth Creationists."[2]

An even larger number of cichlid species have evolved in Lake Tanganyika, among them one that has been particularly well studied by ethologists, though it is known only by its scientific name, *Haplochromis burtoni*.[3]
Haplochromis burtoni belongs to a subgroup of cichlids known as the mouth-
brooders, because of the unconventional manner in which they protect their
young. The eggs, and then the fry, reside in their mother's mouth until they
can swim unencumbered by their yolk sacs. More unconventional still is the
manner in which the eggs are fertilized.

Male *H. burtoni* perform a fairly elaborate courtship, in which they entice
females into their breeding territory. If a male is successful in wooing a
female into his inner sanctum, his courtship intensifies into a series of rapid
body quivers, which gets the female into an egg-laying mood. Using only
her pectoral fins, she begins to slowly rub her belly against the substrate.
After a few passes, she releases some eggs, which she then proceeds to

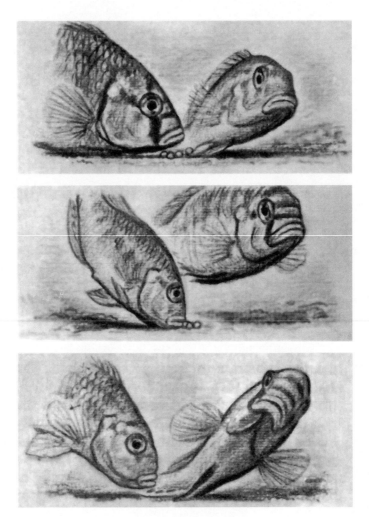

Figure 6.1 Oral conception in *Haplochromis burtoni*
(Adapted from Wickler 1968, pp. 224–25)

gather into her mouth. Now comes the interesting part. Males are equipped
with several orange spots on their anal fin, each of which resembles an *H.
burtoni* egg.[4] In the process of picking up her just-laid eggs, the female tries
to grab the "dummy eggs" on the male's anal fin. Just as she does so, the
male releases his sperm into her mouth. In this way are *H. burtoni* conceived
(figure 6.1).

   Haplochromis burtoni's unconventional manner of conception indirectly
promotes intense sexual competition among the males. In many cichlid spe-

cies, such as the Midas cichlids discussed previously, both parents provide care and protection for their young, but not mouth-brooders. Once the male has ejaculated into a female's mouth, he is off to woo another one. The female alone is left to care for their offspring. Unencumbered by parental responsibility, the males are free to pursue as many matings as possible. But a male's life is not as carefree as it may first appear. Aside from courtship, males must devote considerable time and energy to procuring and defending a breeding territory. And desirable territories, the ones a discriminating female would deign to enter, are in very short supply. At any point during the breeding season, only a relative few members of the male population are able to defend these desirable breeding territories. I will refer to them as territorial males. The rest of the males, which lurk in the interstices of territories, will be referred to as nonterritorial males.[5]

The territorial males and the nonterritorial males look quite different: Territorial males have a prominent black eye stripe, red and orange splashes on their flanks, and prominent egg spots on their anal fins; nonterritorial males lack all of these and generally resemble the rather drab, sandy-colored females. Superficially, the territorial and nonterritorial males may seem to represent alternative reproductive tactics. But the nonterritorial males are not the equivalent of toadfish sneakers, or even of bullfrog satellites. They are, rather, as unrequited as bachelor elephant seals; and, as I discussed in chapter 5, not mating is not an alternative mating tactic.

The nonterritorial males can breed only if they displace a territorial male or fill a vacancy after a territorial male has been removed by a predator such as a kingfisher. The territorial males are easy prey for these birds, especially when they are contesting boundaries with other territorial males or displaying for females. Many territorial males meet their doom while their attention is focused on sex, their eye stripes darkened in anticipation.[6] As is so often the case, one male's misfortune is another male's boon. Once a vacancy has been created, several nonterritorial males immediately try to fill it. The winner quickly begins a dramatic transformation. What was once a skulking, drab-looking nonterritorial male metamorphoses into a strikingly colored and pugnacious wooer of females—a territorial male.

There is ample evidence from laboratory studies that the territorial males suppress reproduction in the nonterritorial males through aggressive behavior.[7] This suppression is obviously beneficial for the territorial males but seemingly bad for the nonterritorial males. Hans Hofmann, Mark Benson, and Russell Fernald have proposed, however, that nonterritorial males also benefit from this state of affairs and that their social suppression is therefore adaptive. More specifically, they reason that the social suppression of reproduction in nonterritorial males by territorial males is an adaptation that promotes the most efficient allocation of reproductive resources.[8] It is not only dangerous to be a territorial male; it is also energetically costly, because the reproductive system takes up valuable resources that could be devoted to other physiological systems and growth. By remaining reproductively quiescent until he can defend a breeding territory, the H. burtoni male does not waste these resources. Though the nonterritorial state is not an alternative mating tactic on this view, it is a conditional alternative reproductive tactic in some extended sense of that term.

We will examine this and related adaptationist hypotheses concerning the social suppression of reproduction in this chapter. This will lead us to a more general investigation of the how and why of this phenomenon, which is widespread among vertebrates. I will relate the social suppression of reproduction to some generic properties of the reproductive system in vertebrates, but, in addition, some new considerations will come into play, considerations relating to social processes. I will argue that the social suppression of reproduction is not just a biological phenomenon, but a sociological phenomenon as well. The sociological implications will be especially apparent when we consider the source or causes of the social inequalities that provide the basis for this social suppression. As such, the social suppression of reproduction can serve as an introduction to a more general research area at the interface of the biological and the social, which I will refer to as biosociology.

The Social Suppression of Reproduction

Before we address directly the thesis of Hofmann, Benson, and Fernald as to the adaptiveness of the social suppression of reproduction in *Haplochromis burtoni*, we need to get a better feel for the phenomenon. As I have indicated, it is widespread among vertebrates, and, in contrast to sex change and alternative reproductive tactics, it is quite common in mammals. Let's briefly consider some examples of the social suppression of reproduction in some mammal species that occupy the terrestrial environments around Lake Tanganyika and elsewhere in Africa.

Musthy Males

Beginning in 1992, rare white rhinoceroses (*Ceratotherium simum*), were turning up dead in Pilansberg Park, South Africa. They were not being killed by poachers for their horns, but rather by rogue elephants for what seemed like no reason at all. All of the rhino deaths could be traced to male elephants, and all occurred while these elephants were in musth, a period of heightened sexual arousal during which male elephants become extremely pugnacious. Captive males must be isolated from other elephants during this period, or they can cause gross bodily harm to their elephant companions and human caregivers. In nature, almost all elephants avoid musthy males—with the exceptions of other musthy males (with whom they fight) and receptive females (with whom they mate). Musthy males do not always confine their rage to other elephants, however, as zoo keepers and white rhinos can attest. But what occurred at Pilansberg—40 elephant-induced rhino deaths in 5 years—was unprecedented. In what way did the Pilansberg elephants differ from typical musthy males?

The problem with the Pilansberg males, it turns out, was their youth. The Pilansberg elephants were all young orphans, survivors of a culling at nearby Kruger National Park. They were released at Pilansberg in the 1980s, in an attempt to reestablish elephants there. Whereas males normally enter musth at age 25, and then for only a matter of days, the Pilansberg males became musthy at 18 years and for up to 5 months. The reason for the precociousness of the Pilansberg males was a lack of older males. Teenage males do not enter musth if larger and older musthy males are present.[9] Moreover, young males can also be forced out of musth by aggressive encounters with larger males.

The rhino-killing problem at Pilansberg was solved by introducing six older males, which curtailed musth in the younger males, and therefore the rhino deaths.[10] Evidently, aggression in the younger musthy males is less targeted and disciplined than in the older males. Much as human teenagers, the young males are psychologically unprepared for the increased testosterone levels. So they "act out," with destructive consequences for everything around them. (I once observed a young musthy male acting out at a salt lick in Kenya. During the course of an afternoon he drove off anything that moved, including a black rhino, two large male African buffalo, several gazelle, and numerous Egyptian geese. He was finally driven off by an older, larger bull, and many animals quickly returned.) Older males keep these young potential hooligans from becoming testosterone-intoxicated.

For our purposes here, the important fact about the Pilansberg elephants is that the social suppression of reproduction resembles that of *H. burtoni* in important ways. Indeed, it has been suggested that the reproductive suppression of younger males benefits them because of the energy savings and avoidance of injury,[11] much as has been claimed for nonterritorial *H. burtoni* males by Hofman, Benson, and Fernald. The plausibility of that argument, however, is much less apparent in other cases of socially suppressed reproduction, such as that found in baboons.

White Fang, Pink Penis

Baboons are among the most successful primates. The social system of the olive baboon (*Papio anubis*) is typical of other baboon species and many other Old World monkeys, including, for example, several species of macaques and langurs. It consists of a group of related females, their female and prepubescent male offspring, and several unrelated adult males. In a baboon society, every member has a unique rank that is apparent not only to human observers but also to the baboons themselves, as is evident in the very particular ways they interact with each other. Rank is easiest to diagnose in aggressive interactions, but it can be expressed in much subtler ways, such as through the ritual form of hygiene known as grooming, in which one individual carefully inspects the fur of another and removes any bugs, ticks, lice, or dandruff it happens by. Though the groomer gets to eat whatever he or she finds, it is clear that the groomee gets much the better of the deal, and higher ranking baboons spend more time as the groomee than lower ranking ones. When rank differences are large, reciprocity can be token at best.

Among female baboons, rank is inherited, and thus rank relations are quite stable. Among males, however, no such conventions exist. Rank relations are dynamic, and the transitions thus typically violent. During periods of stability, displays of rank are confined to threats, both veiled and unveiled, such as eyebrow flashes and ritualized yawns that expose formidable canine teeth of leopard-like proportions. And in one of their more unveiled modes, they use their penis to signal status. If you are a male baboon gazing in the direction of one of higher rank, and he is sitting on his haunches, knees far apart, and his fully erect, electric pink penis is pointing in your direction, you should not mistake it for a sexual overture.[12] When a show of fang or penis fails to impress, things can turn violent, and the fangs—though not the penis—are called upon for more than symbolic use. Severe injuries and even death can result.

Finding oneself on the right side of the grooming asymmetry is only one of the perks of high social status, and a relatively minor one at that. It is certainly not worth risking life and limb over. For male baboons, the payoff from these battles comes in the form of reproductive opportunity, which is generally allocated in accordance with rank.[13]

The social suppression of reproduction is not confined to males. In social carnivores, such as wolves and African wild dogs, all subordinate pack members, whether male or female, are reproductively suppressed. Breeding is typically confined to only the alpha male and alpha female member of the pack, and they suppress reproduction in the other pack members through expressions of their social dominance.[14] Other pack members tend, and even feed, the pups, but do not generate pups of their own.

The situation is somewhat different in spotted hyenas. The alpha female, called the queen, does not prevent other females from breeding. There are, nonetheless, substantial differences in the number of young hyena females will produce in their lifetime, and the difference is directly related to social rank.[15] So, within a few generations the entire clan will consist of the dominant female's matriline. The queen's disproportionate success in producing future clan members is a direct result of her social status. Higher ranking females tend to begin breeding at an earlier age than lower ranking females, and they experience shorter intervals between litters.[16]

No one has yet suggested that the reproductively suppressed baboons, wolves, wild dogs, or hyenas are engaging in an adaptive conditional tactic.[17] Rather, it seems obvious that subordination often involves a restriction in reproductive privileges. In no mammal is this restriction more pronounced than in a fascinating subterranean creature known as the naked mole rat.

The Most Social Mammal in the World

When it comes to sociality, vertebrates are not terribly impressive. The most social animals are termites (order Isoptera) and ants (order Hymenoptera).[18] The degree of cooperative interaction that has evolved in these animals is truly awesome, sufficient to deserve a new label: eusociality. The hallmark of eusociality is that reproduction is restricted to a few individuals.[19] Among the most social termite species, a colony may number in the millions, but only one male and one female, referred to as the king and queen, respectively, are reproductively active. Though no vertebrate approaches the termite standard when it comes to organized social activity, a few species are truly eusocial. The best studied is the naked mole rat (*Heterocephalus glaber*) of eastern Africa.[20] As their common name suggests, these are not attractive creatures. Shaped like late-season yams that have begun to sprout; their wrinkled and virtually hairless pink skin looks at once fetal and mummified. They seem to have been plucked from the womb much too early and then freeze-dried. Upper incisors that appear to emerge from the nose and lower ones that project more outward than upward further dim their prospects in the pet trade. Fortunately, for naked mole rats, their visual sense has atrophied. They spend their entire lives underground in the darkness of an extensive maze of tunnels, where they rely on olfactory and acoustic cues to recognize each other and to communicate.

As in termite colonies, there is one queen for each naked mole rat colony (of 30 to 90 individuals), who functions as the baby producer; the other females, as well as nonreproductive males, labor in other ways. Most occupy themselves maintaining and expanding the intricate burrow system. The largest individuals guard the colony. These guards are stationed throughout the underground maze, at checkpoints through which any mole rat must pass on its way to the inner sanctum, where the queen resides. In contrast to termites, all female mole rats are potential breeders, and the queen looks the same as other female colony members. She is simply the highest ranking female in a linear hierarchy. She maintains her status as queen through her aggressive behavior: primarily nudges and shoves directed at other high-ranking females.[21] There is ample reason to contest for the queen role, and some queens are overthrown. For the most part though, ranks are stable, until, that is, the queen dies. Succession is far from peaceful; vicious, unrestrained combat among the highest ranking females breaks out, often resulting in death.[22] High-ranking males can also become casualties of queen succession, their close association with the former queen having become a liability.

Once a new queen is established, and she no longer experiences the suppressive effects of the former queen on her reproductive physiology, it does not take her long to become reproductively competent. Before becoming the queen though, she and all other female members of the colony could not ovulate.[23] All of that nudging and shoving meted out by the former queen had the effect of shutting them down reproductively. This condition is reflected in their low gonadotropin levels.[24]

Male naked mole rats also form a linear hierarchy. Reproductive privileges are somewhat less restricted in the males than in the females. Up to three males can be reproductively active at any given time, and they can be solicited by the queen. The other, non-breeding, males experience reproductive suppression, though not as completely as do nonbreeding females.[25] Nonbreeding males have lower gonadotropin (LH, or luteinizing hormone) concentrations than the breeding males; they also have lower testosterone levels and smaller testes. The nonbreeders do produce sperm,[26] but their sperm are generally not as abundant as in the breeders nor as motile.[27] When the nonbreeding males are removed from the colony, they rapidly become reproductively active.[28]

Presumably, subordinate male and female mole rats also benefit from the energy savings experienced by nonbreeding elephants and *Haplochromis burtoni*. Nonetheless, no one has suggested that their not breeding is, for this reason, an adaptive tactic on their part. In some relatives of naked mole rats, however, the failure to breed by subordinates clearly is an adaptive tactic, albeit one unrelated to energy savings. Some mole rat species in the genus *Cryptomys* are as eusocial as naked mole rats.[29] For example, in the Damaraland mole rat (*Cryptomys damarensis*) breeding is confined to one male and one female.[30] But in this species, the sperm are as abundant and motile in subordinate males as in the dominant breeders.[31] The failure of the subordinate males to breed appears to reflect an aversion on their part to mating incestuously. They are quite happy to mate with any unfamiliar female.[32] This then truly is a case in which the lack of breeding among subordinates is adaptive on their part, but it is obvious as well that in this case the subordinates are not reproductively suppressed by the presence of more dominant individuals, and they are not failing to reproduce in order to divert energy resources elsewhere.

Adaptation or Byproduct?

Evidence of some benefit that the social suppression of reproduction confers to the socially suppressed nonterritorial *Haplochromis burtoni* males is not sufficient to warrant calling this social suppression an adaptation. There is no condition, however obviously undesirable, for which we cannot identify some benefit (slaves don't have to pay taxes). The reduced energy expenditures of nonterritorial males notwithstanding, the reproductively suppressed state of nonterritorial *H. burtoni* males seems undesirable from an evolutionary point of view. At a minimum, it must be demonstrated that the energy savings benefit outweighs any associated disadvantages. And, from an evolutionary perspective, the disadvantage of not reproducing, or even delaying reproduction, is considerable. The onus is on those who claim that being socially suppressed is adaptive to demonstrate that the benefits of that condition surpass the obvious drawbacks. Unless we have compelling evidence to the contrary, we should assume that social suppression serves only the suppresser, not the suppressed. This is obviously true of the social suppression of reproduction in baboons, hyenas, and African wild dogs. Few biologists would be tempted to claim that subordinate hyena females or male baboons are practicing adaptive conditional behavior. Their reproductive suppression obviously results in reduced fitness. They cannot do otherwise than produce fewer offspring than higher ranking individuals. They are just stuck with being reproductively suppressed as a result of their social status.

Recall from chapter 5 that for all true conditional tactics there are a switch-point and a decision rule. For protogynous (female → male) sex-changing fishes, the switch-point was based on relative size, and the decision rule was, "below this size remain female, above this size become a male." Do we have any evidence that territorial and nonterritorial *H. burtoni* males are analogous to male and female cleaner wrasses in these respects? No! The evidence that *H. burtoni* males are following an adaptive decision rule—"breed when you are a dominant territorial male, do not breed when you are a subordinate nonterritorial male"—is reduced to the fact that territorial males, which are socially dominant, breed, but nonterritorial males, which are socially subordinate, do not breed. That is not enough.

In lieu of an objective switch-point, we are reduced to circumstantial evidence of the sort prone to paranoiac abuse. Given the available evidence, we should conclude that the suppressed reproduction of nonterritorial *H. burtoni* males is not an adaptive conditional strategy for the efficient allocation of reproductive effort but rather an evolutionary byproduct of generic properties of the vertebrate reproductive system and the way it is hooked up to other physiological systems.[33] It is now time to explore in more detail what these generic properties might be.

The How-Biology of Reproductive Inhibition

We begin by considering how, physiologically, social suppression comes about. First, though, we need to address a common misconception, one that is perpetuated by a false distinction between "behavioral" (or "psychological") inhibition and "physiological" inhibition. This distinction is the misleading residue of an antinaturalistic Cartesian dualism, in suggesting that, in some cases, social stimuli somehow influence an animal's behavior without affecting its physiology. Any social influence on behavior is mediated by changes in brain and hormone states, processes that are no less physiological than digestion. The more useful distinction is between cases in which social suppression causes alterations in the hypothalamic-pituitary-gonadal (HPG) axis and those cases in which the social suppression of reproduction involves neural and hormonal processes less directly related to reproductive physiology.

I will focus here on the social suppression of reproduction through alterations in the HPG axis. Evidence for such can take a variety of forms, including a reduction in sperm counts, gonad size, testosterone levels, gonadotropin levels, and GnRH levels. In determining whether any such changes have occurred as a result of social interactions, we first need to establish whether there is any difference between dominant and subordinate in these measures. In many of the species I have considered here, dominant individuals are better prepared physiologically for reproduction than subordinate individuals, by one or more of these measures. Most of this evidence comes from hormonal assays of circulating sex steroid or gonadotropin levels, or both, which are often lower among subordinate than among dominant individuals. But this correlational evidence does not establish cause and effect. It could be the case, for example, that dominant individuals have high testosterone levels because of their social advantage, or rather that dominant individuals are dominant because of their high testosterone levels. Only if high testosterone is the effect, not the cause, of high social rank, does this correlation suggest socially suppressed reproduction. I chose to make *Haplochromis burtoni* the focus of this chapter because the cause–effect relationships have been clearly established in this species, in part, through my own research.

Reproduction and Social Rank in Haplochromis burtoni

As you might expect, the socially dominant and reproductively active territorial males have larger testes and higher testosterone levels than the reproductively suppressed nonterritorial males.[34] More interesting, territorial males also have larger hypothalamic GnRH neurons than nonterritorial

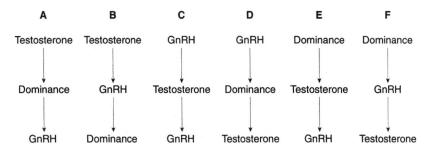

Figure 6.2 Six possible scenarios for the causal relationships among social dominance, testosterone levels, and GnRH neurons

males. So there is an association between social dominance, large testes, and large GnRH neurons in adult males. But there are numerous ways in which these traits could be causally linked. Six possible scenarios are illustrated in figure 6.2. I set out to determine, through experimental manipulations, which of these causal sequences explained these associations in *H. burtoni*.

It could be the case, as represented in scenarios A and B, that the testes or their products (i.e., testosterone) are the primary causal factors in these correlations. That is, territorial males are dominant and have large GnRH neurons because they have large gonads. This would be the case if, for example, testosterone level is the primary determinant of dominance status and males with large gonads have more testosterone than those with relatively small testes (A), or, as in B, testosterone level affects dominance status indirectly through its affect on GnRH neurons.

In order to test for the effects of testosterone on the GnRH neurons, we reduced the testosterone levels experimentally by castrating some territorial males and compared their GnRH neurons with those of intact (sham-operated) territorial males. To our surprise, the GnRH neurons actually enlarged in the castrates.[35] Therefore, testosterone constrains the size of the GnRH neurons; it does not cause them to grow. The territorial males have large GnRH neurons *despite* their high testosterone levels, not *because* of them.

What about the causal relation between testosterone and dominance status? We tested this by pairing castrated territorial males with intact nonterritorial males of the same size. The results were interesting: castrated territorial males continued to dominate the now better-endowed nonterritorial males, despite their seeming handicap. As a reporter for the *New Scientist* put it when she described these results, castrated territorial males "fail to lose their balls."[36] It is also interesting to note that they managed to maintain their dominance despite the fact that their aggressiveness dropped along with their testosterone titers.[37] These results further undermine the

simplistic notion that there is a unidirectional causal relation linking testosterone and social dominance.

So neither the dominance of territorial males nor their large GnRH neurons can be explained by their relatively large gonads and high testosterone
levels. As such, if there is a unifying explanation for the characteristics of
territorial males—their social dominance, large testes, and large GnRH
neurons—it is either the size of the GnRH neurons or social status itself
that is causally primary. I investigated the causal relation between these two
variables in another set of experiments.

From Territorial to Nonterritorial, and Back

Before this study, I had assumed that once a male became a territorial male,
he stayed a territorial male for the rest of his life. I discovered through a
serendipitous observation that I was mistaken. The main animal facility at
the University of Oregon was separated from the laboratories by an incon-
venient distance, so fish that were destined for experimental use were temporarily housed in a holding facility adjacent to the lab. During the course
of an experiment unrelated to this discussion, I transported several territorial males from the main animal rooms to the holding facility; I then
decided not to use one of them. Because it was a Friday afternoon, I placed
this fish in an empty tank in the holding room over the weekend. The
following Monday morning I returned him to his home tank. Once back in
familiar surroundings, he immediately went back to his former territory. In
his absence, however, one of the nonterritorial males had assumed owner-
ship and had begun to transform into a territorial male, as indicated by his
coloration and behavior. The erstwhile (but temporarily exiled) occupant
attempted to evict the usurper but without success. Within days, he lost his
coloration, including the eye bar, and, from the outside at least, came to
resemble a typical nonterritorial male. It seemed that I had inadvertently
managed to convert a territorial male into a nonterritorial male.

This little drama led me to consider what happens to the GnRH neurons
of a territorial male when it becomes a nonterritorial male. I could already
infer what was happening inside the usurper. As a newly minted territorial
male, his testes were beginning to enlarge, as were the GnRH neurons in his
preoptic area (POA). But what about our protagonist? Were his GnRH neurons starting to shrink, or would they retain their territorial male size? What
about his testes—would they regress to a nonterritorial state? To answer
these questions, I needed to find a more systematic way to induce a change
in social status. After exploring some options, I settled on a simple procedure (figure 6.3).

In the main animal rooms, the fish were maintained in stable communities consisting of 3 to 5 territorial males, 5 to 8 nonterritorial males, and

$$T \longrightarrow NT$$

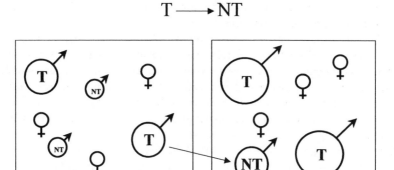

$$NT \longrightarrow T$$

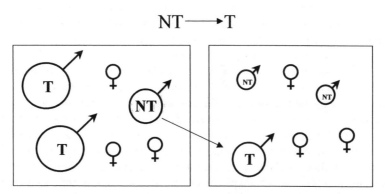

Figure 6.3 Experimental protocols for the conversion of territorial males into non-territorial males (T → NT) and for the conversion of nonterritorial males into territorial males (NT → T)

The boxes on the left represent home tanks; the boxes on the right represent experimental manipulations of the social context.

10 to 15 females. Under these conditions, a nonterritorial male could become a territorial male only if (1) one of the territorial males died or was removed and (2) one of the surviving territorial males did not expand his territory to include the newly vacated one. In order to make the nonterritorial to territorial (NT → T) transition more probable, I simply removed nonterritorial males from their home tanks and placed them in new ones along with lots of females and smaller nonterritorial males. In order to generate the territorial to nonterritorial (T → NT) transition, which rarely if ever occurred in the stable communities, I removed territorial males from their home tanks and placed them in new tanks containing larger territorial

males and females. In this way, I was able to obtain the desired behavioral transition, without fail, within 1 or 2 days.

As expected, the NT → T males experienced marked testicular growth. Their POA GnRH neurons also enlarged significantly. The results for males making the opposite (T → NT) social transition were more noteworthy. Those territorial males that I had caused to become nonterritorial males experienced a reduction in the size of both their testes and their GnRH neurons. This experiment demonstrated that the size of these GnRH neurons is socially regulated and in a symmetrical manner.[38]

It is now apparent that the entire HPG axis in *Haplochromis burtoni* is responsive to—indeed, is regulated by—social interactions, from a juvenile stage through adulthood.[39] In adults, inhibitory social signals cause a reduction in GnRH expression, which is reflected in the shrinkage of GnRH neurons; the reduced GnRH expression, in turn, causes a reduction in gonadal activity. When these inhibitory signals are removed, as occurs in the NT → T transition, GnRH expression increases, and, as a consequence, so does gonadal activity. It was later determined that the inhibitory signals are tactile, not visual or olfactory.[40]

Recall that the basic causal flow in the HPG axis is as follows: gonadotropin-releasing hormone (GnRH) produced in the hypothalamus causes the release of gonadotropins from the pituitary, which causes an elevation in gonadal activity. We can be confident, therefore, that the effect of the reduced GnRH expression on the gonads of nonterritorial males is mediated by a reduction in pituitary gonadotropin levels. These conserved causal relationships within the HPG axis are important to bear in mind, but for the purpose of evaluating the byproduct view of socially suppressed reproduction, we now need to consider the way the HPG axis itself is causally hooked up to the rest of the organism. Of particular importance is the way the HPG axis is hooked up to the physiological system that mediates the stress response in vertebrates. It is these evolutionarily conserved hook-ups that explain why subordinates are reproductively suppressed, whether being suppressed saves them energy or not. First, though, a few words about social stress.

SOCIAL STRESS

Social life has some obvious advantages: A pride of lions can kill very large animals, such as the formidable African buffalo; a solitary lion, though, must avoid this dangerous animal. The reverse is also true: A solitary ungulate, even an African buffalo, is much more likely to become food for lions, hyenas, or wild dogs than is a buffalo that is a member of a herd. But life in groups is a sword with two edges. With social living comes social stress.

Most of what we read about stress concerns its pathological effects, but without an effective stress response no animal could long survive. The stress response is a systematic reaction to any event in the environment that threatens to disrupt an animal's internal equilibrium. Energy is mobilized from storage and diverted to muscles; sensory capacities increase, though the perception of pain is dulled. The stressed animal is generally in a heightened state of preparedness for dealing effectively with the stressor, whatever it may be. The stress response enhances the impala's chances to escape the leopard and the likelihood that a cornered rat can fend off its attacker. But what are obviously adaptive responses to acute challenges become pathological when the stressor is chronic. The diversion of resources to deal with the immediate threat comes at the expense of digestion, growth, immuno-competence, and, yes, reproduction. If the stressor does not go away, the physiological toll becomes a problem. Because social interactions are among the most chronic stressors, chronic stress is a price that many social animals have to pay for being social.

The Stress Axis

The stress response, like reproduction, is ultimately regulated by neurons in the hypothalamus. These hypothalamic neurons, as do the GnRH neurons, project to the pituitary, but they produce a different peptide hormone, cor-ticotropin-releasing hormone (CRH), and project to a different population of pituitary cells. These pituitary cells manufacture a hormone called cor-ticotropin rather than gonadotropins. Corticotropin, like gonadotropin, cir-culates in the blood; it is bound by receptors in the adrenal glands, which are adjacent to the gonads. In response to the corticotropin, the adrenals release steroid hormones known as glucocorticoids. One of the common glucocorticoids in many vertebrates is cortisol. By way of summary then, in response to a stressor, CRH from the hypothalamus causes the release of corticotropins from the pituitary, which in turn causes the release of cortisol from the adrenals. This is called the stress axis, or the hypothalamic-pitu-itary-adrenal (HPA) axis (figure 6.4).[41]

Cortisol is related to CRH in the same manner as sex steroids (e.g., tes-tosterone) are to GnRH. And, as are sex steroids, cortisol is much easier to measure than its corresponding peptide hormone. For this reason, cortisol levels have been the measure of choice in studies of the stress response. In one such study, Robert Sapolsky monitored cortisol levels in baboons.[42] This study is noteworthy for a number of reasons, not least because the cortisol levels were measured in the field, a considerable logistical feat. The results of this study provide some important insights into the relationship between social status and social stress.

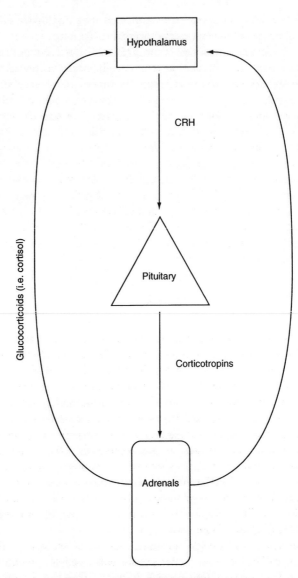

Figure 6.4 The hypothalamic-pituitary-adrenal (HPA) axis

Stress in Baboon Society

In the absence of an obvious stressor, high-ranking male baboons have lower cortisol levels than their lower ranking counterparts, which means they experience less (social) stress. High and low ranking males experience different social environments, and the social environment of higher ranking individuals is better, stress-wise. In response to a stressor, cortisol levels rise at a much faster rate in high ranking males than in their lower ranking peers, but they return to baseline levels more quickly.[43] These results suggest that the stress response of high-ranking baboons is largely confined to acute stressors and hence minimizes the many potential negative effects. Low-ranking males, though, appear to be under constant stress, no doubt related to the presence of higher ranking males, and must experience some of those deleterious side effects of a chronically activated stress response. A repressed reproductive system is among these deleterious effects.

The influence of stress on reproductive physiology results from the intimate connections between the stress (hypothalamic-pituitary-adrenal) axis and the reproductive (hypothalamic-pituitary-gonadal) axis, which exist in all vertebrates (figure 6.5). For example, in addition to causing the release of corticotropins, and therefore cortisol, CRH also causes the release of various other neurochemicals, including opiates produced naturally in the brain. Among the effects of these endogenous opiates is a lowering of GnRH levels in the hypothalamus, resulting in a reduction of gonadotropin release from the pituitary, which, in turn, can lead to reduced sperm production and lowered testosterone levels.[44] Cortisol itself reduces testosterone production by reducing the sensitivity of the testes to gonadotropins.[45] Because of these hook-ups, any vertebrate that experiences chronic stress will be subject to a depressed reproductive physiology as a sort of default condition, whether it is adaptive or not.

Consider again Sapolsky's baboons. Though the basal testosterone levels of high- and low-ranking males do not differ during periods of social stability, there is a marked difference in the way they respond to acute stressors (including social conflicts). Whereas testosterone levels drop in the low-rank males, they actually increase in high-rank males during periods of acute stress, as their cortisol levels also increase. So, during periods of social instability in which there is frequent fighting, testosterone levels are higher in high-rank males than in low-rank males. These results indicate that the chronic socially induced stress experienced by low-rank males renders their reproductive system more vulnerable to acute stressors such as fights. This is exactly what we would expect, given the way the stress (HPA) axis is hooked up to the reproductive (HPG) axis in vertebrates. The lower circulating levels of sex steroids and gonadotropins in low-ranking baboon males in response to acute stressors, is simply an unavoidable consequence of their low social status, rather than an adaptive conditional response. Barring compelling evidence to the contrary, we should assume that the same is true of nonterritorial Haplochromis burtoni males.

What sort of evidence should we look for in determining whether the social inhibition of reproduction in nonterritorial H. burtoni males is adaptive? We should look for a dissociation between the effects of inhibitory

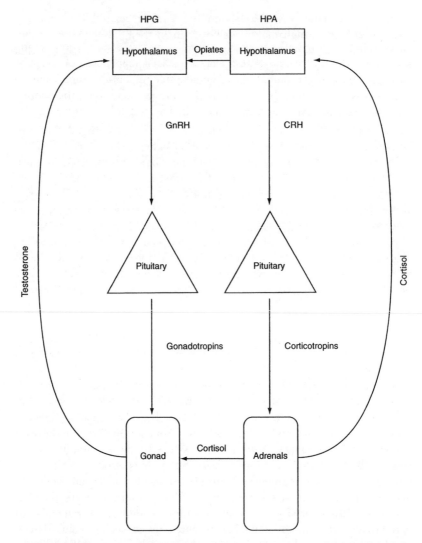

Figure 6.5 Interactions between the hypothalamic-pituitary-gonadal (HPG) and the hypothalamic-pituitary-adreanal (HPA) axes

social stimuli on reproduction and the stress response system. Evidence, that is, that the effects of social subordination on reproduction are not mediated by the same final common pathway as, say, hunger-induced stress. Put another way, we need to find evidence of a mechanism through which inhibitory social stimuli are transduced into diminished HPG function that is independent of the HPA axis, as is the case in some of the most social mammals.[46] In fact, however, the nonbreeding nonterritorial *H. burtoni* males have higher basal corticosteroid levels than the reproductively active territorial *H. burtoni* males, suggesting an HPA-mediated social suppression.[47] This is not what we would expect were the social suppression of reproduction in subordinate *H. burtoni* males an adaptive response to subordination.

Natural selection has obviously played a role in establishing the connections between the HPG and HPA axes. It does not follow, however, that these connections were established through selection for the social regulation of reproduction in *H. burtoni* or any other vertebrate species. The stress response can be stimulated by diverse stressors, including hunger, injury, cold temperatures, predators, and social bullying, to name a few. In each case, we should distinguish between acute and chronic effects. The acute effects are most likely to be adaptive, and the chronic effects are most likely to be unavoidable byproducts of the adaptive response to acute stressors. (An acute elevation in corticosteroid levels actually has a stimulatory effect on the HPG axis in many animals.[48]) Pathological responses to chronic stress in humans are well known and include cardiovascular disease, malfunctions of the gastrointestinal system, reproductive malfunctions of various sorts, depression and other diseases of the central nervous system, as well as a reduced immune response. It is probable that each of these pathologies is simply an unfortunate byproduct of an adaptive response to acute stress that is essential for survival. We should not assume that these pathologies are confined to humans and only by virtue of recent changes in our "way of life." The inhibitory effects of low social status on reproductive physiology in *H. burtoni* could be pathological in the same way.

In the three previous chapters we have seen how how-biology considerations altered our understanding of evolutionary phenomena ranging from sexual reproduction itself to alternative reproductive tactics. In those cases, the interaction of how-biology and why-biology explanations lay somewhere between the complementary and replacement poles. With respect to this specific putative conditional tactic in *H. burtoni*, however, the how-biology explanation should be viewed as a replacement for the why-biology explanation unless it can be demonstrated that subordinate males exhibit a true conditional tactic.[49] The how-biology may be all we need to understand the social suppression of reproduction in nonterritorial males. Actually, that is not quite right. The "how" of social suppression may be all we need here, but in this case the "how" extends beyond the how-biology. With a social phenomenon such as this we must extend our how-considerations beyond biology to sociology as well. For the remainder of this chapter, I will address some misconceptions that constitute barriers to developing a science of "sociobiology" worthy of the name.

From Sociobiology to Biosociology

Though I was pleased with my experimental demonstration that alterations in the social status of *Haplochromis burtoni* males cause changes in GnRH neurons, I was unprepared for the popular attention this research garnered.[50] It took me some time to realize that this popular attention was based on common misperceptions as to the way biological factors are related to social factors. I had set out to identify the mechanism for the social suppression of reproduction in *H. burtoni*, but what attracted the attention of the popular media was the fact that a social state (dominance status) determined a biological state (size of GnRH neurons). But that should not be at all surprising; we are everywhere surrounded by evidence that biological phenomena are influenced by social phenomena—but clearly it must go largely unrecognized by many. For the remainder of this chapter I will explore the barriers to this recognition, which, I will argue, can be traced to two fundamental conceptual errors.

I have already alluded to the first of these, which I will refer to as "anemic materialism." The distinction between "psychological" and "physiological" suppression is an example of anemic materialism, as is the false dichotomy between drug therapy and psychotherapy. Most people who adhere to these dichotomies are not really Cartesian dualists who believe in radically separate psychological (immaterial, mental) and material (physiological) realms; they just fail to appreciate the full implications of materialism. Certainly, no full-blooded or robust materialist should be surprised that social interactions have physiological effects; ultimately, physiological effects are the only kind of effects that social interactions could possibly have on individual organisms. Changes in the size of GnRH neurons are not particularly noteworthy, in this respect.

But anemic materialism was actually only a minor factor in the response to my research. The second fundamental misconception, which I call "misguided materialism," played a far more important role.

Misguided Materialism

The second misconception concerns the way social phenomena are causally related to biological phenomena. Misguided materialists believe that being a materialist means that, because social systems are composed of biological entities, and all biological entities are composed of chemical entities, and so on, we should explain the social in terms of the biological and the biological in terms of the chemical, and so on, but not the reverse. It is generally assumed, therefore, that given a biological fact and a related social fact, the biological fact is the cause and the social fact is the effect. Given this assumption, finding that social status determines the size of the GnRH neurons is more than noteworthy.

Misguided materialism is often referred to as reductionism, but that term covers a multitude of sins (and some virtues as well), so I want to be more specific about what I have in mind here. The form of reductionism at issue here is steeped in a venerable metaphysical perspective known as atomism. There is a strong form of atomism and a weak form. The weak form is vacuous,[51] so I will focus on the strong version, which alleges that the biological entities of which social systems are composed, and the chemical entities of which biological systems are composed, are not themselves substantively altered by their interactions. They can be affected by their interactions, to be sure, but not in their essence. The essential properties of an individual organism, which are related to its biology and, ultimately, to its genes, are "intrinsic." "Accidental" properties of an individual, on the other hand, which do not result from its essential biological nature, result from its interactions with the environment, including other individuals. These accidental properties are no less the properties of individuals than are its essential properties, but they are not intrinsic. When individuals, so construed, interact, it is their individual attributes qua intrinsic properties that are presumed to determine the results. Hence, any group properties are truly epiphenomenal, effects but not causes.

This attitude permeates sociobiology, a discipline whose goal it is to Darwinize social phenomena. Heretofore I have focused on the adaptationism in sociobiology and allied disciplines; here I examine a second component of the sociobiological worldview: its atomistic reductionism. I focus in particular on the ways in which this view limits our understanding of social animals, with respect to both their sociality and their biology. For sociobiology provides an impoverished perspective not only on social phenomena; it provides an impoverished view of biological phenomena as well. In treating social phenomena—including social systems—as epiphenomenal consequences of biological processes, we must ignore a rich vein of causal factors in biology. Social phenomena qua social phenomena influence biological phenomena from brain states right on down to gene states. Once we acknowledge this neglected category of causal factors, we will not find it at all surprising that social status determines the size of GnRH neurons in *H. burtoni*. Moreover, we will be even less likely to mistake the territorial and nonterritorial states of *H. burtoni* for conditional tactics. The "how" is all we need to explain the fact that nonterritorial males do not reproduce, but this "how" includes social causes as well as biological causes.

Causes of Social Inequalities

Some obvious problems with the sociobiological approach to social phenomena can be illustrated through a consideration of social hierarchies, which are a fairly common result of the interactions of social animals. Peck-

ing orders in chickens are paradigmatic.[52] One chicken, the alpha, dominates all others. Beta dominates all but alpha, and so on, in a linear order.
There are two social facts that need to be explained. First, Why do individual chickens come to occupy their particular rank in the hierarchy? And
second, Why is this social order linear?

The answer to the first question might seem obvious: the social ranks are
the result of intrinsic individual differences in some biological trait or traits
that determine social dominance. Each chicken can be ranked according to
these traits, and that rank will correspond to its position in the social hierarchy. It is these intrinsic biological chicken traits that determine a
chicken's social status. Assume that testosterone-induced aggressiveness is
the salient intrinsic factor. Alpha would have to exhibit the highest levels,
beta somewhat less, and so on, until we reach omega, who has the least.
Unfortunately, it is an empirical fact that we rarely get such nice correlations between a biological variable and social rank. Hence, more biological
variables must be considered, such as body weight, or claw size, or the levels
of serotonin in the hypothalamus, or the number of neurons in a particular
brain area. But biological variables such as these, even when combined in
statistically ingenious ways, are often still poor predictors of social rank.[53]
Moreover, as we saw with H. burtoni, even when biological factors are
highly correlated with social rank, we cannot assume that the cause is the
biological factor and the effect is the social rank. In fact, it is often the case
that biological factors such as testosterone levels are the effect, and not the
cause, of an individual's social rank.[54] Put another way, testosterone level is
not a simple intrinsic biological factor but rather a trait that is also socially
conditioned.

When biological traits such as testosterone levels turn out to be not so
intrinsic, the most common response is to look for another trait that is more
so, and genes often provide the last refuge in this regress toward the intrinsic. But the limitations of this approach have been dramatically demonstrated in some simple experiments conducted on some of H. burtoni's cichlid cousins, Pseudotropheus zebra from lake Malawi. In these experiments,
fish of roughly the same size, and which could be identified individually,
were assembled at random. They soon formed a stable hierarchy, after which
they were separated for a brief period. They were then placed back together
in order to monitor their social interactions. Not surprisingly, the reas-
sembled fish quickly formed a linear hierarchy, just as they had before their
separation. What is surprising, however, is that the ranks of particular individuals had changed substantially.[55] Hence, though linear hierarchies are a
robust result of these social interactions, individual ranks are not. In this
case, it is clear that social rank cannot be reduced to intrinsic individual
differences of any kind. Other factors, including highly contingent social
interactions, must be considered as well.

We have ample evidence of some very generic properties of contingent social interactions in animals; these are easiest to recognize when the social interactions are experimentally restricted in various ways. The most minimal social process results from the interactions of two individuals. With regard to social dominance, we can ask what sort of rank relationships result from a series of dyadic, or pairwise, interactions. When the contestants are chosen at random, the victors of an encounter tend to win the next encounter as well, and the losers tend to lose the next encounter. One obvious interpretation is that whatever attributes of, say, chicken A that caused it to defeat chicken B also made it more probable that it would subsequently defeat chicken C. Winners tend to continue to win, because of some intrinsic biological properties, and losers tend to continue to lose, for the same reason. If the pairwise encounters were structured like a round-robin tournament, in which every contestant plays every other contestant, we would expect the ultimate rankings to reflect the intrinsic individual differences in prowess. This ranking, in turn, should be identical to the hierarchy that results when all contestants are simultaneously introduced and allowed to sort things out all at once. In fact, this outcome is so obvious that there has been virtually no effort to confirm it experimentally.

What little research has been directed toward this issue deeply undermines this commonsense view: the two experimental procedures can result in two very different rankings of the very same individuals.[56] The winner of the round-robin tournament will not necessarily acquire the alpha rank in the group context. One salient difference in the two situations is the order in which individuals interact with each other, but that should not matter if the intrinsic qualities of the contestants determine their ultimate rank. What if, however, the probability that chicken A will defeat chicken C is affected by the outcome of its prior encounter with chicken B? That is, what if chicken A is more likely to defeat chicken C if it had previously defeated chicken B, independent of any intrinsic properties of chicken A or chicken C? In that case, we would have to say that the contingent experience of winning or losing affects subsequent interactions, and, as such, the contingent order of interaction will play a large role in determining rank. Such experiential effects are extremely important in determining rank relations in a wide range of animals, including chickens, mice, cichlids, and primates.[57]

My own research on social dominance in paradisefish (*Macropodus opercularis*) revealed pronounced experiential effects of this sort.[58] In addition to these experiential effects on social dominance, I also attempted to isolate some relevant "intrinsic" factors. Because intrinsic influences on social states are ultimately traced to gene states, I artificially selected for the capacity or propensity to win initial paired encounters with opponents of the same size. After a few generations of selection, there was a marked separation in the high-dominance (HD) and low-dominance (LD) lines, such that HD fish were consistently more successful than LD fish in contests against un-selected naive opponents of the same size.[59]

But this was true only of naive HD and LD fish. The effects of selection could be completely obliterated by simple manipulations of their experience in prior dominance encounters. I could predetermine the outcome of any contest simply by choosing significantly larger or smaller opponents for the subjects. The "psychological" effects of losing such staged encounters were particularly dramatic.[60] Hence, though naive HD fish were more likely than LD fish to win a contest against a naive opponent, the HD advantage was eliminated if both fish had experienced one staged defeat. Moreover, naive LD males tended to dominate HD males that had experienced defeat. Hence, any intrinsic propensity for dominance resulting from artificial selec-tion was swamped by experiential factors.[61]

Paired encounters constitute the simplest and most basic social interac-tions. More complex social forms that exist in larger groups are even less likely to be amenable to analyses restricted to intrinsic individual differ-ences. For example, consider the linearity of social relationships, which is a common outcome of group living. Clearly, this linearity does not require stable rank relations resulting from individual attributes that existed before the interactions. Rather, linearity is a generic property of a large class of social interactions that emerges through the social interactions themselves, no matter the preexisting individual differences, whether biologically intrin-sic or accidental.

In a linear hierarchy, all dominance relationships among individuals must be transitive; that is, if chicken A is dominant relative to chicken B, and chicken B is dominant relative to chicken C, then chicken A must also be dominant relative to chicken C. Any intransitivities among group members will preclude the formation of a linear hierarchy. What ensures this tran-sitivity, if it does not result from transitive individual differences? It turns out that the way social experience is structured during hierarchy formation may be sufficient to guarantee across-the-board transitivity.

Ivan Chase, who is unique among sociologists in his use of diverse animal species, has proposed a general model for the dynamics of hierarchy forma-tion, one that predicts linearity even when no individual differences exist.[62] In his "jigsaw" model, linear hierarchies are self-organized assemblies of so-cial order that generically result from the effects of the experiential histories of the interactants during hierarchy formation. On this view, neither rank nor linearity need be prefigured in the properties of the interactants; in-stead, they can result entirely from the interactions themselves. It should be noted, however, that the jigsaw model does not imply that individual differ-ences do not matter—only that they need not. The jigsaw model is compat-ible with a significant association between rank differences and individual differences in, say, body size. But the jigsaw model does predict that hierarchies will be linear whether individual differences exist or not. It suggests, moreover, that even when there are preexisting individual differences, social rank will be influenced by contingent social variables, such as the order in which opponents are encountered.

Spotted hyenas illustrate a completely different sort of limitation on the influence of intrinsic biological individual differences on social rank, one that also permeates human societies. Hyena societies are well ordered and stable. In large part, this stability results from the linearity of rank relations and the conventional nature of rank acquisitions. All adult females outrank all adult males. Among females, there are a number of distinct lineages, or matrilines, which are themselves linearly ordered. Within a matriline, rank is entirely a function of age; so the alpha female (sometimes called the queen) is the oldest, and hence the highest ranking, female of the highest ranking matriline; and the lowest ranking female of the highest ranking matriline outranks the highest ranking female of the matriline that is next in rank. Hence, female rank is socially inherited and depends entirely on when you were born and who your mother is. This is not a meritocracy. In fact, hyena society has a caste-like quality.[63]

Social rank in spotted hyena males is also determined by convention. Male spotted hyenas emigrate from their natal groups when they approach maturity. When they join their new clans as adults, they enter at the bottom of the social ladder. Their subsequent rank is entirely a function of time in residence, or seniority.[64] There is remarkably little aggression among these males and no challenging for higher rank. In fact, despite the enormous fitness consequences of social rank in both sexes, social conventions, not intrinsic individual differences, determine social status.

From the Social to the Biological

Once we recognize that social processes are not simply biological but have "a life of their own," we have access to a whole new class of causes in explaining biological states, including reproductive physiology. The social suppression of reproduction is but one category of social influences on biological phenomena, one example in which social factors are the cause and biological factors the effects. And the fact that social status determines the size of GnRH neurons in one species of fish should not be considered terribly noteworthy, except in explaining how social stimuli influence the gonads.

These alterations in the size of GnRH neurons involve changes in gene states. If, in fact, it is established that the GnRH gene itself is socially regulated, the news of that discovery will create a much bigger stir than the demonstration that the GnRH neurons are socially regulated. Again, however, the excitement will largely reflect a fundamentally mistaken assumption concerning how biological processes—in this case, at the genic level—are causally related to social processes. It is only in this context that such a discovery should be cause for any excitement at all. For we should expect social influences on gene expression to be commonplace, given what we already know about gene regulation in development. Most genes, those DNA molecules that function in development, are not constitutionally active; they are reactive. Whether a gene is generating messenger RNA transcripts or just sitting there doing nothing depends entirely on the chemical milieu within a cell, which is itself influenced both by chemical interactions between neighboring cells and by neural and endocrine signals from more remote parts of the organism. The neural and endocrine states, in turn, are responsive to many aspects of the environment, including the social environment. So there is nothing mysterious about social influences on gene states. What is noteworthy is that such effects are considered noteworthy.

BIOSOCIOLOGY

This chapter extends the critique of sociobiology and other adaptationist evolutionary accounts of social behavior by noting a second failing, above and beyond the excessive teleology. This second failing is an impoverished view of the causal arrows connecting biological and social processes. We began with a putative case of an adaptive conditional tactic in *Haplochromis burtoni*. It was proposed that the suppressed reproductive axis in nonterritorial males is a strategy to save energy and resources. We saw that this explanation did not pass muster. Rather, given the existing evidence, we should assume that socially suppressed reproduction in *H. burtoni*, African elephants, baboons, hyenas, and naked mole rats is just a byproduct of the way the reproductive axis is hooked up to the stress axis in vertebrates. Given these generic hook-ups and the social stress that attends subordinate status, suppressed reproduction is simply something that cannot be helped.

The notion of social suppression itself led to a consideration of the causal relations between social and biological processes, the second failing of the sociobiological perspective. I have attempted to demonstrate the value of viewing social processes as more than just the effects of biological processes. For biologists, the value in this view is its making explicit a class of causal factors that would otherwise be ignored. The misguided view that the causal arrow always points from the biological facts to the social facts impairs our grasp of both biology and sociology, and it certainly reduces the prospects of any successful interdisciplinary endeavor to research the processes at their interface.

Consider research at another interface, that between biology and chemistry. It is called biochemistry. Biochemistry is not merely the study of how chemicals influence biological phenomena, but also the study of how biological processes qua biological processes influence chemistry. Indeed, the biochemicals that constitute the objects of study in biochemistry come to exist only in biological systems. By analogy, we should refer to the study of how social phenomena influence biology as "sociobiology." Unfortunately, that name has already been adopted for an antithetical enterprise.

We will better understand those fascinating phenomena that occur at the interface between biological and social processes when the current propensity to simply biologize social phenomena is recognized as inadequate, not only for a science of sociology but for a science of biology as well. Only when the causal significance of social processes, with respect to biological processes, is recognized will we have a science of "sociobiology" worthy of the name. In the meantime, I will call the research described in this chapter "biosociology," in order to avoid confusion.

7

Why Does the Mockingbird Mock?

ONE CONSEQUENCE of our domination of nature is that our experience of wildlife has become increasingly mediated and indirect. Televised nature programs provide vicarious access to wild things in wild places, and zoos and museums provide a direct experience of wild things, but only in simulated environments. Our direct experience of wildlife is primarily of pests: the flies, termites, cockroaches, and spiders that invade our houses and the aphids, thrips, earwigs, weevils, and mites that plague our gardens. There are exceptions of course—butterflies spring immediately to mind—but most of our direct experiences of wildlife are not happy ones. I believe this is why birds are so important. Among the wild vertebrates, birds are exceptionally accessible to us; they are everywhere, from inner cities to the most remote human outposts, and they are not at all reclusive. We all have an opportunity to directly experience these wild creatures, and only the most curmudgeonly would begrudge it.

Of birds' many salutary qualities the most celebrated is their song. All birds make noises, but only some of those noises are considered songs, and not all song-singers are true songbirds. The anatomical feature that makes a songbird a songbird is the syrinx, a vocal organ. The true songbirds, or oscines, constitute but one suborder of the perching birds (Passeriformes). This suborder is, however, extremely species-rich and contains most of the birds with which we are familiar, including sparrows, finches, thrushes, wrens, blackbirds, larks, buntings, vireos, and warblers—to name a few common types. Each species sings characteristic songs that can be distinguished by such qualities as pitch, color, phrasing, and length. Expert birders can identify many songbird species by their songs alone. For the less discerning, a visual printout of the song, called a sonogram, can help identify the salient features.

The sonogram is also useful in assessing the size of a songbird's repertoire, a trait known as versatility, though there is some disagreement as to how to measure it. Everyone agrees that the marsh wren is a more versatile singer than the white-crowned sparrow and that a California thrasher is more versatile than a marsh wren, but quantifying these differences can be somewhat tricky.[1] What are the relevant song units? There are several possible candi-

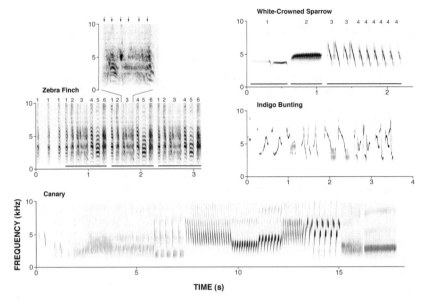

Figure 7.1 Representative sonograms for the zebra finch (A), white-crowned sparrow (B), and canary (C)

Numbers above the sonograms identify distinct syllables, each of which consists of one or more notes. Syllable number 3 in the zebra finch sonogram (a) is blown up to reveal six notes, as indicated by the arrows

(Adapted from Brenowitz et al. 1997, p. 497)

dates (figure 7.1). A note is the basic element of song and is defined as a continuous trace on a sonogram. A syllable consists of two or more notes; and a song consists of either one syllable that is rapidly repeated or, more commonly, two or more syllables. By either of these criteria, marsh wrens have larger song repertoires than white-crowned sparrows. But, because some of the most adept songsters, such as thrashers, sing more or less continuously without the sort of punctuation typical of wrens and sparrows, the most suitable measure of versatility is the number of syllable types sung within some specified period of time.[2] By this criterion the most versatile singers in North America come from a family (Mimidae) that includes the thrashers.

In general, the best time of day to judge a bird's versatility is just before sunrise. It is at dawn that thrashers and most other birds are particularly vocal, their combined efforts resulting in what is known as the dawn chorus. Members of some species, however, do not respect this convention and instead opt to get a jump on the sonic competition. It is these birds that are most likely to be classified as pests, even by those like me, who are generally sympathetic listeners. One such convention violator is a thrasher relative, and one I can enjoy from my own backyard—the mockingbird (*Mimus poly-glottus*). Generally, I consider myself fortunate to live in the company of mockingbirds, but not on those summer nights when I must keep my bedroom windows open. Mockingbirds, unfortunately, are among that minority of birds that do not respect the dawn chorus convention. In fact, they seem most motivated to sing during the midnight-to-dawn time slot. Regrettably, their varied songs are not nearly as enjoyable from bed as they are from the deck.[3]

Aside from the sheer variety of their vocalizations, and their occasional sleep-negating effects, mockingbirds are noteworthy for their ability to imitate non-mockingbird sounds. Mockingbirds—along with the thrashers—belong to a family of birds commonly referred to as the mimic thrushes, and their capacity for imitation is truly impressive. More than once, I have been duped into wondering why a Pacific treefrog (*Hyla regilla*) was calling from the top of a telephone pole, and during the middle of the day, only to realize the call was coming from a mockingbird. Mockingbirds can do various insects, squirrels, and squeaky doors as well. But it is their imitations of the songs of other birds that are the best documented. One mockingbird reportedly included the songs of 32 other songbird species in its repertoire.[4] I certainly would not trust my ear to evaluate the quality of the imitation. Fortunately, we can use sonograms to make precise comparisons of the original and the copy. Such a comparison was made of a whip-poor-will (*Caprimulgus vociferus*) song and a mockingbird's imitation. The whip-poor-will is named for its well-known three-syllable call. To the mockingbird, it must sound like five syllables because, according to the sonogram, that is how many are included in its imitation. But then the sonogram indicates that the whip-poor-will original is actually five syllables as well.[5] The mockingbird must also get other details right, because imitator and imitated cannot be distinguished on a sonogram. That is an aptitude beyond perfect pitch. When mockingbirds sing to each other, we are obviously hearing only a fraction of what they hear. How do we explain this remarkable ability? This would seem to be one of those occasions when it is not just paranoia that compels us to ask "Why"?

Adaptationists have offered a variety of teleological explanations for the mockingbirds' capacity for mimicry.[6] The assumption that undergirds all of them is that vocal mimicry in birds is a capacity "designed" for the purpose of imitating the songs of birds of other species. (As yet, there have been no attempts to explain the adaptive significance of the mockingbird's imitation of a Pacific treefrog). According to one such hypothesis, the mimicry of other bird species evolved because mockingbirds compete and interact aggressively with birds of a variety of other species. The ability to produce their songs confers a competitive advantage.[7] Other adaptationists have proposed that copying the songs of other songbird species is a way to generate a large song repertoire,[8] and large repertoires are thought to be desirable for male mimics because females are attracted to the most versatile singers.

But none of these adaptationist explanations, it turns out, help us understand mockingbird mimicry. They are simply off the mark. The remarkable ability of mockingbirds to imitate ambient sounds and songs is just a byproduct of their versatility, rather than an adaptation of any kind. In exploring the byproduct explanation of mimicry, we will need to delve more generally into the how-biology of birdsong, including its neural substrates. This fascinating phenomenon will in turn provide our entree into two topics, sexual selection and the brain, which I have heretofore approached only indirectly. Birdsong is ideal in this regard for several reasons. First, ethologists have provided abundant behavioral data; second, neuroethologists have provided extensive information on the neural substrates for this behavior; and finally, the evolution of both the behavior and the neural substrates is considered to result primarily from sexual selection.

DARWIN'S SECOND BIG IDEA

Darwin did not rest on his laurels after *The Origin of Species*. He continued to write prolifically for his own pleasure and for publication. One of these post-*Origin* efforts, entitled *Sexual Selection and the Descent of Man*, would itself have made the reputation of any lesser scientist. In this book, Darwin, in addition to addressing the particular case of human evolution, broached the idea of sexual selection. Darwin posited sexual selection to meet a particular sort of challenge to the idea of evolution through natural selection. Natural selection was Darwin's very successful attempt to explain the origin of adaptations, conceived as those traits that help their bearer survive. Certain sorts of characters, however, though obviously influenced by selection, were difficult to categorize as adaptive in this sense. For example, the feathery finery of peacocks (*Pavo cristatus*), although certainly attractive, seemed to make their lives more difficult in a number of ways. Because of those feathers, peacocks are more conspicuous than peahens, so they are more likely than peahens to be preyed upon by predators of peafowl; furthermore, peacock ornamentation makes escape from predators difficult. Nor is their conspicuous coloration adaptive in the manner of the striking pelage color of a striped skunk. A skunk's bold stripes reduce the likelihood that harm will come to it, because potential harmers learn to associate the stripes with noxious sickening odors.[9] The gaudy coloration of the peacock actually increases the likelihood of his becoming prey.

Darwin felt the need to distinguish between the kind of selection that increases fitness by increasing the chances of survival (natural selection) and the kind of selection that causes peacocks to be so colorful and ornamented. It is the latter that he termed sexual selection. According to this principle, peacocks are highly ornamented because females prefer to mate with highly ornamented males. Darwin considered sexual selection exceptional in not resulting in adaptation.[10] Some biologists, following Darwin himself, prefer to view natural selection and sexual selection as distinct nonoverlapping categories, but the majority of evolutionary biologists now view sexual selection as a subtype of natural selection, albeit one with special properties. Sexual selection refers to natural selection for traits related more or less directly to reproductive success. An important characteristic of such selection is that it results from interactions with other members of the same species.

There are two categories of sexual selection, distinguished by the nature of these interactions: *Intrasexual* selection results from interactions with other members of the same sex; *intersexual* selection results from interactions with members of the opposite sex. The natures of intrasexual interactions and intersexual interactions are qualitatively different. Intrasexual interactions are typically aggressive, as, for example, those between male elephant seals. Intersexual interactions, such as those between peacock and peahen, fall under the category of courtship. Hence, the two components of sexual selection result in different kinds of traits: Intrasexual selection promotes traits that enhance social status, such as size and strength; and intersexual selection promotes traits that enhance the attractiveness of members of one sex to members of the other sex.

Either component of sexual selection can result in sex differences, or "sexual dimorphisms." Moreover, differences in the intensity of selection on males and females can result in sex differences as well. Males typically experience more intense sexual selection than females, simply because there is generally more variation in male reproductive success. One indicator of this sex difference is the proportion of successful breeders among males and females. In many species, the proportion of females that successfully reproduce exceeds that of males. This variation in male reproductive success is the engine for sexual selection; the greater the magnitude of this variation, the greater the intensity of sexual selection. This variation is especially high in those polygynous species in which a relative few males monopolize matings. For monogamous species, on the other hand, there is much less variation in male reproductive success and hence less opportunity for sexual selection. Therefore, we would expect to find greater sexual dimorphisms in species with polygynous mating systems. If, as in elephant seals, sexual selection is primarily intrasexual, we would expect sex differences in size and weapons such as horns, antlers, and teeth. If, however, sexual selection is primarily in the form of intersexual selection, we would expect sex differences in coloration and bodily adornments. Indeed, the most spectacularly adorned male birds are those, such as peacocks, for which males experience especially intense intersexual selection. Peacocks do not bother fighting amongst themselves; they just attempt to position themselves so as to show their impressive feathery finery to maximum effect. Male-male competition is confined to getting in each other's way or screening a competitor from the view of a female, much as bodybuilders do in a flex-off. The female chooses her mate solely on the basis of his appearance and demeanor.

Why should a peahen choose to mate with the most beautiful peacock? What is the evolutionary advantage for the peahen? One adaptive rationale for female choosiness would be to identify a mate that could protect and provision her and her offspring. The female preference for big gorgeous tails would make sense if the peacock's tail were an indicator of what kind of a father he will be. But peacocks are not at all paternal. In fact, the peacock's contribution to his offspring ends with fertilization. After mating, the peahen ambles off, alone, to lay her eggs and incubate them. The peacock returns to his preening in the quest for more copulations—the more, the better.

Thus the peahen's preference for elaborate peacock tails presents something of a conundrum for adaptationists: it is not apparent how mating with a well-adorned peacock can enhance the peahen's fitness, given the peacock's minimal role as a father. Even R. A. Fisher, something of an iconic figure for adaptationists, despaired of finding a function for the peahen's preference for, and the peacock's possession of, a gaudy tail. He proposed instead that both resulted from a physical linkage between a gene that causes the enhancement of the tail in males and a gene that causes females to prefer males that possess these adornments. This linkage can result in a feed-forward or runaway process in which the female preference for males with enhanced adornments results not only in an increase in the frequency of adorned males but also in an increase in the frequency of females who prefer adorned males, which further increases the fitness value of such adornments and subsequent further adornment enhancements. The fitness enhancement that results from the female choice, it should be noted, has nothing to do with adaptation in the original Darwinian sense. This non-adaptive explanation of peacock tails is still the one preferred by many biologists who approach the problem from a genetic perspective.[11]

Adaptationists do not find this a very satisfying explanation, primarily because it provides a how-answer for a why-question. They want the choosiness of a peahen and the peacock's feathery finery to be adaptively linked, not simply a matter of contingent genetic covariation. They assume that peahen choosiness serves some greater good, so peacock tails must be an indicator of the degree to which the peacocks possess that greater good. Because the peacock provides only his sperm to the peahen, this greater good must reside in his genes. The peacock's tail must therefore be an indicator of the goodness of his genes.[12] On this view, the most beautiful peacocks are the ones that are genetically blessed, so, in choosing a mate on the basis of beauty, the peahen is not merely selecting males with beauty alone. Beautiful peacocks are also strong, physically robust, and virile pea-cocks—all-around studs—and they are manifestly studly because they are genetic studs.

Good Genes and Sensory Exploitation

Originally, sexual traits such as peacock tails were conceived, by good-genes enthusiasts, to be "honest" signals that directly reflected a peacock's genet-ically endowed studliness.[13] But there have been a number of proposals as to what, precisely, the genetic substrates of studliness are. One proposal was that robust sexual traits such as peacock tails reflect the overall genetic diversity, or heterozygosity, of their bearer, not any particular genes or com-bination of genes.[14] Heterozygosity has been found to be associated with developmental stability[15] and increased viability or general health.[16] So a female who mates with the most heterozygous males would produce more heterozygous and, hence, more viable offspring. But the good-genes-as-het-erozygosity hypothesis failed to garner much enthusiasm among good-genes enthusiasts,[17] partly because it departs from the notion that males can be ranked on an absolute scale for their genetic value in a way that females can directly assess.[18] Instead, the best mate for any given female will depend on her own genetic makeup.[19] Much more popular is the somewhat related view that a quality known as *fluctuating asymmetry* is the good-genes indicator. Fluctuating asymmetry, which refers to random deviations from bilateral symmetry—the symmetry of the left and right halves of the body—has long been of interest to evolutionary biologists as an indicator of developmental stability,[20] but it has only recently been put to use by good-genes enthusiasts. In the case of the peacock's tail, fluctuating asymmetry may be measured, for example, in terms of differences in the number of feathers on the left and right halves. Large deviations from symmetry indicate developmental insta-bility, whereas low fluctuating asymmetry indicates developmental robust-ness, an ostensible source of studliness in males.[21] The theory of fluctuating asymmetry as a good-genes indicator of studliness had an initial wave of faddish enthusiasm, but skepticism has begun to mount on a number of fronts, and its advocates are on the defensive.[22]

Aside from the fluctuating asymmetry bubble, good-genes enthusiasts have become less preoccupied with straightforward indicators of fitness; in-stead the emphasis is on "costly" status "badges" that actually compromise the fitness of their bearers but nonetheless render them more attractive to females. This is known as the handicap principle,[23] which is meant to ad-dress an important omission in the original good-genes hypotheses. The problem is cheating: What is to prevent a wimpy peacock from growing a large beautiful tail and thereby fooling the females? One way to keep things honest is to make the tail costly to maintain, such that a wimpy peacock could not grow a large tail without risking death. Only the true studs could maintain and flaunt the wedding gown–like feathery train in the face of its fitness-reducing costs. The large, ornate peacock tail is the equivalent of a wooden leg, and those males that possess such a tail are the ones over which the discerning female is most likely to swoon, because they manage to be studly despite their handicap.

But what, ultimately, are the costs that make the studly males' good genes good? There have been a number of proposals, but the one that has garnered the most attention focuses on resistance to the malign effects of parasites.[24] Perhaps the peacock's feathery finery signals prospective mates of his vigorous health in the face of this ubiquitous curse. If this resistance is heritable, the adapted female should prefer those males with the biggest and brightest tail as fathers for her offspring. The logic is sound, but to date the evidence for the parasite version of the good-genes hypothesis for mate choice is decidedly mixed.[25]

The latest version of the good-genes account of intersexual selection, called the immuno-competence handicap hypothesis (ICHH), combines the parasites theory and the handicap principle. On this view, the size of the peacock's tail is directly related to his testosterone level; the higher the testosterone titer, the bigger the tail. But the high testosterone levels are burdensome in a number of ways. In particular, high levels of testosterone can compromise the immune system.[26] Immuno-handicappers infer from this fact that testosterone—not the large tail itself—constitutes a handicap; the more testosterone, the greater the handicap. But testosterone levels cannot be directly observed by the peahen, of course, so the peacock's tail functions primarily as a testosterone handicap indicator or as an indicator of the degree to which this handicap has been surmounted by the peacock's studliness-promoting genes, or both. A peacock that can maintain a large tail— and vigorously display it—in the face of his testosterone-compromised immune system should be recognized as a truly deserving father by the discerning peahen. If this all sounds rather rococo, you are perhaps not a sufficiently committed adaptationist. But there are reasons other than esthetics alone to resist the immuno-handicap principle.[27] Most fundamentally, the immuno-handicap principle, as do all other handicap explanations of sexual traits, suffers from a lack of one critical piece of evidence—evidence that these traits are in fact costly.[28] For a sexual trait to qualify as costly in the relevant sense requires (1) that the trait be demonstrated to have negative fitness consequences and (2) that the negative fitness consequences are condition-dependent; in particular, the negative fitness consequences must be greater for individuals in poor condition than for those in good condition. To date, there is but circumstantial evidence that any sexual trait meets those criteria.[29]

Recently, a third player has joined the arena to compete against the runaway-selection and good-genes accounts of mate choice; it is called the sensory exploitation hypothesis.[30] Michael Ryan, to whom we owe this idea, proposed that all animals have sensory biases of various sorts, sensitivity to light of particular wavelengths (colors), for example. But, just as important, there are also response biases to certain stimuli within the detectable range. These biases may be the byproducts of selection for species recognition or of selection for detecting and responding to particular sorts of food items, for example. As far back as the 1950s, the ethologist Niko Tinbergen recognized the importance of such biases, which he incorporated into his concept of the "supernormal stimulus." A supernormal stimulus, according to Tinbergen, is one that exceeds, in one or more dimensions, the stimulus that normally triggers a particular response. The "egg dummies" on the anal fins of male *Haplochromis burtoni* are supernormal stimuli in this sense. They are both larger and more brightly orange than actual *H. burtoni* eggs, but *H. burtoni* females, as we have seen, find them irresistible; indeed, the egg dummies, more than the actual eggs themselves, stimulate the females to attempt to place them in their mouth; it is this response that makes oral copulation possible in this species.

James Gould was perhaps the first to connect sensory biases of this sort to sexual selection,[31] but it was Ryan who worked out the full implications of this connection. He suggested that in many types of courtship, males, in advertising themselves, play to the particular sensory biases of the females of their species, whatever the source of this bias.[32] If for example, members of a species are particularly sensitive to red colors, the males are more likely to employ red colors in their courtship rituals. This is precisely what occurs in the guppy *Poecilia reticulata*. These denizens of the rivers of Trinidad are attracted to orange things because it is the color of the fallen fruit that is their primary food source. Female guppies are also attracted to male guppies with orange spots, not, as was previously supposed, because orange spots indicate good genes, but because of a sensory bias for things orange.[33]

From cases such as these, Ryan derives some surprising conclusions about the nature of sexually selected traits, which contrast in important ways from both the runaway selection and the good-genes views. Sensory exploitation differs from runaway selection because it does not require genetic covariation in the genes that contribute to the male trait and those that influence female choice. More fundamentally, however, sensory exploitation resembles runaway selection, but contrasts with the good-genes idea, in providing a nonteleological explanation for mate choice.

The sensory exploitation hypothesis is particularly compelling in providing an explanation for the results of some comparative studies that just don't make any sense from a good-genes perspective. For example, Ryan and his associates have identified a sensory bias within a group of Neotropical frogs (family Leptodactylidae).[34] Males in several species of the genus *Physalaemus* emit a call with a distinct "whine" component. Some species have evolved elaborations of the basic whine, thereby adding complexity to the call. *Physalaemus pustulosus*, for example, has added "chucks" to its whine. Not surprisingly, *P. pustulosus* females are more attracted to male mating calls that contain chucks than to those that consist only of whines. What *is* surprising is that females of the closely related species *P. coloradorum* are also more attracted to male calls that contain chucks. This fact is surprising because *P. coloradorum* male calls consist only of the whines; they are completely chuck-less. *Physalaemus coloradorum* females have a distinct preference for a sound that *P. coloradorum* males cannot produce. We can predict that some lucky future mutant *P. coloradorum* male that produces chucks may spend the rest of his waking hours in amplexus, and not because the chuck is a handicap or because the chuck is any indicator of his good genes.[35]

Another nice piece of evidence for sensory exploitation comes from yet more studies on members of the poeceliid family of teleost fishes, which includes the Amazons and platys discussed in previous chapters. Swordtails belong to the same genus (*Xiphophorus*) as platys. Unlike platys, the males have sword-like projections from the bottom of their caudal (tail) fins. As you would expect, female swordtails prefer males with these extensions to those that lack them. But so do female platys; they find male platys whose tail has been artificially augmented with a sword-like extension more attractive than normally endowed male platys.[36] From this finding we can conclude that there is a preexisting bias for these extensions, which has been exploited by male swordtails but not by male platys.

One of the considerable virtues of the sensory exploitation idea is that it considers mate choice in the context of the properties of the nervous systems of the animals that are doing the choosing, their how-biology. Another (related) virtue is that it considers mate choice in the context of evolutionary history or phylogeny.

Sexual Selection and Birdsong

Birdsong is thought to be primarily the product of sexual selection. As we would expect of a sexually selected trait, it is sexually dimorphic. Singing is primarily a male activity.[37] Songs are employed to intimidate rivals and stake claims for territories (intrasexual selection); they are also used to attract mates (intersexual selection). It has been proposed that the two components of sexual selection promote different singing attributes. If intrasexual selection (male-male competition) predominates, song repertoires are small and consist of simple songs; whereas, if intersexual (mate attraction) selection predominates, the songs are more complex and the song repertoires are larger.[38]

Polygynous songbirds in which intersexual selection predominates should therefore be the most versatile singers. Adaptationists presume that females prefer to mate with the most versatile male singers, because their versatility is an indicator of their fitness, the goodness of their genes.[39] If a male's good genes are good because they help him resist parasitic infections, we should expect parasite loads to be inversely related to versatility within a species.[40] We would also expect stronger intersexual selection for versatility in species with high rates of parasitic infections than in those with low rates of infection. To date, however, there is little evidence of such correlations.[41]

Of course, even if this particular good-genes hypothesis is universally deemed falsified, an adaptationist can always come up with another variation on the theme.[42] Perhaps, though, we should seriously consider some alternatives to the good-genes explanations for female preferences, such as sensory exploitation. Consider, for example, the case of the common grackle (*Quiscalus quiscula*), a member of the blackbird family (Icteridae). Female grackles prefer males that sing a variety of songs. Unfortunately for them, all male grackles have a repertoire of only one song.[43] Why would female grackles prefer large repertoires, if male grackles cannot deliver them? This preference is not as mysterious as it may at first seem. It is a generic trait of many animals, completely unrelated to ecological factors or mating systems, that they habituate or become desensitized to a stimulus the more it is repeated. Female grackles may prefer large repertoires simply because they find large repertoires more stimulating than small ones.[44] So female grackles are doing what many animals would do under the circumstances. Their preferences do raise an interesting why-not question: Why don't male grackles have larger song repertoires? Why do they bore their prospective mates by singing the same song over and over and over again despite the females' preference for variety?

Mockingbirds pose another problem for traditional sexual selection accounts of birdsong, because these most versatile of singers are monogamous. As a monogamous species, they should not be subjected to the sort of intense intersexual selection experienced by peacocks.[45] So why do they have the vocal equivalent of a peacock's tail? Perhaps mockingbird versatility is not at all analogous to the peacock's tail. Indeed, from the perspective of the sensory exploitation hypothesis, it might behoove us to look to factors other than selection in explaining variation in repertoire size—genealogy, for example. Perhaps there is considerable evolutionary inertia with respect to repertoire size.[46]

Adaptationist explanations of the variation in birdsong versatility focus, of course, on the selective milieu, especially the mating system. Yet, attempts to identify the ecological correlates of variation in versatility have met with only very limited success.[47] On the other hand, there is ample evidence that genealogy matters. For instance, it is a safe bet that a species of wren (family Troglodytidae) or thrush (family Turdidae) chosen at random, will be a more versatile singer than a species of New World sparrow (family Emberizidae) chosen at random, irrespective of ecological considerations.[48] Moreover, many of the most versatile singers belong to one of two families: Mimidae (mimic thrushes) and Sturnidae (starlings and mynahs), which are closely related.[49] It would seem that the time has come to look beyond ecology. Considering birdsong in the context of its development will help to understand why there might be an ecology-defying inertia in repertoire size.

THE DEVELOPMENT OF BIRDSONG

All songbirds learn their songs by copying the songs of other members of their species. To do so, a young songbird must not only remember heard songs; it must also match its own vocalizations to this auditory memory. First a song model is acquired in the form of an auditory memory; second, the vocalized song is acquired as a motor skill. I will refer to the first step as *sensory* learning and the second as *sensorimotor* learning. The sensory phase is always initiated before the sensorimotor phase, but in most species the sensorimotor phase commences considerably before the end of the sensory phase. Hence, the young bird is both laying down new song memories and attempting to sing at the same time. But in some species, such as the song sparrow, there is no overlap of sensory and sensorimotor phases; young birds do not attempt to sing at all during the sensory phase, and in some cases there may be a several-month lag between the offset of the sensory phase and the onset of the sensorimotor phase. Young swamp sparrows (*Melospiza georgiana*) are maximally prone to memorize the vocalizations of other birds from about 20 to 50 days of age.[50] This is often referred to as the sensitive period.[51] The youngster does not begin to sing itself, however, until months later. During the sensory phase, it is all ears (figure 7.2).

Ethologists have identified a number of distinct behavioral landmarks in the sensorimotor phase of song learning. During the first, or *subsong*, stage, the initial efforts result in a cornucopia of almost random-seeming sounds, albeit at low volume. The sheer variety of sounds is impressive, but there is an obvious lack of discipline and structure. Think of this as the John Cage stage of song development. Next comes *plastic song*. Now there is some species-typical syllable structure, but the songs are still continuous, and they have an improvisational quality about them. There is no recognizable repertoire. Think of this as the Keith Jarrett stage of song development. At some

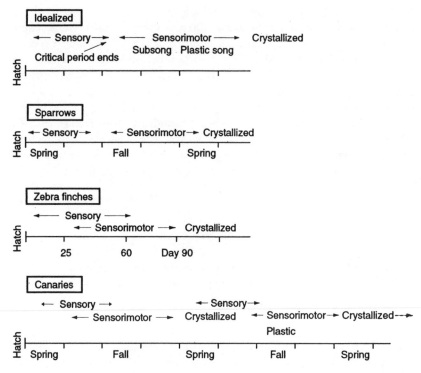

Figure 7.2 Stages of song learning in an idealized songbird, two age-limited learners (the zebra finch and the white-crowned sparrow), and an open-ended learner (the canary)

(Adapted from Doupe 1993, p. 105)

point late in the Jarrett stage, distinct songs start to precipitate out of the unstructured music matrix, a process known as *crystallization*. The number of crystals that precipitate out varies widely between species. White-crowned sparrows (*Zonotrichia leucophrys*) usually end up with one crystal, northern juncos (*Junco hyemalis*) 3 to 5, and song sparrows (*Melospiza melodia*) with over 10.[52] In most songbird species, all the non-crystalline material is lost, never to be sung again. Only the crystallized songs are sung—over and over and over again, once this period of full song or *final song* commences. Call this the Philip Glass stage of song development.

Tutoring and Instrumental Learning

On the traditional ethological account, song learning is viewed as a tutored process, the older birds acting as instructors.[53] This can't be the whole story, however. It does not explain the winnowing process that takes place toward the end of plastic song, during which the repertoire is drastically reduced. Moreover, some of the songs that remain are further fine-tuned. The tutoring model breaks down here, because these later changes clearly result from sexual and aggressive social interactions. For example, during its first breeding season, the young male often ends up far from his natal nest. And, as do humans, many bird species speak regionally distinct dialects, often on very local scales.[54] There is ample evidence that it is advantageous during territorially motivated singing for the males to match dialects.[55] Males that can match are more successful in securing a territory than those that cannot. Hence, a young male will drop his initial dialect in favor of the local one, solely on the basis of its relative efficacy in producing the desired effect.

Similar changes in song repertoire result from male-female interactions, as demonstrated in some wonderful studies on brown-headed cowbirds (*Molothrus ater*) by Meridith West and John King.[56] A generic example will suffice for my purposes here. Say a male is courting a female. He has been singing a song, *tu weet wee*, to which the female has so far been indifferent. Now, out of frustration perhaps, he puts more energy into the *wee*, so the end becomes a trilled *weeeeeeeee*. The female then immediately goes into a copulatory posture. Our male responds by putting even more emphasis on the final trill. The female becomes even more ardent, and so on. Our male has experienced behavioral modification and has himself a new song. The singer makes song modifications on the basis of the reaction of the receiver. It is not surprising that such interactions would change a bird's tune.[57]

The Song Circuits

Song learning has been enabled by some evolved alterations in the forebrain of songbirds. Of particular importance are two distinct circuits of interconnected brain nuclei (figure 7.3). One circuit (HVC → Area X → DLM → lMAN) functions in laying down song memories during the sensory phase of song learning. I will refer to this as the song acquisition circuit. If this circuit is damaged, the bird will fail to develop a normal song because, it is thought, it cannot form a proper template or model. The other circuit (HVC → RA → n12ts) regulates song production during the sensorimotor phase. I will refer to this as the song production circuit. If a significant number of neurons in any of these three nuclei are damaged or inactivated, the bird will not be able to produce any song, though it may assume the singing posture and even move its beak appropriately.[58]

The HVC, which participates in both song circuits, and hence in both song acquisition and song production, has received more attention than any other song nucleus from birdsong researchers. And it will be my primary focus throughout the rest of this chapter.

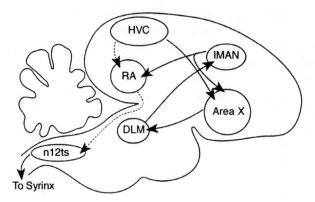

Figure 7.3 The song nuclei of songbirds. Dotted lines connect the nuclei in the song production circuit; solid lines connect the nuclei of the song acquisition circuit

(DLM = medial portion of the dorsolateral nucleus of the thalamus; lMAN = lateral portion of the magnocellular nucleus of the anterior neostriatum; HVC = high vocal center; nXIIts = tracheosytingeal part of the hypoglossal (twelfth cranial nerve) nucleus; RA = robust nucleus of the archistriatum; X = area X)

(Adapted from Doupe 1993, p. 105)

Age-Limited and Open-Ended Learners

Songbirds have been divided into two broad categories, based on their capacity to acquire new songs. Those species, such as zebra finches and swamp sparrows, that can acquire new songs for only a limited period during the first year are called age-limited learners. Those species, such as mockingbirds and thrashers, that can acquire new songs throughout their lifetime, are called open-ended learners. Recent investigations have begun to reveal some interesting differences in the development of the neural substrates for song learning in age-limited and open-ended learners.[59] This research is part of a broader effort to relate changes in the song circuits to song learning, which has become a central project for an offshoot of ethology, called neuroethology.[60]

The zebra finch (*Taeniopygia gutatta*), has long served as the primary experimental model of age-limited song learning for neuroethologists. Native to arid areas of Australia, it breeds opportunistically, any time during the year when there is sufficient rainfall. Even heavy rains only briefly meliorate the desert enough for plants to grow and insects to proliferate, so zebra finches must proceed through the breeding cycle with alacrity. Young zebra finches must also grow up and learn to sing quickly. In fact, young zebra finches can begin to memorize songs before they are 25 days old,[61] and they begin to sing sometime around 25 to 30 days of age. Even more impressive, the song learning is virtually complete by 60 to 70 days, and song has crystallized by day 90.[62] So by 3 months of age these birds have learned the only songs they will ever sing. For, like some age-limited learners, zebra finches stop learning after they have acquired one song.

The basic pattern of song circuit development in open-ended learners such as mockingbirds, red-winged blackbirds (*Agelaius phoeniceus*), and various thrushes, is, initially, much the same as in the zebra finch. But open-enders maintain the ability to learn throughout their lives, and this requires a somewhat different pattern of neural development. The canary, beloved of miners and pet fanciers alike, has been the experimental model for this group of species. Compared with zebra finches, young canaries are slow learners, but then, they have a lot more to learn, and less urgency. Their first-year songs are not fixed until sometime between 180 to 300 days (6 to 10 months) of age, which means they can still be fiddling with the repertoire during their first breeding season (spring), which commences around day 240.[63] Canaries make their first Cage stage (subsong) attempts to sing around days 42 to 45. The transition to Jarrett (plastic song) singing occurs at around day 60, and they gradually polish their songs for the next several months. After its first breeding season, the canary begins to sing less but experiment more. During the fall, it returns to the plastic song phase again. This seasonal transition from crystallized song to plastic song and back again occurs throughout the canary's life. Fernando Nottebohm, a pioneer in the field of neuroethology, noticed seasonal changes in the song system that paralleled these behavioral changes.[64] The HVC, in particular, shows remarkable seasonal plasticity, reaching its largest size during the spring breeding season and its nadir in late summer, when canaries are reproductively quiescent.

Brain Space for Learned Song

Ever since Fernando Nottebohm discovered the relationship between seasonal changes in HVC size and song production in canaries, there has been a concerted effort to link variation in HVC size with variation in repertoire size. In assessing the relationship between the size of the HVC and repertoire size, the most straightforward comparisons are within species. It is therefore noteworthy that those canaries with the largest repertoires also tend to have larger HVCs than their peers who are less versatile.[65] Subsequent studies on other songbird species have yielded conflicting results, however;[66] so at this point we can only keep an open mind as to the nature of the association between HVC size and versatility within species.

In any case, evolutionary biologists are primarily interested in how HVC size differs among populations and species. The most compelling study on population differences in HVC size was done on a member of the wren family (Troglodytidae). Wrens are drab little ground-dwelling birds that are generally excellent songsters. The marsh wren (*Cistothoras palustris*), a species with a wide range across the United States, is an outstanding singer with an impressive repertoire. But some birds have larger repertoires than others, and the size of their repertoire is largely a function of where they live. For some reason, California birds are much more versatile than New York birds.[67] The more versatile California birds also have larger HVCs than their smaller-repertoired counterparts from the Empire State.[68]

Timothy Devoogd and his colleagues adopted a much more broadly comparative approach in their investigation of the relationship between HVC size and versatility.[69] They found a significant association between HVC size and repertoire size among 45 species—including the mockingbird—of songbirds, representing a wide variety of families. This is certainly an interesting result. It should be noted, however, that variation in HVC size accounts for only about one-third of the variance in song repertoire size. A lot of the variation in repertoire size is therefore not explained by variation in HVC size.

Sex Differences in Singing Behavior and Song Nuclei

Adaptationists have been eager to link variation in both versatility and HVC size to sexual selection. To this end, the first order of business is to document sex differences. Because singing is primarily, if not exclusively, a male endeavor, there has been a concerted effort to identify sex differences in the song circuits, and particularly in the HVC. This research is significant in providing the benchmark for research on neural sex differences in general—including that conducted on humans. Hence, it is instructive to examine what is known about sex differences in the song circuits and how they are related to singing behavior.

Most of the studies of sexually differentiated song circuits have been conducted on zebra finches. As it happens, zebra finches are among the most sexually dimorphic bird species with respect to singing behavior. In fact, zebra finch females do not sing at all. The degree of sexual dimorphism in singing behavior varies greatly from species to species, however. Though the female mockingbird does not sing at all during the breeding season, she holds her own during autumn and winter, at which time she may engage in lengthy song duels with her former mate, whom she now treats like any other territorial rival. The wren family contains examples of every degree of female singing. Carolina wrens (*Thryothorus ludovicianus*) resemble zebra finches in that the females almost never sing. In contrast, two Neotropical species, the bay wren (*T. nigricapillus*) and the buff-breasted wren (*T. leucotis*) sing duets to which males and females contribute equally.[70] There are all manner of intermediate conditions as well. The rufous and white wren (*T. rufalbus*) is another duetting species, but the males contribute about twice as many syllables to these joint efforts as the females.[71]

Neuroethologists have looked for correlated differences in the neural song system of these birds. As expected, zebra finches, in which only the males sing, show the greatest dimorphism in the song circuit. At the neural level, the HVC in female zebra finches is borderline vestigial, and females entirely lack a definable Area X.[72] The zebra finch is exceptional in this respect, however. In most species, the entire song circuit is present in both males and females. But among wrens, at least, the sex differences in HVC size are roughly correlated with the sex differences in the amount of singing. Neither bay wrens nor buff-breasted wrens, two duetting species, evidence any sex differences in HVC volume.[73] The HVC is larger in males than in females in another duetting species in which the male has the larger role in the duets. On the basis of the foregoing, we would expect an even greater sex difference in the song nuclei of the Carolina wren, a male-only singer. And a recent study does confirm that the sex difference in the song nuclei of this species approaches that of zebra finches.[74]

The standard assumption in neuroethology is that there is a deterministic, indeed linear, relationship between the amount of brain space and the behavior it subserves. The more neurons there are, the more of the related behavior or capacity.[75] As a working hypothesis, this is unobjectionable. The data on singing behavior and song nuclei in wrens are consistent with this working hypothesis, but there is enough counterevidence available to warrant caution. The research of Manfred Gahr and his colleagues on sex differences in the song nuclei of the East African bush shrikes (*Laniarius funebris*) indicates that the relationship between brain space and birdsong is more complex.[76]

However complex the relationship between HVC size and singing behavior turns out to be, there is no doubt that there are robust sex differences in both. We are still a long way from understanding the role of sexual selection in the evolution of these sex differences.[77] But we do know a great deal about the how-biology of sexual differentiation of the HVC and other song nuclei. In fact, we probably know more about the how-biology details of the sexual differentiation of the HVC and the song circuits than about any other comparable neural substrates for a complex behavior.[78] It is for this reason that the sex differences in birdsong brain-behavior patterns provide something of a standard by which to judge the claims for adaptive sex differences that I will discuss in subsequent chapters.

Genealogy, Development, and Evolution of Birdsong

The sex differences in brain and behavior described above are at least broadly consistent with the view that birdsong has been influenced by sexual selection. But we have yet to establish a link between sexual selection and versatility. Indeed, the most broadly comparative studies conducted to date indicate that this link is tenuous at best.[79] It is therefore important to consider some additional factors that might influence versatility.

One such general factor to consider is genealogy of course. Substantial alterations in the size of a species' repertoire require alterations in sometimes intricately coordinated events during the development of the song circuits. To the degree that changes in this circuitry involve substantial developmental reorganization they will be resisted. If this resistance, or evolutionary inertia, is substantial, we would expect versatility to vary more with genealogical factors than with the intensity of sexual selection.[80] Consider, for example, what would happen if a population of swamp sparrows were subjected to intense intersexual selection for increased repertoire size. The first thing to note is that, if versatility is to be increased beyond a certain point, there must be a shift from age-limited learning to open-ended learning. But consider the fact that whole families or subfamilies of songbirds tend to be one or the other. All New World sparrows, for example, are age-limited learners; all thrushes are open-ended learners. This fact alone suggests that transitions between age-limited and open-ended learning are not trivial. We should consider the possibility that age-limited and open-ended learners constitute two fundamentally different developmental patterns.

Nonetheless, there is substantial interspecies variation in versatility within both age-limited and open-ended learners. So even if the transition from age-limited to open-ended song learning is a difficult one, we might still find much room for sexual selection in explaining variation within each category. For example, among open-ended learners, some, such as common grackles and red-winged blackbirds have small repertoires, while others, such as mockingbirds, have very large repertoires. But even here, the fact that the virtuoso mockingbirds are monogamous and the much less versatile red-winged blackbirds are polygynous should give us pause. At a minimum, this fact indicates that more than sexual selection is involved in determining versatility.[81]

Why Mimic?

There is an obvious taxonomic factor in the evolution of vocal mimicry, parallel to that which we observed for versatility. Most of the best vocal mimics belong to one of two closely related families of versatile singers, the one that includes the mockingbirds and thrashers (Mimidae) and another that includes the starlings and mynahs (Sturnidae). Indian Hill mynahs are perhaps the most famous mimics of all; their popularity as pets is related to their ability to mimic even non-bird sounds such as human speech. Starlings are also good mimics.[82] Like mynahs, but unlike mockingbirds, starlings can be raised in captivity. And it is under captive conditions that their mimicry is most evident. Starlings are particularly good at imitating human sounds when they are reared alone, that is, in the absence of other starlings. When other starlings are present, young starlings prefer to imitate them rather than humans. In fact, humans, it seems, are imitated only as a last resort. In nature, starlings, like mockingbirds, do imitate the songs of other birds, but starlings primarily copy the vocalizations of other starlings, just as any other songbird imitates its own kind. Their non-starling vocalizations are essentially mistakes.[83]

We might ask then, why do they make so many mistakes? But if we look at the big picture, neither mockingbirds nor starlings make all that many mistakes in nature. The vast majority of what starlings sing is starling songs, and the vast majority of what mockingbirds sing is mockingbird songs.[84] Moreover, older mockingbirds are less prone than younger ones to use phrases copied from other species, and this age difference is especially noteworthy given the way the mockingbird song repertoire is constructed. Mockingbirds are always making new songs by rearranging bits of older songs or adding and deleting bits of song from their current repertoire. When a phrase is dropped from the repertoire, it is not because it has been forgotten. Sometimes a mockingbird will resuscitate a phrase it has not used for years. Hence, if anything, mockingbirds are especially prone to remove heterospecific bits as they become more experienced.

It is true that starlings and mockingbirds make more mistakes than your average sparrow, and this fact, ultimately, is what invites speculation about the function of vocal mimicry. But the increase in mistakes made by mockingbirds relative to sparrows is simply a byproduct of their greater versatility. Assume for the moment that there is selection for large repertoires. In order to have a large repertoire, a bird must retain the ability to learn new songs into adulthood. This requires that both the period of sensory learning and the plastic song stage of sensorimotor learning extend into adulthood. Most of the remembered sounds come from parents or other adult conspecifics in the vicinity, especially if, as in zebra finches, the sensory phase is confined to a sensitive period that ends before the young are fully fledged. In this case, the juvenile's undivided attention is directed toward its parents and the food they come bearing. Even if the sensory phase extends beyond this period, most of the sounds to which the juvenile attends will be the songs of other members of its species, which, after all, constitute the most salient stimulus objects. There is, however, more opportunity to pick up extraneous bits from birds of other species and to practice them. Chances are, however, those elements will not elicit the desired response from either male or female conspecifics, and they will be dropped when song crystallizes.

If, as for the mockingbird, sensory learning is lifelong, there is ample opportunity for all kinds of extraneous sounds—not only the songs of other species, but non-bird and even inanimate sounds as well—to enter the vocal repertoire. From an evolutionary perspective, these should be viewed as harmless mistakes.[85] Selection for large song repertoires requires that mock-ingbirds and starlings learn more songs, and learning more songs requires an extended period of vocal plasticity, which makes them vulnerable to copy-ing extraneous stuff, which, fortunately for them, is of little consequence.

Mimicry has been reported for a wide variety of songbirds in natural set-tings,[86] but, as you would expect, it is much more pronounced in laboratory settings, with their abnormal array of salient stimulus objects. Among adult birds, vocal mimicry is largely confined to the most versatile singers,[87] but all songbirds experience a developmental period, during plastic song, in which they can mimic the sounds of other species and many other sounds as well. During plastic song, all birds match their own vocal outputs to remembered sounds, not all of which were produced by members of their own species. All songbirds mimic sounds with uncanny precision during plastic song. Even the swamp sparrow, one of the most variety-challenged singers on the earth, can mimic nonstandard sounds during this stage of its development.[88] Age-limited learners such as swamp sparrows limit their mimicry to the songs of their elders, primarily because the plastic song stage and the sensory learning phase are completed early in development. Open-ended learners still manage to copy the vast majority of their songs from conspecifics, and this habit too bears the hallmarks of selection. But open-ended learners mimic a lot of other sounds as well, not because they were designed to do so, but because doing so is of little consequence. And, initially at least, it just can't be helped.

MIMICRY, VERSATILITY, AND SEXUAL SELECTION

By way of summary then, the ethology of song learning is all we need in order to explain mockingbird mimicry. Mockingbird mimicry is merely a byproduct of how mockingbirds come to acquire their songs; we don't need to tell an adaptive story about it. As such, we should consider this how-explanation an outright replacement for any of the proposed why-biology explanations. In this case, the causal explanation competes directly with the teleological explanation and wins the contest outright.

The issue of interspecies variation in versatility is a more complex matter in this regard. Here we can make a more compelling case for the teleological approach, and sexual selection would be the obvious principle on which to focus. As we have seen, however, to date, this focus on sexual selection has not taken us very far in explaining species differences in versatility. The mockingbird, for example, is one of the most versatile singers, the vocal equivalent of a peacock; yet, unlike the highly polygynous peacock, the mockingbird is more or less monogamous and hence much less subject to intersexual selection. This combination of versatility and monogamy led us to consider the how-biology of changes in repertoire size, which, in turn pointed to alternative, genealogy-based, explanations for variation in versatility among songbirds. We should consider the genealogy-based explanation for variation in versatility a rival to the sexual selection explanation but expect both to play an important role in any adequate evolutionary account of this phenomenon.

But sexual selection itself, as we have seen, is not a monolithic process. Moreover, our discussion of intersexual selection, or mate choice, illustrates a point I made in chapter 1: we can embrace selection without going teleological. The teleological, or "good-genes," interpretation of sexual selection *is* the one favored by adaptationists, but there are two nonteleological accounts of sexual selection as well. One of these, the feed-forward, or "runaway selection," account, focuses on genetic processes; the other, sensory exploitation, focuses on the how-biology of sensory processes. The sensory exploitation hypothesis is particularly attractive, from the perspective advocated throughout this book, in embedding our thinking about intersexual selection in both the how-biology and genealogical contexts. Indeed, the contrast I described above, between intersexual selection and genealogy as explanations of versatility, is much less apparent for intersexual selection as sensory exploitation than for intersexual selection as good genes.

Finally, whatever difficulties we had in connecting singing behavior and its neural substrates (e.g., HVC) to sexual selection, we did identify real sex differences in both. Indeed, the evidence for sex differences in HVC and birdsong represents something of a gold standard in evaluating putative sex differences in brain and behavior among vertebrates generally, as we shall see in the following two chapters.

Brain Ecology

SONGBIRDS, LIKE most wildlife, have been adversely affected by human activities. Human environmental alterations are not detrimental to all songbirds, however. The house sparrow is now abundant throughout the world, inhabiting virtually every location that humans have managed to colonize. As are certain cockroaches and black rats, house sparrows are human symbionts, preferring our carpentry to anything a forest has to offer. Another songbird that benefits from human activities—though much less directly—is the brown-headed cowbird (*Molothrus ater*). Its preferred habitat is open grasslands interspersed with occasional clumps of trees or bushes; it avoids extensive forests, but patchy forests interspersed with open areas are ideal. This happens to be precisely what results from many human–forest interactions. Hence, cowbirds are thriving as never before. Cowbirds, however, are the bane of other songbird species, because they prefer to leave the care of their young to others, and not to other cowbirds. As do the Old World cuckoos, from whence the term *cuckold* derives, cowbirds parasitize the parental care of other songbirds, ranging from wrens to orioles. This is no mere inconvenience for the host species. Parental care requires considerable time and energy, so boom times for cowbirds mean bad times for other songbirds.[1]

In order to appreciate the hardship that cowbirds impose on their hosts, we need to consider not only the time spent incubating the eggs but also the time and energy songbirds spend on the nestlings once they hatch, because when songbirds hatch they are quite unprepared for independent life. Not all birds are so unprepared; the young of many species, including chickens, ducks, and geese, hatch in a remarkably mature state, nerves and muscles already well integrated; and, after an hour or so of stretching and flexing, they are ready to forage for themselves. For species with such "precocial" young, parental duties are limited to guidance and defense. In marked contrast, songbirds hatch in an extremely inchoate state—naked, skeletal, and blind, their oversized eyes seemingly sewn shut. They are not a pretty sight. In fact, as babies go, these are among the ugliest. They are also quite helpless. Baby songbirds cannot even stand up, much less walk, much less forage for themselves. It takes a lot of work to get a baby sparrow to a developmental state equivalent to that of a newly hatched chicken. Typically, the concerted efforts of both parents are required if the chick is to survive and thrive. This parental care is both energy-intensive and risky. Cowbird parents manage to avoid these considerable costs of parenthood.

Given the costs of being parasitized, potential cowbird hosts are understandably on their guard. Hence, if a cowbird is to succeed in its cuckoldry, it cannot just dump a load of eggs willy-nilly. It must be sneaky and stealthy or the eggs will be rejected. Though a female cowbird may lay more than a dozen eggs, she lays only one egg in a given nest. Hence, she must monitor many nests within her range. She must remember where each of these nests is, which ones she has already bequeathed with an egg, and which ones are suitable hosts for her next egg dump. When both parents are away, she covertly deposits one of her own eggs into the host nest; then, to add insult to injury, she often removes one of the host's own eggs. The male cowbirds seem to help somewhat as spies or in creating diversions, but the females do the bulk of the work. Given this division of labor, some adaptationists have reasoned, we should expect female cowbirds to have a better spatial memory than males. Given the greater demands on the spatial memory of female cowbirds, they also conclude that we should expect to find a sex difference in the part of the brain that supports spatial memory. The hippocampus has long been suspected to be such a structure, so reports that female cowbirds seem to have larger a hippocampus than male cowbirds was considered a compelling example of an adaptive sex difference in the brain.[2]

The adaptationists who have identified this adaptive sex difference in the cowbird brain participate in a research program that I call "brain ecology."[3] Brain ecology bears a superficial resemblance to evolutionary neurobiology, but, as I will emphasize in this chapter, the aims and organizing concepts of these two programs differ markedly.[4] The putative sex difference in the hippocampus of cowbirds and some mammals will serve as an exemplar in this regard, a way to highlight the differences between brain ecology and evolutionary neurobiology. First, though, some historical background will provide some insight into the conceptual foundation of brain ecology.

From Ethology to Brain Ecology

It is ironic that at the same time that the ethological perspective was exerting its widest influence—extending to research areas as diverse as child psychology and robotics[5]—ethology began to decline as a distinct discipline. In large part, the decline of ethology was the result of the success of two of its offspring: neuroethology, which I discussed in chapter 7, and behavioral ecology, which I will discuss here. To my mind, one of the great strengths of the ethological perspective was its thoroughgoing explanatory pluralism embodied in Tinbergen's four questions: (1) By what mechanism is a behavior caused? (2) How does it develop? (3) What is its evolutionary history? and (4) What is its function? This combination of questions is distinctly ethological.[6] But neuroethologists and behavioral ecologists have become specialists relative to their ethological predecessors. In neuroethology, as we have seen, the focus is on Tinbergen's first question, with a particular emphasis on the neural substrates for behavior. In behavioral ecology, on the other hand, attention is exclusively focused on function or adaptation.[7]

The original behavioral ecologists were former ethologists who had become increasingly attracted to the adaptationist program, especially after the publication of Edward O. Wilson's *Sociobiology* in 1975. Wilson's book galvanized many ethological functionalists into seeking a new identity, one more directly linked to adaptationist evolutionary biology. Their goal is to explain behavior by relating it to ecological factors. Little or no attention is paid to development and or physiology—Mayr's "how biology."[8] From the outset, one of the central projects in behavioral ecology concerned the means and methods by which animals acquire adequate nutrition. It was assumed that, having been shaped by natural selection, an animal's foraging behavior, its food-acquiring habits, is probably the best it possibly can be under its present circumstances. Behavioral ecologists sought to test this assumption for particular cases by constructing a model of what the optimal solution to the foraging problem would be, then comparing the animal's actual foraging behavior with this model. This area of research is called optimal foraging theory.[9]

The first ever international conference on optimal foraging theory was held at Brown University in July 1984. Though I had little interest in the subject at the time, I happened to accompany a friend to the conference. As a fairly disinterested outsider, I was struck by the shared sense in the attendees of being on the vanguard of something important. It was if they had discovered not only a more productive way to analyze behavior but also a framework that allowed for a "deeper" understanding. There was also a sense of unanimity and bonhomie that I had not experienced since I attended a religious retreat in my youth. As was true of the earlier retreat, my own reaction to the presentations was much less enthusiastic than the reactions of the other attendees. So I was both pleased and, like the rest of the audience, somewhat stunned, when a brave graduate student from New Zealand, Russell Gray, presented a scathing critique of the whole enterprise.[10] His criticisms were wide-ranging, but they generally centered on the naive adaptationism of most practitioners of optimal foraging theory.

Gray's presentation was much less well received by the rest of the audience at Brown that day, but ultimately his critique, along with a number of subsequent critiques of optimality thinking in general, greatly reduced the messianic zeal so evident at Brown. Though it remains an active research area, optimal foraging theory is attracting fewer and fewer new researchers.

Moreover, some of the key pioneers in optimal foraging theory largely abandoned this research and turned their attention, surprisingly, to the brain. I
say, "surprisingly" because, as adaptationists, behavioral ecologists have traditionally resisted looking under the skin. But as we shall see, the departure
is only apparent. For better or worse, they have retained their thoroughgoing adaptationism; in fact, they have attempted to extend its reach.

Remembrance of Things Cached

The behavioral ecologists cum brain ecologists have retained their focus on
foraging behavior. They have also retained many of the same animal
models, most notably birds of the family Paridae, which includes North
American chickadees, as well as their European counterparts, called tits.
Now, however, they are not concerned so much with how these birds choose
what to eat as with how well they remember where they have stored excess
food during times of bounty. A number of bird and mammal species cache
food, which they then retrieve, as needed, over a wide range of time intervals.[11] Some types of birds are quite good at remembering, most notably
several parid species, as well as many members of the crow family (Corvidae), which includes, in addition to those clever birds, jays, magpies, nutcrackers, jackdaws, ravens, and rooks.[12]

This capacity to store food and then retrieve it days, or even months,
later has some obvious advantages, especially where food abundance varies
seasonally as it does throughout the temperate zone. Of the bird species that
do not migrate out of the temperate zone during the winter, a sizable portion
are either parids or corvids. Migration is one of the most energetically demanding and hazardous endeavors imaginable for a bird; hence, it would
seem a good thing to avoid, prospects of a tropical interlude notwithstanding. But in order to avoid migration, a bird must somehow manage to en-
dure not only the cold of winter but also a dearth of food. Food caching is
one way to cope with these harsh conditions.

The black-capped chickadee (*Parus atricapillus*) is at home in a variety of
habitats throughout much of North America. Remarkably, these little birds
are not deterred by either cold or snow and manage to thrive during the
severe Midwest winters. In January, when the warblers, vireos, tanagers, and
flycatchers are seeking sustenance in the tropics, the little chickadees are
out and about in a much less hospitable environment. Their ability to re-
cover stored food items is essential in this regard. Seeds can still be had in
midwinter, and insects as well. Chickadees harvest dormant insect larvae
nestled inside the bark of trees. They are especially adept at finding larvae at
the ends of small twigs. But when conditions are blizzard-like, foraging becomes impossible, and it is during these periods that their food-storing propensity becomes important. Chickadees have a very high metabolic rate:

just a relatively short period without food during the cold of a Wisconsin (or Manitoba, for that matter) winter and they are doomed. So, on relatively balmy days, when they can harvest more than they can immediately eat, they stuff the excess into various holes and nooks, which they can then access on those particularly nasty days.

Some of the corvids have even more impressive memories. Clark's nutcrackers (*Nucifraga columbiana*) inhabit the subalpine and alpine zones of the mountain ranges of western North America. Even in midsummer, this environment is harsh, though for a few weeks, a profusion of wildflowers transforms the desiccated, wind-battered and virtually soil-less landscape into an Oz-like state. Few birds even visit this zone, but the nutcrackers manage to survive and thrive there, retreating to lower elevations only under exceptionally bad conditions.[13] These birds survive such harsh conditions only because they can remember, over a period of months, where they separately cached thousands of pine nuts.

The nutcrackers in the Sierra Nevada rely almost exclusively on the nuts of whitebark pines, which are generated only during the summer. In order to sustain themselves until the next crop becomes available, they must put away an 8 or 9 month supply of the nuts during the brief harvest. The pine nuts cannot all be put in one place lest some lucky squirrel happen upon the cache and clean it out in preparation for its hibernation. The depredations of other nutcrackers must be avoided as well. Therefore, nutcrackers scatter the nuts widely to avoid being robbed. They must also somehow remember where they put them even after the landscape has been dramatically changed by snow cover. Fortunately, their memory is prodigious. During the middle of winter and into the spring they can remember with remarkable accuracy where they put thousands of pine nuts during the course of the summer.[14]

Behavioral ecologists have demonstrated that the memory required for cache and retrieval is not characteristic of birds in general, it only occurs in those species that depend on this capacity for their survival. Under controlled laboratory conditions, black-capped chickadees, which cache food, outperform slate-colored juncos (*Junco hyemalis*), which do not, in tests for spatial memory.[15] And the food-storing marsh tits (*Parus palustris*) outperformed the non-storing blue tits (*P. caeruleus*) and great tits (*P. major*).[16] Other laboratory studies, however, have yielded results at odds with the picture painted by behavioral ecologists with respect to the association between cache and retrieval in nature and spatial memory.[17]

Whatever we make of these studies, the research I have described to this point is pretty much mainstream behavioral ecology: locate a behavioral difference, then explain the behavioral difference by referring to a difference in some aspect of natural history, on the assumption that the behavioral differences reflect the action of natural selection. But if that were the whole story, the cache-and-retrieve behavior, interesting as it is in its own right, would not have generated so much excitement. This research is deemed important primarily because of its connection to neuroscience. This is a story about how some behavioral ecologists ventured into neurobiology and created the new discipline of "brain ecology."

Brain ecology is considered by its practitioners the logical extension—from behavior to brain—of Darwinian thinking. This extension is logical because the brain is the ultimate material source of behavior. Hence, selection cannot modify behavior without first modifying the brain. Brain ecology represents a purely teleological approach to neurobiology, in which brain parts are explained by reference to their function in an ecological context. The brain part on which the behavioral ecologists cum brain ecologists focus is called the hippocampus, which is known to play a critical role in memory. Brain ecologists, however, argue that the hippocampus subserves only very particular kinds of memories, those related to spatial locations. Non-spatial memories are presumed to be processed elsewhere. They take their inspiration from the HVC song nucleus, the organ of song memory, and the paragon of adapted brain parts.[18]

Brain ecologists have looked for an association between the size of the hippocampus and the amount of cache-and-retrieve behavior along the lines of that between the HVC and song complexity. And, in a flurry of research conducted during the last 15 years, they seem to have found it. The food-storing marsh tits and willow tits (Parus sp.) have larger hippocampi than the nonstoring blue tits and great tits.[19] Similar results have been obtained for corvids as well. Most corvids store to some degree, but some species store more than others. Those that do the most cache and retrieval have an especially well-developed throat sac, an obvious specialization for that behavior. Nutcrackers, which are among the most accomplished long-term memorizers, are endowed with the largest throat sac; they also have a considerably larger hippocampus than rooks and jackdaws, which do not cache food.[20] These results would indeed seem to vindicate the effort to relate brain parts to ecological circumstances.

But I will argue that there is a better explanation for these results than the one posed by the brain ecologists, one that emerges naturally from the framework of evolutionary neurobiology. Though both brain ecology and evolutionary neurobiology are evolutionary frameworks, they are quite different evolutionary frameworks. Brain ecology is thoroughly imbued with a teleological mindset inherited from behavioral ecology. That is, brain ecology is primarily an exercise in evolutionary adaptationism and only secondarily a neurobiological endeavor. In contrast, in evolutionary neurobiology the emphasis is on the material how-biology, and neurobiological considerations are not subordinated to evolutionary considerations, teleological or otherwise. As a consequence, evolutionary neurobiology is more sensitive to, and constrained by, what we know about neurobiological processes in general and what we know about the hippocampus in particular.

From the evolutionary neurobiology perspective, brain ecologists make three problematic neurobiological assumptions. First is the assumption that the hippocampus functions exclusively or primarily in spatial memory. We can call this "the hippocampus specialization assumption," or the "hippocampus as HVC" assumption. Second is the assumption that relatively small differences in gross volume should translate straightforwardly into performance differences. We can call this "the bigger is better assumption." Third, and perhaps most important, is an assumption about cause and effect—that the size of the hippocampus determines the amount of spatial learning, and not the reverse. We can call this the "causal arrow assumption." I will now consider each of these in turn.

Is the Hippocampus Dedicated Solely to Spatial Memory?

In the view of brain ecologists, the hippocampus is to spatial learning what the HVC is to song learning. It is certainly understandable that the brain ecologist would see the HVC as a model, because the HVC is the paragon of adapted brain parts. The size of the HVC, as we have seen, roughly corresponds to the versatility of songsters, and sex differences in HVC volume seem to roughly correspond to sex differences in the amount of singing engaged in by males and females of a given species. But even if the hippocampus of cowbirds can be shown to meet HVC standards in these regards, there is an obvious disanalogy. There is a much more particular and direct connection between the HVC and the singing of a mockingbird than there is between the hippocampus and the nest parasitism of a cowbird. The hippocampus is only indirectly connected to nest parasitism or food caching, by way of the construct "spatial learning." But spatial learning itself is a much broader and more heterogeneous category than birdsong. Moreover, it is not at all clear that the hippocampus is devoted specifically to spatial learning in the way that the HVC is devoted to song learning. Rather, there is ample evidence that the hippocampus functions in a much broader cognitive category—memory. If the hippocampus is not devoted specifically to spatial memory, but is essential for all sorts of other kinds of memories as well, the analogy with the HVC breaks down, at least in the eyes of an evolutionary neurobiologist.

Much of what we know about the hippocampus derives from the extreme misfortunes of a man known only by his initials: H. M. From an early age he suffered recurrent and extremely debilitating bouts of epilepsy, as a result of which he was not able to live anything approximating a normal life. He received various kinds of treatments but none were effective. H. M.'s neurologists determined that the focus of his seizures was in the medial portion of his temporal lobes. As a last resort, they decided to remove the blighted region.

He made what seemed to be a complete recovery from the operation and his seizures never recurred. His personality was unaltered, and his IQ, already in the normal-to-high range, actually seemed to improve somewhat after the operation.[21] Nonetheless, H. M. paid a high price to be rid of his seizures. Though his memories from his earlier life remained largely intact, he could no longer consolidate new ones. He could not remember, from one day to the next, the names or faces of his caretakers. In fact, he could not remember anything, no matter how salient, minutes after the experience. As a result, he became completely helpless, and his continued survival depended on the constant surveillance of others.

What happened? It turns out that what his surgeons took to be a selective removal of the damaged part of his temporal lobes was, by today's standards, none too selective. Subsequent surgeries on similarly afflicted patients revealed that memory loss was directly related to the amount of damage to a part of the limbic system known as the hippocampus. Unfortunately for H. M., the damage to this area resulting from his operation was total (figure 8.1).

H. M. was not alone in inadvertently providing very useful information about the hippocampus. Other men and women received similar operations. But the bulk of the research inspired by H. M. has been conducted on nonhuman primates and laboratory rats. The hippocampus is now one of the most thoroughly studied parts of the brain. Although this research has generated lots of neurobiological information about the hippocampus, there is no consensus as to what the function of this structure is. Instead, there are a number of competing hypotheses.[22]

The particular hypothesis that provides the foundation for much research in brain ecology is that the hippocampal formation is devoted primarily, if not exclusively, to spatial memory. On this view, the hippocampus functions as a cognitive map, by means of which the subject represents the geometrical relationships of salient parts of its environment.[23] This view is most popular among those researchers who use the laboratory rat as their animal model. It is much less popular, however, among those who study humans and other primates.[24] The view of the hippocampus as cognitive map certainly does not help make sense of H. M.'s experience. It is true that H. M. would easily get lost after his operation. But he also failed to remember faces, pictures, names, or situations. Clearly H. M.'s problems extended beyond his spatial memory.

If, in fact, hippocampal processing is not restricted to spatial tasks, the significance of the observed differences in hippocampal volume in storing versus nonstoring birds is much less apparent. The interpretation of these

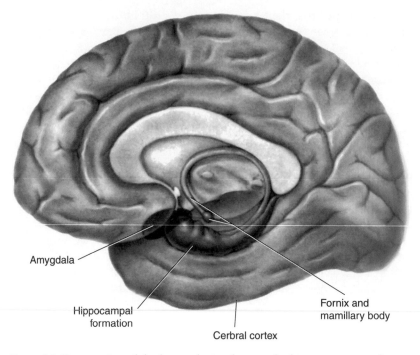

Figure 8.1 Cross section of the human brain, showing the hippocampus in relation to other limbic structures (amygdala and mamillary body) and cerebral cortex (Adapted from Martin 1989, p. 380)

size differences as adaptive depends critically on the view that the hippocampus is the HVC of spatial memory. If the hippocampus is not the HVC of spatial memory, we would need to look beyond ecology to explain the size variation—which brings us to the next neurobiological assumption in brain ecology.

Does Bigger Mean Better?

Over some range of size differences bigger must be better. Take total brain size. We would expect a brain consisting of 2 billion neurons to have greater computational capacity than one consisting of 2 thousand neurons. The same must be true of brain regions that perform particular sorts of computations. The cerebellum, for example, subserves most motor skills. Birds have proportionately much larger cerebellums than sloths; and the size of their cerebellum goes a long way toward explaining why birds can maneuver so rapidly and with such coordination in three dimensions, whereas the sloth's deliberate and much more restricted movements seem almost palsied. These are the sorts of gross comparisons that Jerison had in mind when he formulated his "principle of proper mass": "the mass of neural tissue controlling a particular function is proportional to the amount of information processing involved in that function".[25] Jerison himself was primarily interested in the evolution of animal intelligence and how it is related to the size of the cerebral cortex and other large-scale structures.

Consider variation in the mass of the neocortex among mammals. Most of what we consider intelligent behavior is mediated by this structure. So it is not surprising that the cortex is much larger in primates than in rodents, for example, or that the human cortex is substantially larger than that of other primates.[26] We are obviously smarter than monkeys, and they in turn can perform mental feats that rats could not begin to understand. The difference in cortex volume between humans and even our closest relatives is extremely large. But what do much smaller differences indicate with regard to intelligence? For brain ecologists invoke the work of Jerison to warrant much stronger claims about the relationship between brain structure and brain function than would an evolutionary neurobiologist. In particular, brain ecologists use Jersion's principle to warrant adaptive claims about fairly small differences in hippocampal size among closely related species.[27] At this scale, it is much less obvious that a bigger-is-better principle should hold, given our current understanding of neurobiological processes. Performance differences of the sort reported by brain ecologists need not be manifest as volume or mass differences in gross brain regions. In fact, myriad previous research indicates that it is at least as likely that such differences can be accounted for by differences in connectivity or synapse number, neither of which is detectable by the crude measurements heretofore employed on tits and jays. Moreover, small differences in the gross size of particular brain areas can have a variety of causes, most of which are not related to computational power; the amount of space between neurons, the proportion of non-neural tissue, water content, and even the size of the genome, all influence such measurements.[28]

To better understand the difficulty in interpreting such small differences, we need only examine variation in we humans ourselves. There is substantial variation in the size of the cortex and of total brain volume among humans. But it is not at all clear what it indicates regarding intelligence. Einstein's brain, for example, was found to be of about average size for a man of his physical stature, not at all what was expected of a man widely regarded as the prototypical genius. Whatever aspect of Einstein's neural endowment contributed to his being able to formulate the theories of special and general relativity, it apparently was not related to the size of his cortex.[29]

The bigger-is-better assumption is especially problematic when comparisons are made using gross measures of heterogeneous regions such as the hippocampal complex. The hippocampus consists of numerous cell types

TABLE 8.1
Tits and Jays

Summary of associations between food storing, spatial memory, and hippocampal
volume in tits and jays

Species	Food Storing	Spatial Memory Tests	Hippocampus Size
Blue tit	No	+ + + +	+ +
Great tit	No	+ + + +	+ +
Marsh tit	+ + +	+ + + +	+ + +
Willow tit	+ + + +	+ + + +	+ + + +
Mexican jay	+ +	+ +	+ +
Scrub jay	+ +	+ +	+ +
Pinyon jay	+ + +	+ + +	+ +
Nutcracker	+ + + +	+ + + +	+ + + +

The number of (+) symbols indicates relative magnitude

organized in discrete zones. Neuroanatomists recognize a number of hippocampal regions distinct enough to warrant different names. We cannot just
assume that each cell type and each area of the hippocampal formation
contribute equally to spatial memory.[30] We should expect, rather, that some
types of neurons and some types of projections are more significant than
others in this regard. So it should not be surprising that recently the relationship between gross hippocampal volume and spatial learning has become increasingly muddied in both tits and jays (table 8.1).

Among corvids, for example, the relationship between hippocampal volume and spatial memory is not as straightforward as first suggested, as is
evident from a study of four members of this family that inhabit the moun-
tainous regions of the southwestern United States. Hippocampal volume
was not correlated with species differences in cache-recovery aptitudes as
measured under controlled laboratory conditions.[31] The parid story has gotten equally messy from a brain ecology perspective. Blue tits and great tits
do not store food and have a much smaller hipppocampus than marsh tits,
which do. Yet blue tits and great tits perform as well as marsh tits in laboratory tests for memory of spatial cues.[32] Thus, the large differences in hippocampal volume do not predict performance in spatial memory tasks. Even
among the food-storing tit species, there are marked differences in food-
storing behavior. Marsh tits store food for only a few hours or days; willow
tits, however, store food for several weeks to months. As would be expected
on the brain ecology account of such differences, the willow tits have a
considerably larger hippocampus than the marsh tits.[33] But again there are
no clear species differences in their performance on spatial tasks in controlled laboratory tests that require shorter- or longer-term memory.[34] In gen-

eral, hippocampal volume seems much more closely related to the tendency to cache food than to performance differences in spatial skills. This conclusion brings us to the third and most important problematic assumption in brain ecology.

Cause and Effect in Brain and Behavior

Throughout this book, I have critically examined some common assumptions about cause and effect in biology. In this case, we need to examine the assumption that hippocampal size determines the amount of cache and retrieval and not the reverse. This assumption is convenient, given the adaptationist agenda of brain ecologists, because it makes ecological determinism plausible; but the assumption seems dubious to an evolutionary neurobiologist, given what we know about the material properties of neural tissues. A raison d'être of the brain is its responsiveness to experience, which enables future effective action. Inherent in the brain's being responsive to experience is its being physically altered by that experience.[35] This alteration can take many forms, from a modification in patterns of coordinated neuron firing, to changes in synaptic connections, to changes in the allotment of resources to one part of the brain at the expense of others. Some of these experiential effects may result in alterations of mass.[36] Brain ecologists have placed insufficient emphasis on these experiential effects because such effects undermine their main thesis.

Many of the observed differences in hippocampal volume that are functionally related to spatial memory may themselves *result* from the experience of spatial learning and the behavioral propensity to store and retrieve. If this is so, an enlarged hippocampus does not enable cache and retrieval; it is the byproduct of cache and retrieval. That interpretation would mean that the brain ecologists have gotten cause and effect backward and, in so doing, have misidentified the trait on which natural selection is actually acting. The evidence for this interpretation comes from developmental studies conducted by the brain ecologists themselves. Two European parids have been the species of choice for these developmental studies, the marsh tit, which stores, and the blue tit, which does not.

The brain ecologist Nicola Clayton invites us to make the following distinction in order to help us think about the relation between hippocampal growth and experience.[37] On the one hand, the brain could change in preparation for future cognitive demands. This is called the experience-expectant hypothesis. An analogy would be the increase in gonadal size in many birds before the spring breeding season.[38] The alternative hypothesis is that the brain changes in response to experience—the experience-dependent hypothesis. An example would be the increase in muscle mass in response to mechanical resistance. Is the hippocampus more like a gonad or a muscle?

Brain ecologists need it to be more like a gonad. If the hippocampus is more gonad-like, we would expect the size differences in the two tit species to emerge before the marsh tits start storing. They start storing quite early, so a difference at birth would not be too much to expect. If the hippocampus is more muscle-like, we would expect the size differences to emerge after the marsh tits start storing, no matter how late in development they start.

The first important finding was that the hippocampal size difference of marsh tits and blue tits does not exist at birth. Instead, it emerges only gradually after the birds have fledged and become independent.[39] Hippocampal volume increases in both species as the young mature, but it increases more rapidly, relative to the rest of the brain, in the food-storing marsh tits.[40] If marsh tits are deprived of the opportunity to store seeds, however, their hippocampus enlarges at the slower rate typical of blue tits.[41] There is no sensitive period for this experiential effect. Marsh tits deprived of the opportunity to store seeds for up to a year achieve only typical adult hippocampal size once they have been given the opportunity to store.[42]

It seems obvious, therefore, that the hippocampus is more like a muscle than a gonad. Clayton, however, opts instead for the (experience-expectant) hippocampus-as-gonad view.[43] The fact that she wanted it to turn out that way cannot be ignored as a factor in her conclusion. If the size difference merely results from experiential factors, there is no need to invoke natural selection. Clayton's most substantive argument is that it takes relatively few experiences of cache and recovery to trigger the growth in the hippocampus of the marsh tit, but this claim is based on paltry data.[44] Even if we accept her interpretation of the data, it does not make the hippocampus gonad-like. It is gonad-like only if the growth is anticipatory. At best, she has demonstrated a difference in the muscle-like responsiveness of marsh tit and blue tit hippocampi to a given experience. Even that claim requires more evidence than she provides.[45]

Brain ecologists frequently use elliptical and hedged language when interpreting their results,[46] but they clearly promote the view that there has been selection for increased hippocampal volume in order to facilitate cache recovery or memory for the nests of suitable host nests. On this view, size differences in hippocampi are adaptive per se. There is an alternative hypothesis that accords well with the developmental data provided by the brain ecologists themselves, according to which selection acts only on food-storing behavior, and hippocampal volume varies as a consequence of differences in this behavioral propensity.

At the very least, the developmental data suggest that the brain ecologists should be looking beyond the hippocampus to understand the neural substrates that function in cache and retrieval. For example, there is suggestive evidence that food storers are more attentive to spatial cues than are non-storers.[47] This is not an innate difference, however. Rather, it has been demonstrated that food-storers only gradually learn to attend primarily to spatial cues through something like an operant conditioning process. Ultimately, the interspecific differences in hippocampal volume may be a passive consequence of species differences in the propensity or motivation to store food.[48] If this is so, the neural differences between storers and nonstorers is merely a secondary consequence of this more fundamental behavioral propensity and not itself the target of natural selection.[49]

This concludes our examination of the three problematic neurobiological assumptions in brain ecology: (1) that the hippocampus is solely concerned with processing spatial information, (2) that small differences in hippocampal volume translate directly into performance on spatial skills, and (3) that the size of the hippocampus determines spatial learning and not the reverse. The neurobiological critique also applies to the putative sex differences in the hippocampus of cowbirds. But these putative sex differences reveal a further problematic inference in brain ecology, this one concerning the way in which sex differences are conceived. First, let's explore the evidence for these sex differences in a bit more detail.

THE BRAIN ECOLOGY OF HIPPOCAMPAL SEX DIFFERENCES

The research on chickadees and nutcrackers is directed toward the variation among species in spatial learning and hippocampal volume. But some of the most attention-grabbing research in brain ecology concerns sex differences in these attributes such as were reported for cowbirds. Most of the research on sex differences in the hippocampus has been conducted on rodents, particularly voles (genus *Microtus*).

Voles are unprepossessing critters with which you are probably unfamiliar unless you live in a rural environment and happen to own a cat. In that case, you may have had one of them deposited in your living room, partly mutilated, by your beloved pet. If you are lucky and have actually witnessed the drama, you can lock the cat in the bathroom and direct the vole out of the house so you don't have to watch it die. If you are unlucky and discover the vole only because of the odor emitted by its corpse, do not, under any circumstances, place it in your garbage can! Proceed, instead, into the woods with your charge, leave it, and let nature take its course. If it is not too ripe, you might take a few seconds to examine it. Look at its ears. Are they relatively large? Then, it is probably a mouse. Are they seemingly nonexistent? Then, check the eyes; no, the cat did not remove them. You are holding a dead mole. Feel its fur to confirm. It should be very dense and soft. If the ear size is intermediate relative to mouse and mole, the dead creature in your hands is probably a vole.

Voles are among the most abundant mammals in the Northern Hemisphere, yet, because they are nocturnal, small, and not particularly cute, they are given short shrift by nature programs and have managed to live largely anonymous lives. They have long been familiar to ecologists, however. The population dynamics of voles constitutes one of the paradigms within population ecology. Recently, they have come to the attention of behavioral ecologists and brain ecologists as well, because of some interesting variation in hippocampal volume among species.

In promiscuous (and/or polygynous) breeders such as meadow voles (*Microtus pennsylvanicus*), the males must range farther in their search for mates than do males of monogamous species such as the pine vole (*M. pinetorum*), so male meadow voles have larger home ranges than male pine voles.[50] All else being equal, the less area a vole has to cover in its quest for food or mates, the better off it is. There are many ways for voles to die prematurely—death by fox, snake, owl, and, of course, domestic cat, to name a few—and most of these risks are increased in direct relation to the amount of its activity above ground. So a male vole whose relative fitness is not compromised by settling for one mate should wander less than one who must find multiple mates. Female voles, on the other hand, whether monogamous or polygynous, do not seek out mates. Therefore, both meadow vole and pine vole females retain the same home range throughout the year.

So there is a sex difference in the home range size of meadow voles, but not of pine voles. Members of both species were taken into the laboratory to test their spatial learning skills, as indicated by their ability to master maze learning.[51] The logic underlying these tests was simple and straightforward: if male meadow voles have larger home ranges than female meadow voles, then male meadow voles have more need than female meadow voles for the spatial learning skills demanded by the maze. The males should perform better. And they did. For the monogamous pine voles, on the other hand, in which the size of the home ranges of males and females are similar, we would not expect a sex difference in the ability to master this task. This prediction too was confirmed.[52]

The respective hippocampi of these voles were also examined, and, in harmony with the behavioral results from both lab and field, those of male meadow voles were larger than those of female meadow voles. No such sex difference existed in pine voles.[53] It appears, then, that meadow voles, in which males must range farther to mate than male pine voles, develop as a consequence a dimorphism in both spatial skill and its ostensible neural substrate (table 8.2).

A More Congenial Rodent

Kangaroo rats (genus *Dipodomys*) are much more likely than voles to appear on television nature shows. Named for their style of locomotion, these ro-

TABLE 8.2
Voles

Summary of species differences among pine voles and meadow voles with respect to sex differences in home range, maze learning, and hippocampal volume.

Species	Home Range	Maze Learning	Hippocampus Size
Pine vole	Male > Female	Male > Female	Male > Female
Meadow vole	Male = Female	Male = Female	Male = Female

dents are almost irresistibly cute, especially when their cheek pouches are stuffed with seeds. Their temperament is also much more attractive than that of voles. Whereas voles appear extremely stressed under captive conditions and have an unfortunate tendency to kill one another, kangaroo rats quickly become relaxed and tame.

Kangaroo rats inhabit the arid regions of western North America. They exhibit some spectacular physiological adaptations to desert conditions, including a kidney that is so efficient at recycling water that the bladder is virtually superfluous.[54] Members of some species never drink water, deriving all of their moisture from their food, though the food itself would be considered quite dry. They spend the hot daylight hours in their burrow, emerging at night to forage when the temperature is cooler.

Several species of kangaroo rat coexist at one study site near Portal, Arizona. One species, Merriam's kangaroo rat (*D. merriami*), stores seeds in scattered caches. The bannertail kangaroo rat (*D. spectabilis*), on the other hand, utilizes only one central cache, which it vigorously defends from would-be interlopers. Bannertail kangaroo rats tend to spend their entire adult lives in one burrow, from which they venture out each night. They also tend to exploit richer food sources than Merriam's kangaroo rats. Consequently, bannertails don't venture as far from their burrows as Merriam's kangaroo rats.[55] Based on these facts, it was predicted that Merriam's kangaroo rats should experience greater selection to remember spatial relationships and to recall with accuracy many particular locations than their sympatric congener. It was further predicted that Merriam's kangaroo rats should have a larger hippocampus than bannertail kangaroo rats, given the role of the hippocampus in spatial memory.

The results were as expected (table 8.3). The authors of this study also observed sex differences in hippocampal size in both species, which they attributed to their polygyny.[56] As with polygynous voles, kangaroo rat males of these two species range farther during the breeding season than their female counterparts, a fact that, according to these brain ecologists, explains why males have a larger hippocampus than females.[57]

TABLE 8.3
Kangaroo Rats

Summary of species differences among bannertail kangaroo rats and Merriam's kangaroo rats with respect to sex differences in home range and hippocampal volume

Species	Home Range	Maze Learning	Hippocampus Size
Bannertail kangaroo rat	Male > Female	??	Male > Female
Merriam's kangaroo rat	Male > Female	??	Male > Female

Remembrance of Eggs Passed

The comparative method has been applied to cowbirds as well in order to demonstrate adaptive sex differences in the hippocampus. Several cowbird species, which vary in the degree to which they practice nest parasitism, are native to Argentina. The shiny cowbird (*Molothrus bonariensis*) greatly resembles the brown-headed cowbird in its habits. Screaming cowbirds (M. *rufoaxillaris*), on the other hand, though also nest parasites, specialize in duping a single host species. Screaming cowbirds also differ from brown-headed cowbirds in that both male and female do the nest inspections. And a third species, the bay-winged cowbird (M. *badianus*), is not parasitic at all. As expected, shiny cowbirds exhibited sex differences in hippocampal volume resembling those of the brown-headed cowbirds. But neither the screaming cowbird nor the bay-winged cowbird, in which there are no sex differences in spatial behavior related to nest parasitism, evidences any sex difference in hippocampal volume.[58]

Taken together, the sex differences observed in cowbirds, voles, and kangaroo rats seem to make a nice story from the perspective of brain ecology. But from the evolutionary neurobiology perspective, there is even more reason for skepticism about the brain ecology explanations for sex differences in hippocampal volume than there was for the brain ecology explanations for interspecies differences in hippocampal volume. In addition to the three problematic assumptions discussed above, there are two more problems specifically related to brain ecology interpretations of hippocampal sex differences.

SEXUAL SELECTION AND SEX DIFFERENCES IN THE HIPPOCAMPUS

From the evolutionary neurobiology perspective, the first problem with the brain ecologists' claims about adaptive sex differences is the lack of an adaptive rationale. The second is that, even given an adaptive rationale, the brain ecologists fail to appreciate the amount of work required by evolution in order to bring about such sex differences.

Is What's Good for the Goose Bad for the Gander?

Assume that an enhanced capacity for spatial memory may benefit the female cowbirds more than the males, or that it would benefit male voles more than female voles. Assume also, for the moment, that this spatial learning capacity is a direct function of hippocampal volume. We cannot assume therefore, that a sexual dimorphism in hippocampal volume will ipso facto evolve. That this assumption is mistaken follows from the fact that male and female voles or cowbirds are members of the same species, and hence share species-wide development. Hence, selection for a trait in one sex will result in a correlated response in the other, unless, that is, there is selection against this correlated response. Call this "intergender hitch-hiking."[59]

The brain ecologists must demonstrate that there is a force that overrides this natural tendency for intergender hitchhiking. They must demonstrate, in other words, that what's good for the goose is bad for the gander in cowbirds, and vice versa in voles. That is, they must demonstrate that selection *for* increased hippocampal volume in female cowbirds is accompanied by selection *against* an increase in hippocampal volume in the males. Otherwise, we should expect selection for an increased hippocampal volume in females to produce a comparable increase in the hippocampal volume of males, even if the males don't benefit thereby.

Most functionally significant sexual dimorphisms occur when either what's good for the goose is bad for the gander or vice versa—what's good for the gander is bad for the goose—usually the latter.[60] For instance, flashy coloration benefits many male birds when it comes to attracting mates. The females, however, cannot benefit in this way. On the other hand, flashy coloration increases the likelihood of being eaten by a predator. So there is selection for conspicuousness in male birds but selection against conspicuousness in female birds.[61] The magnitude of the dimorphism is not simply a function of differences in the degree to which the two sexes benefit; if there were no downside to increased coloration both sexes should be of the same coloration, even if there is selection for increased coloration only in males.[62]

When only one sex benefits from a particular trait or character, it is the magnitude of the downside for that trait in the opposite sex that largely determines the magnitude of the sex difference in that trait. Take antlers and horns, for example.[63] Antlers are confined to the deer family (Cervidae). These are physiologically costly structures that must be generated anew each year. The "velvet" that covers the developing antlers each spring is packed with blood vessels carrying nutrients that could have been used for muscle or some other vital tissue. This expensive adornment reduces the bearer's prospects for survival. The benefits of antlers that accrue to the male deer outweigh their costs, however, especially in contests with other males, on which their breeding prospects ride. Female deer derive no such benefit. Nevertheless, they would possess male-like antlers were it not for their cost.[64]

The likelihood that female deer would have antlers were it not for this cost becomes more apparent if we compare antlers with horns. Horns benefit male antelope, cattle, sheep, and goats in much the same way that antlers benefit male deer. They also generally do not benefit female antelope, cattle, sheep, and goats, just as antlers do not benefit female deer. But horns are permanent structures that do not have to be generated anew each year; they are much less costly than antlers. This cost difference largely explains why females with antlers are the exception, whereas females with horns are the rule.[65] That female antelope generally have somewhat smaller horns than males indicates that horns are not cost free.[66]

Now consider birdsong. Males typically sing more than females, primarily to attract mates and repel competitors, two very important functions. The downside to singing is that it takes a lot of energy, and while so engaged the bird cannot feed. Moreover, most songbird predators have ears too. This is a considerable cost. But what is the cost of remembering spatial cues? There is no obvious behavioral downside. It does not make an animal more vulnerable to predators; nor is there a behavioral energy cost. So even if we grant that female cowbirds require a higher aptitude for spatial learning than other female blackbirds, given their idiosyncratic egg-laying habits,[67] it is not at all apparent why male cowbirds need to be kept dumb relative to female cowbirds. Selection for spatial learning in female cowbirds should result in a correlated increase in the capacity for spatial learning in males as well. Given these considerations, what could be the adaptive rationale for the sexual dimorphism in the hippocampus?

Brain ecologists will appeal to the energy cost of maintaining this brain structure. The brain as a whole is a major energy consumer, and perhaps for those animals living on the energetic edge, even a slight increase in the hippocampus could have a significant associated energy cost.[68] Is this the downside we are looking for, one that would explain why male cowbirds and female voles must be kept dumber than their mates? In order to assess the force of this argument, we need to consider some generic properties of the development of the nervous system in vertebrates, the most important of which is the "use it or lose it" principle.

It is a generic property of the nervous system that the neurons that are chronically activated proliferate and become integrated into active circuits; those that do not become integrated into active circuits die. Of course, to the extent that this use-it-or-lose-it phenomenon explains the observed sex differences in hippocampal volume of voles and cowbirds, we cannot argue that the sex difference is a special adaptation to make the females smarter while keeping the energy costs for the males down.[69] Rather, the sex difference would merely reflect a generic property of the nervous system. On this account. the sex differences in hippocampal volume have no evolutionary function.

The primary reason brain ecologists fail to appreciate the force of the intergender hitchhiking argument is that they ignore how-biology considerations. Once an adaptive rationale for a sex difference has been identified, the how-biology is assumed to take care of itself. Because they ignore the how-biology, they grossly underestimate the amount of evolutionary work that goes into creating an adaptive sex difference. In evolutionary neurobiology, however, the how-biology is the primary consideration. It is because of this how-biology that conditional traits, such as these putative sex differences, require more, not less, evolutionary work than species-wide traits. Let's now consider this matter in more detail, through a focus on the sex differences in voles.

The evolution of any adaptive sex difference in any behavioral or neural trait in voles or any other creatures requires alterations in that component of sexual development called sexual differentiation, the process through which members of a given species become either males or females. During the first phase of sexual differentiation the gonad becomes either a testis or an ovary. In voles and other mammals, the fate of a gonad is determined by a gene on the Y chromosome called SRY.[70] It initiates a cascade of events that result in testicular development; if it is absent, ovaries develop by default.

The ovaries and testes produce steroid hormones (estrogens and androgens, respectively) that stimulate the development of the secondary sexual characteristics, ranging from genitals to behavior. The primary androgen, testosterone, is often considered something of a proxy or agent for SRY: when testosterone is present, the individual is masculinized in body and behavior; when it is absent, feminization ensues as a sort of default condition. But this account is grossly simplistic on a number of levels, perhaps most fundamentally in the degree of agency ascribed to this chemical. This simplistic interpretation of the role of testosterone and other gonadal steroids is not limited to brain ecology, but it does play a prominent role in their program, because it makes their adaptive stories about sex differences, including spatial cognition, more plausible. A fleshed-out version of the how-biology of testosterone-mediated sex differences will make us more skeptical of the brain ecologists' claims for adaptive sex differences in hippocampus and spatial learning.

Testosterone Rex

Of all the factors relevant to sexual development, steroid hormones, such as testosterone, are perhaps the easiest to measure, manipulate, and monitor, so they tend to be accorded more explanatory weight than other developmental factors. Consider a young male chicken. When he is castrated, his testosterone levels drop, as you would expect; he also fails to develop the comb and wattle typical of roosters. From these effects of castration, we can conclude that some testicular product is necessary for the normal development of those features. If, soon after castration, the chick is provided exogenous testosterone, he will develop normal comb and wattle despite lacking testes; other hormones are not as effective, evidence that testosterone in particular is the salient factor. Naturally, the less we know about the other developmental factors, the more significance we will ascribe to testosterone. In lieu of any information about other factors, it is even tempting to assume that testosterone is the only relevant factor.

But the same chick could be chemically castrated, by treating it with one of several known anti-androgens, and with the same results. These anti-androgens are not called anti-androgens because they destroy or annihilate androgens. Instead, they block androgenic effects by interfering with testosterone's ability to bind with a protein known as the androgen receptor.[71] If testosterone does not bind with the androgen receptor, it is effectively inert. The pathology known as androgen insensitivity syndrome (AIS) provides dramatic evidence of this effect. In an affected individual, whether it be mouse or human, there is little if any development of male secondary sexual characteristics of any kind, including the genitals. Though genetically male, these individuals are indistinguishable from females. There is no shortage of testosterone in those with AIS, only a dearth of androgen receptors.

Androgens and other steroid hormones influence cell processes by acting as a precursor for a transcription factor. A transcription factor is any chemical that influences gene expression. Transcription factors influence gene expression by physically binding to the DNA. In this case, the testosterone-bound androgen receptor undergoes a conformational change that allows it to bind particular bits of DNA. So without androgen receptors, there is no way for testosterone to influence traits such as comb and wattle, no matter how much androgen levels are artificially boosted. This is why the focus of much recent endocrinology is on the receptors and how they come to be distributed in particular parts of the body.

In the mammalian brain, as it happens, testosterone exerts its masculinizing effects even less directly than elsewhere in the body. In most neurons, androgens do not bind directly to androgen receptors; they are attacked instead by an enzyme called aromatase and are immediately converted to the estrogen estradiol.[72] This "female hormone" is the primary masculinizing influence in the mammal brain. As does testosterone, estradiol binds to a receptor—in this case estrogen receptor (ER)—and the estrogen–estrogen receptor (E-ER) complex binds to particular stretches of DNA, thereby influencing gene expression. Hence, with regard to the nervous system, testosterone functions primarily as a vehicle for the "female hormone" estradiol. Male brains are masculinized because more of this female hormone is available to the neurons in male brains than in female brains.

But that is only part of the story. Even when testosterone (T) binds to androgen receptors (AR), it has different effects on different cell or tissue types, not because it binds to different androgen receptors but because the T-AR complex can potentially bind to a number of different genes, and the genes available for binding vary with cell and tissue type.[73] Different genes are available for binding in muscle cells than in neurons, and different genes still are available in genital tissue. In most cells, no genes at all are available for binding, because all of the genes that the T-AR complex could potentially regulate have been permanently inactivated or turned off during the course of development. Let's briefly consider how this inactivation process works, because it is a key to understanding the process of development, or differentiation, in general.

All of the cells in the body have an identical suite of genes. During the course of development, the vast majority of genes in all cells become permanently turned off. But a different suite of genes gets turned off in different cells. This process of differentiation, whereby a cell lineage gives rise to neurons or blood cells, and the blood cell lineage to red blood cells or white blood cells, and the white blood cell lineage to monocytes or eosinophils, and so on, requires the progressive disabling of different genes with successive cell divisions. There is a stage, early in development, in which all cells can potentially give rise to any kind of adult cell type (totipotency), because none of the genes have been disabled. These are the embryonic stem cells that have received so much attention of late. During successive cell divisions, the options become increasingly narrowed, as the number of permanently disabled genes increases, until a given cell can give rise only to an identical cell, one that has an identical suite of disabled and enabled genes. This disabling of particular genes is selective, not random, and is caused by cell–cell interactions, both direct and indirect. Selective disabling is, of course, only half of the story; of those genes not permanently turned off during development, different ones become activated at different times, also depending on the cell lineage. Whether a gene becomes disabled or enabled depends on the state of the cell as a whole. The state of the cell as a whole is a complex function of both the state of the cell from which it arose through division and the state of the other cells with which it is presently interacting.

This is the basis for the "timing-is-everything" nature of testosterone's influence, especially in the brain. Testosterone's developmental effects on the brain are limited to very specific periods of development called critical periods. Critical periods are not brain-wide. Rather, their onset and duration vary widely from one integrated group of neurons to the next. When testosterone is available during the critical period, it is said to have an organizing effect for that group of neurons, a more or less permanent alteration of their response capacities, one of which is continued sensitivity to estradiol into adulthood. The subsequent effects of testosterone on the suitably organized neurons are referred to as activational effects. It is these more transient activational effects of testosterone with which we are most familiar, because their onset and offset are more directly related to circulating testosterone levels than are the organizing effects. It is important to remember that these activational effects, including those on aggression and muscle mass, can occur only in cells that have already been properly organized earlier in development.

The term *organizing effect*, however, is unfortunate in seeming to ascribe to testosterone more credit that it deserves for the subsequent alterations in its target tissues. Testosterone itself does not organize anything. It merely stimulates or activates the organizing process in the same way it stimulates or activates the events that are called activational effects. The credit for the organizing that gets done as a result of testosterone's organizing effects should go primarily to the target tissues, each of which gets organized somewhat differently. Next in importance, if we are ranking the developmental actors with respect to the magnitude of their causal role, are the androgen receptors. Testosterone comes in a distant third.

The Role of T. Rex in Brain Ecology

Testosterone becomes T. rex when the how-biology of testosterone's effects on the body is ignored. Given the adaptationist agenda of the brain ecologists, T. rex is a much more suitable entity than testosterone. By giving T. rex all of the explanatory credit for secondary sex differences, including that which is due to the target tissues and the androgen receptors, brain ecologists obscure the evolutionary work that is required of any functional sex difference, including that alleged for spatial cognition and the hippocampus. Let's consider, for example, what is required for the evolution of sex differences in the spatial cognition of a vole, given the anemic how-biology favored by brain ecologists, compared with what is required given the robust how-biology of an evolutionary neurobiologist.

On the adaptationist brain ecology account, once sexual selection demands more spatial learning in male voles than in female voles, the males will inevitably evolve a larger hippocampus, because hippocampal development will be linked to testosterone and testosterone levels are always higher in males. This presumed spontaneous linkage of spatial cognition with testosterone is one reason that brain ecologists do not seriously consider interender hitchhiking. Sex differences just sort of happen automatically, given sexual selection.[74]

But consider now what evolutionary work is required, given the robust how-biology perspective on sexual differentiation described above, in which testosterone is the least significant actor in the play. At a minimum, there must be a genetic alteration so that the androgen receptor is expressed in the appropriate target tissue. Actually, because the target tissue is in the brain, it is the estrogen receptor that needs to be present. And any sex-specific change in spatial learning caused by sexual selection will require a change in the distribution of estrogen or androgen receptors or both. But testosterone does not bind to estrogen receptors directly; it must first be converted to estradiol by the enzyme aromatase. As such, there must also be a genetic change (in the regulation of aromatase expression) so that aromatase is also present at the appropriate time in the appropriate neural targets.

And we are not finished yet. At this point we have in place the means to influence gene transcription in the target neural tissue by way of the estrogen–estrogen receptor (E-ER) complex, but that alone will not result in the appropriate response in the target tissue. The appropriate sex-specific increase in cognition will probably require a change with respect to the genes in the target tissue that are available for regulation by E-ER. Genes that had been turned off earlier in development must now be enabled so that they can be turned on when bound by E-ER. This step too does not just happen spontaneously; it takes evolutionary work involving mutations in the genes influencing the developmental disabling and enabling of the genes to be regulated by E-ER. It will be particularly important that the right genes are available during the critical period when the "organizing" events occur.

The differences between evolutionary neurobiology and brain ecology, with respect to how much evolutionary work is required for the evolution of a sex difference in spatial learning, are pronounced. And the differences boil down to how much attention is paid to the material how-biology. From the functional evolutionary perspective of brain ecology, a super-actor, Testosterone rex, takes care of everything, so the appropriate how-biology alterations seem to occur spontaneously given selection for a sex-specific increase in spatial learning. For the evolutionary neurobiologist, however, the biochemical called testosterone is just another actor in an ensemble, not the plenipotent executor of selection's demands. Once testosterone's role is placed in materialistic how-biology perspective, it becomes obvious that the evolution of a sex difference in spatial cognition takes a lot of evolutionary work. In fact, it is fair to say that conditional traits, such as testosterone-linked sex differences, take more evolutionary work, and are therefore less likely to evolve, than species-wide traits, given comparable selective pressure. Therefore the onus is on the brain ecologists to demonstrate that this evolutionary work has actually been done, even if they can come up with a better rationale for why it should have been done. At a minimum, an evolutionary neurobiologist would like to see evidence of either androgen or estrogen receptors in the hippocampus of cowbirds and meadow voles and a lack thereof in closely related species that lack sex differences in hippocampal volume and spatial learning. That, at least, would be a good start.

EVOLUTIONARY NEUROBIOLOGY VERSUS BRAIN ECOLOGY

We began this chapter with a description of an interesting sex difference in the brain and behavior of brown-headed cowbirds. Adopting a teleological evolutionary perspective, the behavioral ecologists cum brain ecologists explained these sex differences as an adaptation related to the cowbirds' peculiar egg-laying habits. We saw reasons to question this explanation from a how-biology perspective. First, we saw reason to question in a general way the adaptationist approach to explaining variation in hippocampal volume, and then even more reason to question their approach to explaining sex differences in this trait. On the basis of how-biology considerations, I proposed instead that variation in hippocampal volume is a consequence, not a cause, of differences in spatial or any other memory-based learning. On this how-biology account, the variation in hippocampal volume is a byproduct of species and sex differences in the propensity to engage in spatial learning or other memory-based tasks. Differences in this propensity may well be adaptive, but their being adaptive does not make the hippocampus an adapted brain part in the same sense as the HVC of songbirds. As such, this how-biology–based explanation should be considered an outright replacement for the why-biology explanations of the brain ecologists.

Our exploration of the hippocampus in brown-headed cowbirds served as an introduction to two disparate programs, brain ecology and evolutionary neurobiology. Though both programs constitute evolutionary approaches to the brain, the similarities pretty much end there. Brain ecology is a direct descendent of behavioral ecology, from which it inherited its adaptationist teleology. Evolutionary neurobiology, on the other hand, encompasses a much broader evolutionary perspective, one that embraces both historically contingent genealogical relationships and current natural selection. From this perspective, brain ecology, as the term implies, is not so much an evolutionary approach to the brain as an ecological approach to the brain.

Ultimately, as we have seen, the differences between these two programs boil down to the fact that brain ecology is a thoroughly teleological enterprise, concerned with why-explanations, whereas evolutionary neurobiology is focused on the how-biology.[75] The evolutionary inferences in evolutionary neurobiology are informed more by what we know about neurobiological processes than are the evolutionary inferences in brain ecology, which are motivated primarily by a teleological evolutionary agenda.

When it came to sex differences in the hippocampal volume of cowbirds, voles, and Kangaroo rats, we saw how a disregard for the how-biology led brain ecologists to erroneously assume that sex differences will evolve whenever one sex benefits more from spatial cognition than the other. This assumption was the direct result of a failure to appreciate the amount of evolutionary work required of adaptive sex differences, which itself stemmed from a failure to appreciate the necessary how-biological modifications. You are more likely to view any hippocampal sex differences as adaptive if you ignore the how-biology of sexual development than you are if you attend to it. When you ignore this how-biology, you are also prone to inflate the causal power and agency of steroid hormones such as testosterone, especially if you have a teleological agenda.

So there are substantial differences between these two superficially similar, because evolutionary, approaches to the study of the brain and behavior. But, though brain ecology and evolutionary neurobiology can be viewed as rival approaches to this research area, they are not antithetical. The antithesis of evolutionary neurobiology is something called evolutionary psychology, which is the subject of the next chapter.

Why Men Won't Ask for Directions

THE HUNTING PARTY, consisting of 10 very tired men, trudges across the drought-parched savanna in the searing afternoon sun. Though uncomfortable, the conditions make their task easier because their prey, clustered as they are at the remaining water holes, are relatively easy to locate. But it's a long way to the nearest water hole, and the men are already exhausted from the heat and thirsty. The leader, call him Fred, grunts sounds of encouragement, and the other men temporarily quicken their pace, but they soon lapse back into a half-stumbling, fatigued gait. Finally, Fred signals a halt near a large acacia tree, and the men gladly retreat into its shade. Some begin to doze. Fred, however, is obviously reflecting. In fact, he is mentally scanning the landscape intervening between him and the water hole, picturing, in his mind's eye, where the long open expanses are, where there is afternoon shade, and the distances between the shaded areas. The image he conjures is not inviting: very few shaded areas, mostly large exposed expanses. Should they wait until the cooler dusk? If they did, would they get to the water hole before it got too dark? And do they really want to be at that water hole after sundown? Though the game will be plentiful, so will those nasty hyenas and lions, who would welcome a meal of human hunters.

Fred is approached by his close friend and confidant, Barney, who suggests an alternative water hole, which, though less reliable during extended droughts, is considerably closer. Fred has already briefly considered this option, but he has not visited that place for over 2 years. He summons his considerable memory for places, and he begins mentally reconstructing, then scanning, that part of his "cognitive map"—including the intervening space between the acacia tree and the newly remembered water hole— "looking" for suitable rest stops. Yes, he decides, they should be able to get there before the sun goes down; and nearby there is a nice rock formation, on the side of which is a cave large enough for all his men. They could sleep there for the night. But will there be any water left?

Meanwhile, back at the main encampment, one of Fred's mates, Wilma, is mashing some sort of roots into a paste. His other mate, Betty, has just returned from gathering acacia pods. Wilma had been watching Betty's two youngest offspring, along with her own toddlers, while Betty was out gathering. Betty settles in next to Wilma, clearly upset. Usually their conversations revolve around domestic issues such as the kids, but not today. Sensing her companion's distress, Wilma waits for Betty to indicate what's bothering her. After composing herself, Betty relates how, while out gathering acacia pods, she became separated from the other women in the group, deliberately on their part, she suspects. Alone, and quite lost, she felt an overwhelming sense of panic. She tried to calm down and find her way back to the encampment, but she ended up walking in circles. After what seemed like hours, she finally stumbled upon the rest of the gathering party. They hadn't even seemed to notice that she was missing, which only increased her distress.

Betty is the first to admit that she lacks Fred's remarkable ability to orient and navigate in space, using only a few landmarks and some kind of sixth sense—an aptitude that arouses more than respect in her. Betty is not alone in this deficiency, of course, or she would not be so open about it. Wilma too tends to get disoriented and lost if she strays too far from camp. All of the women, in fact, seem to understand that the men are superior in this regard, so they leave the navigating, along with the hunting, to them. But even by male standards, Fred has a special spatial sense; and this increases his stature, not only in the eyes of his mates but in those of his peers as well. Barney has a lower aptitude—though still much higher than that of any woman—and perhaps as a result, only one mate: Hortense.

And then there is Milhouse, who lacks any mate at all. He is noticeably lacking in navigational skills; in fact, he is as inept as the women. When he was young, he was always getting lost, and the other boys made fun of him. Milhouse is peculiar in other ways as well. He couldn't hit the ocean with a rock if he was standing in the surf. His ineptitude used to exasperate Fred and Barney, but they long ago resigned themselves to it. Even so, when he first begged off on hunting trips, they suspected he had ulterior motives. Though Milhouse claimed he just wanted to be with the children, Fred and Barney suspected it was their mates he wanted to be with, so they forced Milhouse to accompany them just so they could keep an eye on him. But they eventually realized that he wasn't interested in the womenfolk; he could be trusted to keep his penis sheathed when he was around them. And the women didn't seem particularly interested in him either. Well, actually they were interested in Milhouse, but they tended to treat him as one of their own. By now, everyone in the clan is fond of Milhouse. Wilma and Betty especially appreciate his willingness to lend a hand with child care. And they don't regret the fact that he is not as prone as Fred to those disturbing displays of temper.

THE ORIGINAL DIVISION OF LABOR?

The saga of Fred and Barney highlights some of the key elements of a story oft told about human evolution, according to which the males hunted and females gathered, especially during the formative Pliocene and Pleistocene epochs.[1] The basic outline of the hunter-gatherer story has been around since the late 1950s, and it has become the cornerstone of every attempt to explain the human condition, and "human nature," from an adaptationist perspective. From this simple story, much has been inferred about "human nature," about why we are the way we are: why we are so prone to kill each other and why we die for each other, why we exhibit fidelity to our mates and why we are adulterous, why we court and why we rape, why we are doting parents and why we commit infanticide, why we are ethical creatures and why we are sociopaths, why we have rock stars, why we have art, why we have religion, why we have language, and why we wage war.[2]

All this may seem like a heavy burden to put on Fred and Barney. And it is. But in recent years, the load they have been asked to bear has only increased. Much of the additional explanatory burden concerns the differences between man and woman. Originally, the story of Fred and Barney was used to explain why men are so much more aggressive than women, why they seek to dominate one another, and why they prefer to spend time with other men. These were the primary concerns of human ethologists.[3] Their successors, the sociobiologists, contributed an explanation for the famous sexual double standard: why Fred and Barney have "wandering eyes," while Wilma and Betty keep theirs focused on the kids. Then, as we saw in chapter 2, the sociobiologists decided that Wilma and Betty should have wandering eyes as well, albeit for a different reason. So the double standard was transmuted into a sex difference in mate choice criteria: males are more interested in quantity, females more interested in quality. But, even later, it was decided that males are not as quality-blind as initially believed; they just have different mate choice criteria. Males prefer females with symmetrical breasts, copious hips, and youth—the most nubile. Females, on the other hand, prefer to mate with males of high status, no matter how homely and no matter how old: ergo, May–December romances.

Around the time of the May–December romance insights, the human sociobiologists had to share the adaptationist stage with sometimes friendly, sometimes unfriendly rivals who called themselves evolutionary psychologists.[4] They shared the teleological convictions of their predecessors but, in addition, brought to the table fresh insights gained from the "cognitive revolution" in psychology, as a result of which they were able to increase Fred and Barney's burden still further. They have identified a number of "psychological" sex differences, in addition to mate choice criteria, that could be traced to Fred, Barney, and Co. Of these, the most original fall under the category of cognition.[5]

The focus of this chapter will be on some of their recent claims about adaptive sex differences in spatial cognition, which are alleged to follow from the story of Fred and Barney. More generally, I will use this putative sex difference to develop some minimal guidelines for evaluating the plausibility of any of the claims made by evolutionary psychologists about behavioral or psychological sex differences in humans. Birdsong will serve as something of a benchmark in this regard, the standard by which we measure the warrant for such claims. More generally still, I will use the particular claims about human sex differences in spatial cognition as a concrete way to examine some of the basic tenets of evolutionary psychology. I will then compare and contrast evolutionary psychology with evolutionary neurobiology.

From Hunting to Calculus

The basic outline of the argument is as follows. Because Fred and Barney did the hunting, they had to range much farther than their mates, whose own duties revolved around camp life: watching the kids, gathering foodstuffs in the vicinity, and so on. As a result, the males experienced stronger selection for spatial cognition than did the females, much as has been claimed for voles and kangaroo rats. In addition, hunting required skill in throwing and tracking projectiles of various sorts, another spatial cognitive task. But males not only experienced natural selection for these skills; they experienced sexual selection for them as well. Wilma and Betty were turned on, it turns out, by displays of hunting prowess, and chose their mates accordingly. This story explains why boys are better than girls at solving spatial problems. Moreover, it provides the answer to the question posed in the title for this chapter. A man won't ask for directions because to do so would be the equivalent of taking off his shirt and revealing a sunken chest, sure to dampen the ardor of any prospective mate.

David Geary proposed a novel extension of this logic to account for sex differences in mathematical prowess. According to Geary, the superior spatial skills of human males give them an advantage over females in solving mathematical problems, because spatial representations of these problems provide a superior way to solve them. And males, more than females, are able to effectively utilize these spatial representations.[6]

Evolutionary psychologists have not ignored the distaff staff side of our cognitive inheritance. Though Wilma and Betty were not as adept as Fred and Barney at navigation, they excelled in a different area of spatial cognition: remembering the location of household items and foodstuffs. This was a direct result of their participation in the gathering part of hunting and gathering. Fred and Barney, on the other hand, were hopeless around the camp.

Milhouse is another story. He represents that subset of males who practice an alternative reproductive strategy, in some extended sense of "reproduction."[7] (His alternative lifestyle does not, however, represent an alternative mating strategy.) Though he has low narrow-sense fitness, because he has no offspring of his own, his broad-sense, or "inclusive, fitness" is quite respectable, because of the care he gives to the offspring of his relatives. In essence, Milhouse babysits while the other men are out hunting and the women are out gathering. Presumably, babysitting does not require the spatial skills of either the male hunters or the female gatherers.[8]

The focus in this chapter will be on the alleged sex differences in the spatial cognition of Fred and Wilma, which, evolutionary psychologists allege, have been visited upon us humans to this day.

Sex Differences in Body and Mind

Nature is full of examples of external sex differences, or sexual dimorphisms. For many species, we can tell, at a glance, whether an individual is male or female. The phenomenon is far from universal, however. Male and female geese, ravens, and crows are outwardly indistinguishable, as are many other birds, as well as many fishes, reptiles, and amphibians. Most mammals exhibit some degree of external sexual dimorphism, that of lions, gorillas, and baboons being among the most pronounced. By mammalian standards, humans are moderately dimorphic, about midway between mountain gorillas and cotton-top tamarins, to name two other primate species. This position on the dimorphism continuum will be important to bear in mind.

Though they may look the same on the outside, male and female tamarins are no less fully male and female than their gorilla counterparts. As a biological category, sex as gender refers to a role in reproduction. Males provide sperm, and females provide eggs. Biologically, the essence of maleness—the attribute that makes a male a male—is sperm. That is, any individual that produces sperm is a male, no matter its other attributes, including external genitals, relative size, strength or coloration. Conversely, the biological essence of femaleness is the capacity to produce ova or eggs. Gonads that produce sperm are called testes; gonads that produce eggs are called ovaries.

But gonads, in and of themselves, do not a reproductively competent individual make. Reproductive competence requires, in addition, sex-specific plumbing by means of which the gametes (sperm or ova) are conveyed to the appropriate destination for fertilization. And in vertebrates, there are often further elaborations of the reproductive anatomy. Female mammals develop a uterus within which embryos reside and develop. Male mammals develop pronounced external genitals.

The production and transmission of gametes also require coordination of the actions of both the endocrine and nervous systems, which is actualized in the hypothalamic-pituitary-gonadal (HPG) axis. Of particular significance is a group of neurons in the hypothalamus that manufacture gonadotropin-releasing hormone (GnRH). There is a sex difference in the manner in which GnRH is released from these neurons. In males, GnRH is released more or less continuously, whereas in females GnRH is released in distinct pulses with distinct interpulse periods. This is far and away the best-documented functional sexual dimorphism in the brains of mammals. It is noteworthy that this physiological dimorphism has no known anatomical consequence. It is not possible to distinguish male and female GnRH neuronal populations anatomically.[9]

Males and females can also differ behaviorally. Male mammals tend to be more aggressive than females, maintain larger territories, and engage in stylized courtship rituals. One of the best-documented behavioral sex differences in humans is an activity referred to as "rough and tumble play." Young boys engage in more of it than do young girls.[10] But behavioral sex differences ranging from courtship behavior to rough and tumble play are far from universal and largely a matter of degree. The most widespread and robust behavioral dimorphism in mammals is the act of copulation itself and the events immediately preceding it. Females signal their readiness to mate through various acts, be they vocalizations, olfactory signals, postures, or tactile contacts such as nuzzling. If the male reacts appropriately to this "proceptive" behavior, she then facilitates intromission. At a minimum, as in many mammals (deer, antelope, rhinos, and elephants), she must stand still and move her tail to one side. In some mammals, including rodents, weasels, and cats, the female signals her receptivity by adopting a characteristic posture known as lordosis, in which she supports her weight on her elbows, lifts her rear end skyward, and elevates her tail. The copulatory behavior of male mammals is, in its essence, amazingly invariant. After mounting the female, males of all mammal species perform a variable number of pelvic thrusts until the—with a few notable exceptions—anticlimactic climax.[11]

But no behavioral sex difference in a mammal, whether in courtship behavior or rough and tumble play, has been connected to a sex difference in the brain in the way birdsong has been connected to the HVC and other song nuclei. The reverse is also true. There are anatomical sex differences in the brains of mammals that have no known connection to behavior or any other function. The first sexually dimorphic part of the hypothalamus to be identified in a mammal is called, appropriately, the sexually dimorphic nucleus, or SDN.[12] The SDN is larger in male rats than in females, and the same is true of other rodents.[13] The discovery of the SDN created considerable excitement among neurobiologists; it was anticipated that the sex difference in this population of neurons would be connected to pelvic thrusting in the way the HVC is connected to birdsong. Years of intensive research, however, have failed to uncover SDN's connection to pelvic thrusting or any other sexual behavior.[14] In the meantime, several other sexually dimorphic bits of the hypothalamus have been identified, though none with a robust connection to a particular behavior.[15]

By far the best characterized anatomical sex difference in the mammalian nervous system, for which there is a clear function, is found in the spinal cord, not in the brain. The spinal nucleus of the bulbocavernosus (SNB) innervates the bulbocavernosus muscle, which controls penile erection.[16] It is larger in males than in female rats. The equivalent structure in humans is called the Onuf's nucleus, and it too is larger in males.[17] This sex difference in the human spinal cord is the only one that comes close to meeting the HVC (of songbirds) standard for adaptive sex differences in the nervous system. In contrast to the HVC, the sex difference in the Onuf's nucleus is not obviously the product of sexual selection. Moreover, the cause-and-effect considerations I applied to the sex differences in the hippocampus of birds are especially important to bear in mind here. We cannot assume that a neural sex difference has a biological etiology. Rather they could have been shaped by experiential factors related to socialization. Indeed, with the partial exception of the Onuf's nucleus, a social etiology cannot be ruled out for any existing human sex difference in the brain or spinal cord.[18]

Notwithstanding these considerations and the meager payoffs to date, a concerted effort has been mounted to locate sex differences in the brain that might explain psychological differences in men and women.[19] In contrast to the brain ecologists, researchers looking for the neural basis for sex differences in human spatial cognition have not been particularly concerned with the hippocampus[20] but rather with more global differences in brain organization,[21] particularly brain lateralization—that is, the tendency for one hemisphere of the brain to be more involved in particular tasks than the other.[22] The received view is that men's brains are more highly lateralized than women's brains, and this difference somehow explains the putative male advantage in spatial tasks. The most charitable assessment of the evidence for this view is that it is equivocal.[23] Part of the problem is that the neurobiological considerations that drive this search are based on a conceptual framework that is decidedly shaky and poorly formulated.[24] This gives the whole enterprise, including the search for the neurobiological substrates for sex differences in lateralization, the flavor of a fishing expedition.[25] Actually, it might more closely resemble a snipe hunt, because the object of the chase may be a phantom. Before much more energy is expended in this hunt for the neural substrates of sex differences in cognition, it would seem prudent to make sure that the evidence for these sex differences is compelling.

Sex Differences in Spatial Cognition

The history of research on psychological sex differences is long and inglorious. We can use as a convenient starting point the publication in 1894 of Havelock Ellis's book, entitled *Man and Woman: A Study of Human Secondary and Tertiary Sexual Characters.*[26] One commentator assessed the state of the art as follows: "There is perhaps no field aspiring to be scientific where flagrant personal bias, logic martyred in the cause of supporting prejudice, unfounded assumptions, and even sentimental rot and drivel, have run riot to such an extent as here."[27] This quotation is from 1910; it is at least an open question as to how much things have improved since then. What *is* clear is that Darwinizers and other functionalists have only exacerbated these problems by adding yet one more ideology to the mix, without, however, increasing methodological rigor.[28]

Ideology abounds on both sides of the debate,[29] which has ill served the science. But it is clear that those who want to find sex differences have always driven this research, often ignoring other variables in the process, as indicated by the results of one extensive survey that revealed that far more results from studies on sex differences (25 percent) fall into the zero range than do results from studies on other psychological factors (6 percent).[30] The inordinate amount of attention to psychological sex differences at the expense of other psychological variables would no doubt make for an interesting topic in the sociology of science. What the history of this research area suggests is that researchers are always on the lookout for the slightest hint of a sex difference, which they then tend to inflate. The skeptics then set out to debunk the original claims, often successfully, and the original claims are scaled back or even jettisoned outright. It is in this context that we should consider the ground-breaking book *The Psychology of Sex Differences* by Maccoby and Jacklin, published in 1974.[31] In their review of the literature, Maccoby and Jacklin debunked many of the previous claims for psychological sex differences but did find evidence for male superiority in spatial cognition, as well as less compelling evidence for a female advantage in verbal tasks. Sex difference advocates have focused primarily on these positive results and have made this cognitive sexual dichotomy something of the received view.[32]

Here I will focus on the quality of the evidence for a male superiority in spatial cognition based on the evidence accumulated since Maccoby and Jacklin's review. How good is the evidence that human males have better spatial skills than females? The answer to that question depends, in large part, on whom you ask. According to Irwin Silverman and Marion Eals, the evidence is strong indeed: "one of the most consistent findings [*in studies of human cognition*] has been superior performance for males on tests of spatial abilities."[33] But this comment is more advocacy than a dispassionate interpretation of the literature. In fact, the evidence for sex differences in spatial cognition has long been subject to contentious debate.[34] Part of the disagreement stems from the fact that disparate tests of spatial cognition have been employed.[35] But even if we bracket that factor, much remains in dispute. This much is beyond dispute, however: at one time or another a robust sex difference has been claimed for virtually every test of spatial cognition ever devised. Some of these tests have subsequently been discredited outright; sex differences in other tests have subsequently been disputed or reinterpreted.[36] The exceptions are tests in which the subject must mentally rotate two three-dimensional objects in order to judge whether they are actually different objects or the same objects in different orientations.[37] Males, on average, consistently score higher than females on these three-dimensional rotations. Estimates as to the magnitude of this sex difference vary widely.[38] But by way of perspective, the magnitude of this sex difference, by even the most generous estimates, is not large—less than the sex difference in body size, a sex difference that is widely considered a modest one.[39] Sex differences in mental rotations are such that an equivalent morphological difference in a bird would escape the attention of all but the most obsessive birders.[40]

Now consider the male superiority in throwing-accuracy tests of spatial cognition. These tests are of more recent vintage; they are based on much smaller sample sizes; and they are much less standardized than the mental rotation tests. On the plus side, these tests were devised specifically to test the predictions of the hunter-gatherer story, in contrast to the mental rotation tests. Indeed, the throwing-accuracy test of spatial ability has been touted for its "ecological validity."[41] (But one could just as easily praise them for their "sociological validity," that is, as tests for differences in the socialization of males and females.) The results from these tests to date can be summarized as follows. Males, on average, hurl projectiles more accurately than females;[42] homosexual males, however, score closer to females than to heterosexual males.[43] This latter fact suggests to evolutionary psychologists that there is more to the male superiority than differential socialization, and that something must be biological.

What mental rotations and throwing-accuracy tests have in common is not obvious. But Geary and other evolutionary psychologists find it convenient to lump them together under the rubric of "three-dimensional spatial cognition," solely because both skills would seem to be essential in hunting. This rather ad hoc conglomeration, it should be emphasized, runs counter to the trends in cognitive psychology as a whole, and certainly counter to the trend in evolutionary psychology itself, where there has been a push to differentiate cognitive tasks, rather than amalgamate them.[44]

Finally, we need to consider the spatial task for which *females* outperform males. As were the throwing-accuracy tests, this test was designed with its "ecological validity" in mind. For this test, the subjects view, for one minute, a picture containing an array of common household items; they are next presented with another array, in which many of the items from the first array are present but some have been replaced with different items; they are then asked to identify the changes. Silverman and Eals, who devised this test, reasoned that girls should outperform boys because the remembrance of household items represents the other side of the hunter-gatherer imperatives.[45] If men were out hunting, they would not benefit from remembering household items. Women, on the other hand, should be quite good at remembering these items and their locations, because they experienced selection for their domestic skills.

The household items test of spatial cognition has been employed primarily in a couple of studies by Silverman and Eals themselves,[46] so they are less standardized and the results are based on much smaller sample sizes than even the throwing-accuracy tests of spatial cognition.

Assume for now that the sex differences for all three tests of spatial cognition are equally robust and reliable. We are still a long way from establishing that they represent an adaptive sexual dimorphism such as we observed for birdsong. To that end, we first need to consider the putative adaptive rationale for this sex difference in the context of what we know about how sexual dimorphisms evolve.

THE WHY OF SEX DIFFERENCES

We first need to acknowledge that there may be no "why" of a given putative sex difference, including that for spatial cognition. That is, the sex difference may be, from an evolutionary perspective, epiphenomenal, without functional significance. Consider the following sex difference in finger digits. The ratio of the length of the index finger to the length of the ring finger is greater in human females than in human males. Of what possible functional significance could this sex difference be?[47]

With regard to the sex difference in spatial cognition, there is another reason there may be no why of it, one that does not apply to the sex difference in finger lengths. Any sex difference in spatial cognition may be a function of sexually differentiated socialization processes, in which case, there is no role for a biological—much less an evolutionary—explanation. This idea, however, brings us to the topic of the following section, so I will defer further discussion until then. Suffice it for now to observe that the paranoiac mindset that evolutionary psychology promotes does not allow its practitioners to seriously consider likening spatial cognition to finger lengths, and it compels them to ascribe to socialization a secondary role, in principle.

The Role of Sexual Selection in Human Cognition

Assume for the moment, however, that the sex differences in spatial cognition are functionally significant. How would they arise? Geary assumes that any functionally significant sex difference must result from sexual selection.[48] But functionally significant sex differences have other evolutionary causes as well. For example, the fact that my wife and all other female mammals have a uterus, whereas I and all other male mammals do not, cannot be traced to sexual selection. Uteruses did not evolve in order to attract males or because of some role they played in female-female status competition. The presence of uteruses in female mammals and the lack thereof in males resulted solely from old-fashioned natural selection for uterus-endowed females and non-uterus—endowed males.

The same logic could be applied to sex differences in spatial cognition. Indeed, on the traditional reading of the sexual division of labor, there is no role for sexual selection at all in explaining cognitive sex differences. Evolutionary psychologists such as Geary often confuse this reading with a second, quite different reading. On the second reading, the alleged superiority in male navigation is driven by male—male competition for mates, or female choice. It is only when we can demonstrate such sexual selection that we can equate male navigational skills to deer antlers and explain why men won't ask for directions.

How would we go about demonstrating that the sex differences in spatial cognition result from sexual selection? We have seen how the comparative method was applied to sex differences in birdsong to this end. But evolutionary psychologists, in contrast to all previous adaptationists, eschew the comparative method. They reason that humans have experienced such a unique selective milieu that such comparisons, even with other primates, are unproductive in principle.[49] So what is the evidence that sex differences in cognition result from sexual selection? The evidence that sexual selection shaped human sex differences in spatial cognition consists solely of the story of Fred and Wilma. This sort of evidence has been aptly likened to Kipling's "just so" stories.[50]

The problems for the evolutionary psychology account of the sex difference in spatial cognition extend beyond the weakness of the evidence in its favor. There are also compelling reasons *not* to expect sex differences in spatial cognition to evolve, even when Fred and Barney benefit more than Wilma and Betty from navigational skills: what I referred to in chapter 8 as intergender hitchhiking. In general, directional selection on one sex will lead to a correlated response in the other, except under special circumstances, when what's good for the goose is bad for the gander, or vice versa. Given directional selection for navigation in Fred and Barney, both their male and female offspring should be cognitively enhanced thereby, unless there is countervailing selection in Wilma and Betty. That is, unless enhanced navigation is disadvantageous in some way for Wilma and Betty, there is no reason to expect a sex difference in navigational skills to evolve.

Traits in which what's good for the goose is not good for the gander are called sexually antagonistic. Penises are quintessentially sexually antagonistic traits—a good thing for males, a bad thing for females. Nipples, on the other hand, are not at all sexually antagonistic.

The evolutionary psychologists must demonstrate that navigation is more like a penis than a nipple in order to have any evolutionary rationale for their claims about male superiority in spatial cognition and navigation. But we will see that evolutionary psychologists have even less warrant to assume that selection for enhanced spatial skills in human males will result in a cognitive sexual dimorphism than the brain ecologists had to assume that selection for increased spatial skills in female cowbirds resulted in a hippocampal dimorphism.

Alternative Explanations

It is not as if there are no alternatives to evolutionary psychology's explanations for the existing sex differences in spatial cognition and mathematical aptitude. Moreover, these alternative explanations have the virtues of simplicity and transparency that are utterly lacking in the evolutionary explanations. Consider, for example, some of the contortions Geary must perform in order to Darwinize putative sex differences in mathematical ability.

Recall that his thesis is that males do better at math because of their superior spatial cognition. He relies heavily on standardized tests such as the SAT (formerly Scholastic Aptitude Test, now Scholastic Abilities Test). Males, on average, score higher than females on these tests, and the sex difference appears to be particularly pronounced at the high end.[51] We should note, however, that these sex differences are small, much smaller, even, than those for tests of three-dimensional rotation.[52] Moreover, it is not clear what these tests indicate about mathematical aptitudes. The SAT is designed to predict how well an individual will do in university courses. Ideally, it should accurately predict that individual's grade point average, or GPA. Unfortunately, however, it often fails in this regard. Though males receive higher average scores than females on the mathematical portion of the SAT, there is no difference in the grades that male and female students obtain in their college math courses.[53] SAT scores *underestimate* women's grades in actual college math courses. Most observers would see this as a deficiency in the SAT test, given its stated purpose.[54] But Geary argues instead that mathematics course grades may *overestimate* mathematical ability in females.

According to Geary, the classroom grades are distorted by such extraneous factors as homework completion and perceived motivation. Females benefit from making a better impression on their instructors, the majority of whom are male. But that is only by way of preamble; the truly novel and ingenious part of his argument concerns something called *transfer*. According to Geary, classroom tests involve the reworking of familiar problems that have already been solved in class or assigned as homework—and remember, females are more likely to do their homework. The SAT, on the other hand, tests the ability to transfer mathematical skills to truly novel tasks. It is in this transfer of mathematical knowledge to novel tasks, Geary argues, that males outperform females. Females, it seems, may be able to memorize the answers in the classroom but cannot formulate them on the SAT. Our educational system is stacked against males, and, as a result, their innate superiority in mathematics is masked. The SAT is able to "see through" this female bias in our education system.[55]

Although ingenious, Geary's transference argument has a certain "handwaving" quality about it. It is designed solely to justify the energy he feels compelled to expend on his evolutionary account of the sex differences in SAT scores. Consider that there is a much greater male advantage in standardized tests of such subject areas as European history and political science than in mathematics.[56] Why focus on the math scores? Geary clearly focused on the math scores because he thought they could be more easily related to the story of Fred and Barney.[57]

Given the utter absence of how-biology considerations in evolutionary psychology, it is truly ironic that Geary is compelled to resort to evidence of biological causation in order to justify his evolutionary account. He has to, if only to undermine the obvious alternative sociological explanation. His evidence that sex differences in mathematics are the result of biological factors is of two sorts. First, these sex differences are found in all cultures; second, they are linked to testosterone. Let's consider first the cross-cultural evidence.

The cross-cultural gambit is a standard move in evolutionary psychology, a way to invoke the authority of biology without actually doing any biology: biology by elimination. If a phenomenon exists outside of western Europe and the United States, it is pan-cultural, therefore it is non-cultural, therefore it is biological. Geary claims that there are ample cross-cultural data showing male mathematical superiority. But most of the data Geary cites come from developed Western countries and are heavily weighted toward the United States. Moreover, he neglects to cite other studies that demonstrate marked variation in sex differences for different parts of the world and that clearly show that in most countries sex differences are either much smaller than in the United States, or are completely nonexistent.[58]

But even within the United States, the magnitude of gender differences differs widely among ethnic groups, being higher among whites than among Asians, Hispanics, or African Americans.[59] In fact, for standardized tests other than the SAT, there are either no gender differences or there is a slight female superiority in all but the Caucasian students.[60] Clearly, there are some sociocultural factors worth investigating here. Let's focus for the moment on Caucasians. The first thing to note is that gender differences in standardized math scores are disappearing. There has been a marked decrease in gender differences over the last 50 years.[61] In other words, these scores are coming to better match the grades women obtain in actual mathematical courses, though they still underestimate them.

This trend is what we would expect from a sociological perspective, given the increased educational and professional opportunities women have experienced in the last few decades. It is not, however, what you would expect if the sex differences in math scores were testosterone-driven. This brings us to the second line of argument employed by those seeking to Darwinize sex differences in human cognition. Here the goal is to link cognitive sex differences to testosterone, particularly its "organizing" effects.[62]

Most of the relevant data come from studies of atypical sexual development or congenital hormonal abnormalities. Some of the most widely cited studies concern females with congenital adrenal hyperplasia (CAH). Females so afflicted are exposed, in utero and neonatally, to abnormally high levels of androgens as a result of a malfunction in their adrenal glands. As a consequence, their genitals become masculinized to the extent that some have been raised as boys.[63] Behaviorally, they are often characterized as tomboys; they seem more interested in the activities characteristic of young boys than those of girls and, according to some studies, have a higher incidence of homosexual or bisexual behavior than do other girls.[64]

If spatial or mathematical reasoning were subject to the organizing effects of androgens, we would expect these aptitudes to be enhanced in CAH females relative to typical females. There have been several attempts to find such effects in CAH females, but they have resulted in conflicting claims. In one study, it was reported that CAH females do better than normal females on some tests of spatial reasoning,[65] and in another, that they exhibit a more "masculine cognitive profile" than normal females.[66] But other studies have failed to find an affect of CAH on spatial tests.[67] Moreover, still other studies indicate that CAH females do not exhibit enhanced mathematical ability.[68]

Testosterone, as we have seen, influences events in the mammalian brain only indirectly, through its conversion to estrogens. It is not surprising,

therefore, that in rodents and nonhuman primates, artificially elevated estrogen levels cause some behaviors, including sexual and play behavior as
well as certain forms of aggression, to become masculinized. In humans, a
natural experiment was inadvertently conducted on a large number of fetuses when their mothers were administered the synthetic estrogen diethylstilbestrol (DES) in order to prevent miscarriages. There have been some
reports of an increase in male-typical or homosexual behavior among this
population,[69] but there is absolutely no evidence that DES exposure affected
any aspect of cognition in females.[70]

If androgens exert important organizing effects on cognition, abnormally
low levels should result in a "feminine cognitive profile" in males. There are
several pathologies of male sexual development caused by defective androgen regulation. Of these, the most severe is androgen insensitivity syndrome (AIS), which I discussed in chapter 8. AIS males are often misdiagnosed as females at birth. Many, in fact, have been raised as females and
show a typical female sexual orientation and self-identification.[71] To date,
there is little evidence that AIS affects cognition in any way.[72] One study
identified a decrease in *some* visuospatial abilities in AIS males reared as
females,[73] but even if we were to take these results at face value, there is no
way to separate the contributions of social and hormonal factors. Nor can
these factors be separated in CAH and the other hormonal disorders discussed here. The physical appearance of the neonate significantly influences
the way it will be treated by persons with whom it interacts, beginning with
the parents. If the neonate looks like a male, it will be treated in male-
typical ways, and the reverse is true of female-looking neonates, whether
they be genetically male or female. The differential treatment includes
everything from the toys the child is provided or encouraged to play with, to
the way the child is dressed, to the way the child is treated by peers.

In an extensive review of research on the organizing effects of sex hormones on sex differences in humans, two researchers who are not in any
way hostile to biological explanations of sex differences concluded that the
evidence for a significant effect is most consistent for sex-typical play.[74] They
also found "some evidence supporting a role for androgens in the development of tendencies toward aggression."[75] They characterized the evidence
for such effects on other behavioral sex differences (core sexual identity,
hand preference, language lateralization, and others), including spatial abilities and verbal skills, as "less convincing or even contradictory."[76] But Geary
ignores this deflating assessment of the evidence for hormonal effects on
human cognition.

The evidence for any activating effects of testosterone on spatial cognition is even weaker and more difficult to interpret from an adaptationist
perspective.[77] So neither the cross-cultural evidence, nor the hormonal evidence provides reason to be confident that these sex differences are primar-

ily the result of biology. On the other hand, there is ample evidence that social factors could play an important role in these sex differences. Perhaps it is significant in this regard that American boys and girls both classify math and science as a "male" activity.[78] This fact probably goes a long way toward explaining why, when math courses become elective, girls are less likely than boys to choose to participate in them. This failure to participate in elective math courses itself is no doubt part of the explanation for the fact that gender differences do not emerge until the teenage years, then gradually increase, reaching a peak in the last year of high school.[79]

It is obvious to one and all that boys and girls become increasingly divergent in their experiences with age. One obvious nonbiological explanation for this divergence is that boys and girls are socialized differently. If you have a penis, you are likely to be encouraged to participate in sports, for example; if you lack one, you are likely to be encouraged to knit.[80] Perhaps that socialization difference would explain why men excel at that "ecologically valid" spatial task of throwing objects toward a target and women tend to be superior in skills requiring manual dexterity. It would seem prudent to explore the role of such social processes on mathematical ability before we bother with an evolutionary explanation.

The putative sex differences in human spatial cognition are an interesting case, in that, in contrast to teleological explanations discussed previously in this book, how-biology is invoked to buttress an adaptationist explanation. This move is actually quite common when it comes to human sociobiology or evolutionary psychology. It is attractive to adaptationists because the obvious alternatives to their explanations are often sociocultural.[81] So the how-biology is used to fend off the how-sociology as a first step toward an evolutionary psychology. But if you are going to employ how-biology in this way, you need to bring more to the table than Geary and other evolutionary psychologists have managed. At this point, there is certainly no reason to prefer the how-biology explanations for sex differences to the how-sociology explanations.

Geary does not completely ignore the role of social factors; he just assigns them a secondary role. Secondary, that is, to evolved hormone-mediated sex differences. But we need to seriously consider the possibility that the social factors are the primary determinants of these sex differences or even the only ones. The resistance of evolutionary psychologists to this conclusion is driven primarily by their desire to devise an evolutionary explanation for these sex differences, an evolutionary explanation for which, as we have seen, the only warrant is the story of Fred and Barney.

It is too soon to say with any confidence what the role of the how-sociology will be in explaining sex differences in spatial cognition. If it should turn out that, in fact, the how-sociology is all we need to explain these sex differences, then obviously there is no work for evolutionary psychology to do; we should consider the how-sociology a sufficient explanation, one that preempts any adaptationist account. But even if the how-biology turns out to be important, there is little warrant for an evolutionary psychology explanation, because, as we have seen, any such biologically based sex differences are likely epiphenomenal from an evolutionary perspective. But evolutionary psychologists, the most unreconstructed teleologists to take the stage as scientists to date, do not seriously consider this possibility. They are the truest heirs of Bishop Paley. So the guiding principles of this endeavor deserve some further scrutiny.

EVOLUTIONARY PSYCHOLOGY

I have focused on spatial cognition in this chapter, because it exemplifies what is distinctive about evolutionary psychology, what separates it from the earlier "Darwinian" programs for explaining the human condition, such as human ethology and human sociobiology. Evolutionary psychologists have been particularly concerned to differentiate their framework from that of Darwinian social science, the particular form of human sociobiology that had hitherto held a hegemonic position among those who would Darwinize the human condition.[82] Particularly noteworthy in this regard is evolutionary psychology's claim that Darwinian social scientists had been fishing at the wrong depth. Whereas Darwinian social scientists had sought to Darwinize manifest behavior by demonstrating its adaptiveness, the focus of evolutionary psychologists is on the "deeper" psychological mechanisms to which the manifest behavior is only indirectly related.[83] Where does this emphasis on psychological mechanisms come from?

The Psychology in Evolutionary Psychology

In seeking to understand evolutionary psychology's emphasis on psychological mechanisms over behavior, we must emphasize from the outset that evolutionary psychology is, first and foremost, an attempt to ground, or create a foundation for, the science of psychology. From the perspective of a robust materialist such as myself, psychology does indeed need a new foundation, because psychology has been, from the outset, a problematic science, first, for taking seriously Descartes's fundamental distinction between material stuff and mental stuff, and second, for taking upon itself the task of explaining the mental stuff without reference to the material stuff. Given this task, it has never been clear how psychology could ever join the community of natural sciences.

Indeed, from the outset, psychology has considered itself a "special" science,[84] special because of its preoccupation with the mental, as opposed to the material, sphere. Wilhelm Wundt, who many consider the founder of psychology, was quite explicit in acknowledging this Cartesian legacy. And Sigmund Freud, after a brief and misguided fling with biology, was typical of those who followed Wundt down the mentalistic road. But the problem with most mentalistic frameworks (Freudian and otherwise), it soon became apparent, was that they were empirically opaque. That is, they required methods—such as introspection, the royal road to the conscious, and dream analysis, the royal road to the unconscious—the accuracy of which was impossible to verify independently.

Enter Behaviorism—a revolutionary movement within psychology. Behaviorism was a systematic attempt to make psychology as empirically transparent as the natural sciences. And in this regard, it is universally agreed, it succeeded. In doing so, behaviorists found it necessary to proscribe any reference to mental phenomena, thereby fundamentally changing the very nature of psychology to a science of behavior. The problem with Behaviorism, it turned out, was that it not only threw the baby out with the dirty bathwater of mentalism; it threw out the bathtub as well. Radical behaviorists such as B. F. Skinner advocated a completely atheoretical approach quite unlike anything in the natural sciences.

After decades of hegemony, Behaviorism's puritanical abstemiousness about things theoretical began to wear thin. And Skinner, the most Cromwellian of the behaviorists, came to serve as the bogeyman who inspired a counterrevolution.

The Cognitive Turn

In disavowing mental phenomena as a proper subject for psychological investigation, Behaviorism was not thereby obligated to eschew psychological categories such as emotion, reasoning, desire, and belief. That it did so nonetheless stems from its philosophical Puritanism, which fostered an extremely ascetic mindset, according to which it is a cardinal sin to even talk about that which direct observation does not compel you to say. Its goal was to start afresh in a world with a minimum of furniture, and only Shaker furniture at that. Hence, the negative program, that of removing the existing clutter, was paramount, especially in the early days. And, initially at least, this approach served Behaviorism well, because few natural scientists have much regard for Freudianism or phenomenology, to name two of the more dubious programs to which Behaviorism was opposed. But Behaviorism's austerity inevitably became wearisome to all but the most zealous. Eventually, it seemed to many, that in completely removing the psyche from psychology, Behaviorism was left with too few explanatory resources, especially when it came to certain characteristically human activities, such as devising a scientific theory, or learning a language. The latter, in particular, proved to be something of an Achilles' heel for Behaviorism. But initially the language problem was tackled with characteristic hubris, by B. F. Skinner, in a book entitled *Verbal Behavior*.[85] This book stimulated one of the most devastating and influential critiques ever launched against an entrenched research program.

The critique, in the form of one of the longest book reviews ever penned, was written by the linguist Noam Chomsky.[86] Chomsky attacked Skinner from a variety of angles. He opposed Skinner's extreme environmentalism with an equally extreme nativism: where Skinner saw environmental determinants of language learning, Chomsky saw innate structure. Chomsky also strenuously argued against Skinner's view that language was just another form of behavior. To the linguist, language was a very special form of behavior, one that required the existence of specific deep structures in the mind. Both forms of opposition to behaviorism, but particularly the latter, came to figure prominently in what is known as the "cognitive revolution."

The cognitive revolution within psychology was quite heterogeneous, however, and quickly divided into a number of factions. I will briefly trace the lineage of evolutionary psychology from among these factions.

The first split was between those who emphasized the revival of belief–desire explanations of behavior, often referred to as "folk psychology,"[87] and those who looked to computer technology—both as a model for human cognition and as a way to make talk of mental processes and mental states respectable—whom I will refer to as "computationalists." Computationalism quickly came to predominate and, with it, the construct of "information," whose murkiness is masked by its abstractness. Evolutionary psychology belongs to the computationalist camp, in which the mind is viewed as something that represents things in the world. These representations represent by virtue of encoding information about the world. Cognition is computation over these informationally encoded representations.

But the computationalists themselves soon divided into two primary factions. One camp was composed of reformed behaviorists—reformed in that they were much less abstemious regarding speculations about what goes on inside the head, but not so reformed as to cease worrying about how these speculations cashed out behaviorally. In fact, their emphasis was still on predicting behavior, which for them was what explaining behavior amounted to, a hangover from old-fashioned Behaviorism. Daniel Dennett exemplifies this attitude.[88] Reformed behaviorists are a lot like old-fashioned behaviorists, but with a computationalist veneer, and without the reforming zeal. Dennett's primary criticism of Skinner is of his overzealousness.[89]

The other computationalist camp was much more radical. These more radical cognitivists intended not only to restore the psyche to psychology but to eliminate or at least greatly minimize the study of behavior. For radicals such as Jerry Fodor, the proper subject for psychology is the psyche;

behavior is another thing altogether.[90] Chomsky continues to exert considerable influence within this camp, which some of its adherents refer to as "high-church" computationalism. Of particular importance in this regard is Chomsky's conception of a language acquisition device or language organ, which he views as the mental equivalent of, say, a liver or a kidney.[91] By this analogy, Chomsky invites us to consider the language acquisition device as a distinct part of the mind, with a particular function, a mental "module." A mental module is self-contained, or encapsulated, in that its internal processing is not influenced by whatever else is going on in the mind. It receives prestructured inputs that are appropriate for its function, which it converts into further structured outputs that may function as the inputs for another type of module.

Chomsky was primarily concerned with syntax or grammar, the "rules" by which word or phrase types are put together to make intelligible sentences.[92] Fodor extended the modularity thesis to some nonlinguistic cognitive processes as well.[93] The modularitist computationalists were soon engaged in a spirited debate with the reformed behaviorists, who claimed that any putative modular psychological mechanism could be modeled with equal efficacy as a nonmodular process, given a different conception of the cognitive architecture, which they referred to as parallel distributed processing, or PDP.[94] One of the virtues of the PDP approach, often referred to as connectionism, is its empirical transparency. By contrast, Chomsky-Fodor–modularism is quite opaque, to an almost Freudian degree. Such opacity is inherently suspect for anyone endowed with the conventional scientific ethos.

Evolutionary Psychology as Freudian Phrenology

Evolutionary psychology owes much of its current prestige to its connection with the cognitive science juggernaut. It is most closely connected, however, to the most problematic part, qua science, of this juggernaut: Fodor's computational Freudianism. In fact, with respect to the modularity thesis, evolutionary psychology stands to Fodor as Fodor does to Chomsky, which is to say, as that which makes the other seem like a shrinking violet by comparison. The proliferation of innate modules under the auspices of evolutionary psychology is truly breathtaking. Where once there were only syntax and vision modules, there are now many and much smaller (more specific) ones, including a nubility detector, a Machiavellian manipulator/calculator, our spatial cognition module, and a module for detecting cheaters, to name a few.

The founders of evolutionary psychology, Leda Cosmides and John Tooby, view the mind as a population of modules, each containing a particular problem-solving algorithm designed by natural selection, a "Darwinian algorithm." An algorithm is simply an explicit procedure for solving a problem.

In computer science, this procedure is conceived as a set of instructions or a program that performs recursive computations on data (inputs) and that produces, in a finite number of steps, solutions (as outputs). According to evolutionary psychologists, Darwinian algorithms are domain-specific, by which they mean specialized to perform a specific task. They claim that natural selection would favor such domain-specific, rather than domain-general, mental devices, because specialized mental devices are more efficient in solving the specific cognitive problems that our environment posed to we humans during the course of our evolution.[95]

In its emphasis on special-purpose solutions to problems posed by the environment, evolutionary psychology resembles the brain ecology program discussed in chapter 8. It is also the psychological equivalent of a much older view of the brain known as phrenology.

Franz Joseph Gall is another Viennese "visionary" who, like Freud, is considered by some a trailblazing genius, by others a prototypical quack. He and his student, Spurzheim, often get credit for pioneering the view that the mind is what the brain does, but his conception of brain structure and organization was quite idiosyncratic. On Gall's view, brain functions are highly localized in discrete brain parts. He seemed to use little more than intuition, however, in defining these functions, among which were acquisitiveness, musicality, and propensity to have children (philoprogenitiveness).[96] Gall was most notorious for his claim that individual differences in these capacities and propensities were manifest in the form of bumps on particular parts of the skull, the size of which varied with the strength of the propensity. The philoprogenitiveness bump would be larger, on average, in Mormons, than in the rest of the U.S. population.

Evolutionary psychology advocates a similarly localized view of human propensities, though its modules are in the mind rather than the brain. Furthermore, these mind modules are deep in the same sense as Freud's id, which is why I often refer to the "psychology" in "evolutionary psychology" as "Freudian phrenology."[97] Whereas traditional phrenology's bumps are right there on the surface for everyone to see, evolutionary psychology's modules can be demonstrated only through indirect evidence of a sort that requires great ingenuity to gather, a skill reserved for researchers who know what to look for. And some of the most obvious candidates do not qualify. Straightforward behavioral evidence, for example, is particularly suspect, except under special circumstances of the sort established by evolutionary psychologists themselves. The Freudian phrenology in evolutionary psychology is problematic in shielding it from empirical scrutiny.[98] And things are not improved, in that regard, by linking these modules to the story of Fred and Barney. But this is precisely the move that makes evolutionary psychology special, the new wrinkle proposed by Cosmides and Tooby that differentiates their approach from that of Chomsky and Fodor.

I need to emphasize that the fact that the brain itself is modular in important respects[99] does not lend credence to evolutionary psychology's vision of an extremely modular mind. Even if we conceive of the brain as consisting of very local functionally distinct parts, it is not at all clear that these brain modules would map at all straightforwardly onto the putative innate mind modules of evolutionary psychology, because there is no reason to assume that what is functionally differentiated in the brain maps straightforwardly onto what is functionally differentiated in the mind, however the mind is conceived. This is the reason Fodor so strongly advocates psychology's independence from neurobiology and the other natural sciences. In any case, there is little evidence to date of a relation between brain modules and evolutionary psychology's mind modules. Evolutionary psychologists themselves are of two minds about the importance of such evidence. On the one hand, they want to confine themselves to the mental sphere and, like Fodor, claim complete independence from neurobiology; on the other, they would welcome any support from neurobiology they could get and freely advise neurobiologists on what they should be looking for.[100]

The "Evolution" in Evolutionary Psychology

As it happens, neither Chomsky nor Fodor is much impressed with the evolutionary turn wrought by Cosmides and Tooby. In fact, they are downright skeptical of this elaboration of their program.[101] But it is certainly true that the more modularist your view of the mind, the more tempting it is to look toward a Darwinian explanation of how the mind came to have the particular properties it does. The problem, however, as both Chomsky and Fodor recognize, is that the story of Fred and Barney does not provide a compelling basis for such theorizing; and currently, that is the only basis for such theorizing that we have.

Evolutionary psychologists have not let the absence of an evolutionary warrant for their program deter them from their goal of providing a foundation for their brand of cognitivism, and thereby all of psychology. Darwinism has an apparent credibility, even if this Darwinism does not extend beyond the story of Fred and Barney. Perhaps by way of compensation, evolutionary psychologists bestowed to the story of Fred, Barney, and Co. a needlessly technical name: the "environment of evolutionary adaptedness" (EEA).[102]

All of evolutionary psychology's modules, including Geary's spatial cognition module, are lent credibility, in the eyes of evolutionary psychologists, by virtue of being deduced from the conditions prevailing in the EEA (read: the story of Fred and Barney), notwithstanding the critiques of the modularity thesis emanating from the connectionist camp.[103] I would argue, rather, that to the extent that our belief in any spatial cognition module is based on the story of Fred and Barney, it is suspect. There is certainly no evolutionary warrant for claims about adaptive sex differences in any such modules.

In chapter 8, I compared evolutionary neurobiology with the enterprise that I labeled brain ecology by way of emphasizing the differences. The differences between evolutionary neurobiology and brain ecology, though substantial, pale by comparison with the differences between evolutionary neurobiology and evolutionary psychology. Though both are evolutionary endeavors, these two frameworks represent completely different visions for how to go about the project of understanding the human condition in general and how to go about understanding the differences between human males and females in particular.

Ultimately, the differences between evolutionary psychology and evolutionary neurobiology reduce to two dimensions. First, where evolutionary psychology provides a teleological framework for understanding the human condition, evolutionary neurobiology provides a causal framework. Second, where evolutionary psychology is mentalistic, or at the very least nonmaterialist, at its core, evolutionary neurobiology is robustly materialistic. Teleology and mentalism are mutually reinforcing. It is the very nature of these attitudes to seek each other out. Not only does teleology always look more attractive when nature is conceived nonmaterialistically, but mentalism always seems more attractive when nature is conceived teleologically.

Though they are intimately related, it will be helpful, initially at least, to discuss the causal-teleological and the materialist-mentalist dimensions separately in comparing evolutionary neurobiology and evolutionary psychology. Consider first the causal-teleological dimension, as it applies to the term *evolutionary* in these two projects.

There is no dispute between evolutionary neurobiology and evolutionary psychology as to the value of an evolutionary perspective on the human condition. For an evolutionary neurobiologist, an evolutionary perspective would be a good thing for any science of psychology to have. But the "evolution" in evolutionary psychology is so distorted and simplistic, and so overwhelmingly likely to be discredited, that it will ultimately make this goal more difficult to attain, when, inevitably, all evolutionary perspectives get tarred with evolutionary psychology's sins. First and foremost among these sins is evolutionary psychology's excessive teleology.

For evolutionary psychologists, an evolutionary approach to human psychology begins with "why." This is the teleological mindset that I have criticized throughout this book, as it was applied to other animal species. The teleology in evolutionary psychology is actually much worse, in this regard, than that of sociobiology or brain ecology, because evolutionary psychologists have given themselves unprecedented room for paranoiac maneuver. In contrast, the first question an evolutionary neurobiologist asks is "how?"; teleology has only a small role to play as a heuristic tool. This difference has several important ramifications.

First, as I discussed in chapter 8, the focus on causes and causal mechanisms that comes from first asking "how?" provides a different sort of evolutionary context for considering any trait than that which results from asking why-questions. Most significantly, there is more interest in genealogical or phylogenetic factors in evolutionary neurobiology than in evolutionary psychology, and less on ecological considerations. Once we take the proximate how-biology seriously, we are inevitably struck by the constraints it imposes on any ecologically stimulated reconfiguring of the organism. Once we come to appreciate the amount of evolutionary work required for any such reconfiguration, we are less likely to assume, as a matter of course, that given an ecological imperative, the how-biology can be changed in an adaptive way. And we begin to notice highly conserved causal structure underlying sex differences, especially in their development. So, for example, an evolutionary neurobiologist is less likely than an evolutionary psychologist to assume that any biologically based sex difference in spatial cognition is a specific adaptation to a particular ecological milieu and, hence, is less preoccupied with the story of Fred and Barney. Instead, the evolutionary neurobiologist recognizes that any such sex difference might well be an unselected by-product of a more general, evolutionarily conserved process, for which the story of Fred and Barney can provide little insight. More fundamentally, the evolutionary neurobiologist will be much more skeptical of the evidence that such sex differences actually exist.

The second important ramification of the focus on "how?" over "why?" concerns the role of extrabiological factors in the production of sex differences. As we have seen, there is ample room, in evolutionary neurobiology, for sociocultural causes of cognitive sex differences. Evolutionary psychologists can only construe such factors as secondary, however, given their goal to connect any such sex difference to selection. Natural selection can shape traits only to the extent that they have a biological—and ultimately a genetic—etiology. For the evolutionary neurobiologist, on the other hand, there is absolutely no reason to assume, a priori, that sociocultural factors are secondary to, say, hormone levels, or any other biological factor, in the production of a cognitive sex difference. Indeed, we saw ample evidence in chapter 6 that social processes can be causally primary with respect to the biology. Hence, as a result of the focus on the how of things, evolutionary neurobiology exhibits a welcome humility, with respect to the social sciences, that is utterly lacking in evolutionary psychology.[104]

The focus on how-issues leads inevitably to consideration of the "what," which brings us to the material–mental dimension in the contrast between evolutionary neurobiology and evolutionary psychology. At bottom, evolutionary neurobiology, like all other natural sciences, is robustly materialistic. From the perspective of a robust materialist, the entire enterprise of psychology has been poisoned through its Cartesian roots. When the antimaterialistic stance of its founders proved untenable, psychology, under the behaviorists, evolved into an immaterialist, or nonmaterialistic, endeavor. In this respect, at least, most cognitivists are the direct descendents of the behaviorists. But unlike the behaviorists, evolutionary psychologists sought to revive the category of mental phenomena; moreover, in practice, they treat mental phenomena as something independent of material nature, including the brain. For an evolutionary neurobiologist, this is a recipe for disaster. What makes this lack of neurobiological mooring all the more problematic is that evolutionary psychology, unlike Behaviorism, is so empirically opaque. Mental modules are deep structures in that they are not directly accessible to observation. If these mental modules are not brain modules, it is much harder to cash out these theoretical entities in a principled empirical manner. Rather, the temptation is always to take the unprincipled Freudian route. Indeed, from the perspective of an evolutionary neurobiologist, evolutionary psychology represents the wedding of Darwinian paranoia with Freudian paranoia, a particularly unfortunate way to build a foundation for psychology.

10

A Textbook Case of Penis Envy?

IMAGINE SOME alien creature that is highly social and forms complex societies. These societies are far from egalitarian; social stratification approaches levels found in their human counterparts. But in these societies it is the females, who, by virtue of their greater size and strength, control all of the important resources. Not surprisingly, these are matriarchal societies within which social position is inherited from the mother. Highborn females experience a life of relative ease. For the lowborn females, however, life is a struggle, so much so that they find it very difficult to raise a family. Most of their children starve to death before they reach adolescence. And all males of this particular life form are worse off than the lowest-ranking females. They generally keep a low profile unless summoned by a female to procreate. Their capacity as sperm purveyors seems to be the only reason the males are tolerated at all.

Not surprisingly, the males behave deferentially toward the females. What *is* surprising is that the males also show little aggression toward each other. Whereas the females engage in constant squabbles and vicious fights, the males find very little to get exercised about. Though the males too can be ranked, their rank is based strictly on seniority. Even when it comes to reproductive privileges, the males fail to compete, leaving it to the libidinous females to do the choosing.

Strange creatures indeed—but not as fantastic as they may seem. In fact, these creatures, though certainly alien, are not imaginary at all. They dwell right here on earth, throughout much of Africa, and they are called spotted hyenas (*Crocuta crocuta*). Despite some notable recent attempts to rehabilitate them in the public's eye, hyenas are little respected, much less admired. In large part, this negative image is based on their reputation as craven and gluttonous scavengers, purloiners of the noble lions' hard-earned kills. In any conflict between hyenas and lions, few human observers would root for the hyenas. The truth about the relationship between lions and hyenas is much more interesting and complex. Spotted hyenas often kill more of what they eat than do lions, and when it comes to purloining, hyenas are as often the victims as the victimizers. It may come as a surprise to learn that hyenas, not lions, are the most successful large predators in Africa.

Even after the misperception concerning their eating habits has been cor-
rected, hyenas will continue to have an image problem. I have often won-
dered if hyenas would be accorded more respect if their hind legs were
longer. As it is, they always seem to be dragging their butt, even in full
sprint. Moreover, their tail seems to be perpetually tucked between their
hind legs, a seemingly submissive posture. But it's not. Hyenas actually express submission in a unique way: they lift a hind leg and display an erect
phallus. (I suspect that this behavior, even more so than their butt-dragging,
will keep them from a starring role in a Disney film.) I say phallus, and not
penis, because females use this posture as well. What the female erects, in
lieu of a penis, is an engorged clitoris. The similarity to male genitals does
not end there. She also has, at the base of the clitoris, a structure that looks
remarkably like the scrotum of the male, but which consists of her fused
labia.

It is not surprising then, that throughout much of written history, rumors
have abounded that hyenas are hermaphrodites.[1] Most people find the existence of two sexes in one body somewhat disconcerting, so these rumors did
nothing to enhance the hyena's already inglorious reputation. In truth, hyenas are no more hermaphroditic than you or I. All spotted hyenas are male
or female, not both. But even as pseudo-hermaphrodites, spotted hyenas,
and particularly their genitals, have piqued the interest of biologists. As is so
often the case, the adaptationists got the jump on the competition. But in
this case, they were met, early on, with vigorous opposition from the how-
biology–centered developmentalist camp. This opposition resulted in a rich
and complex history of moves and countermoves as the adaptationists and
developmentalists maneuvered to claim victory. As such, spotted hyena genitals provide an ideal case from which to explore and summarize some of the
main themes of this book. We begin with the initial attempts of the adaptationists to explain this strange phenomenon.

A PHALLUS COUP?

The most compelling and direct evidence that the male-like genitals of
female spotted hyenas were shaped by the invisible hand of natural selection
would be evidence that these male-like genitals exist because they conferred
a competitive advantage to ancestral spotted hyenas over those not so endowed. Obviously, this sort of direct evidence of natural selection is difficult
to obtain. The next best sort of evidence of natural selection for male-like
genitals in females is evidence that such genitals currently confer a competitive advantage to female spotted hyenas so endowed relative to those with
the more typical mammalian endowment. But as far as we know, all female
spotted hyenas have the male-like genitals, so this sort of comparison is not

possible. It is in lieu of this direct sort of evidence for natural selection that the teleological approach has proved especially attractive to adaptationists. It provides a way to indirectly implicate natural selection. The goal is to identify the function of the male-like genitals, which, for an adaptationist, means identifying something about the spotted hyena's environment that makes male-like female genitals a good thing to have. One time-honored way to identify this environmental factor is called the comparative method, of which we saw the brain ecologists make so much use. There are two components to the comparative method, which we can call "convergences" and "divergences."

Let us look first for divergences. Here, the goal is to identify an ecological factor that spotted hyenas *do not* share with their nearest relatives that lack male-like female genitals. The spotted hyena belongs to a small family of carnivores (Hyenidae) that includes only three other species: the brown hyena (*Hyaena brunnea*), the striped hyena (*H. hyaena*), and the aardwolf (*Proteles cristatus*). The aardwolf is a specialized termite eater that occupies its own subfamily, so we can confine our comparisons to the other two hyena species. Of the three extant hyena species, only spotted hyenas have male-like genitals. So what might be the difference that makes a difference in the spotted hyena's environment compared with that of the other two hyena species? Begin by comparing, in the broadest way, the habitats of the three hyena species. Of the three species, only spotted hyenas inhabit the savanna. What sets the savanna environment apart from the more arid environments of the brown and striped hyenas? Perhaps it is the fact that the African savanna has an abundance of antelope and other large ungulates. And, indeed, among the three hyena species, spotted hyenas are the only one to prey on large ungulates. But what is it about hunting large ungulates that would make male-like female genitals a good thing to have? The connection is not obvious. Perhaps convergences will be more illuminating.

Here we first need to identify more distantly related species that occupy ecological niches equivalent to that of spotted hyenas, so we look for a savanna predator that preys on large ungulates. Two species meet our criteria, the African lion (*Panthera leo*), a member of the cat family (Felidae), and the African wild dog (*Lycaon pictus*), a member of the dog family (Canidae). Both lions and wild dogs are savanna-dwelling carnivores that take large prey; in fact, they both compete with spotted hyenas for such prey. If female lions and wild dogs have male-like genitals we might be on to something. Alas, neither female wild dogs nor female lions have male-like genitals, just the typical mammalian endowment. We seem to have reached a dead end.

But wait, perhaps all is not lost. Though we have not identified an ecological divergence or convergence that we can directly connect to the male-like genitals of female spotted hyenas, perhaps we have come across an indirect connection. Spotted hyenas have diverged from other hyena species in being highly social cooperative hunters, and they have converged on both wild dogs and lions in this respect. Their sociality is clearly related to hunting large ungulates. Despite their impressive teeth and jaws, a solitary spotted hyena's diet would be restricted to smaller prey items. In order to effectively capture such large prey as wildebeests, zebras, and African buffalo, spotted hyenas must hunt in groups. Perhaps we can connect the male-like female genitals to the savanna indirectly, by way of spotted hyena sociality. Making this connection is, in fact, what adaptationists have attempted to do. But now, it should be noted, we have entered much less principled territory and are at increased risk of Darwinian paranoia.

Greetings and Salutations

In order to indirectly connect the female spotted hyena's male-like genitals to the savanna by way of their sociality, we now need to connect these genitals to their social proclivities. We need to identify the social function of the male-like female genitals. One such social function was proposed by the ethologist Hans Kruuk,[2] who noted that when hyenas greet each other after a period of separation they do a pretty thorough genital inspection, with much sniffing and licking. Moreover, during this ceremony hyenas tend to have erections. This erection is clearly not a signal of sexual interest, because it occurs without respect to the sex of the partner. Moreover, spotted hyena erections are not, as in most mammals, confined to males; females too can extend their phallus in greeting. In truth, female hyenas are somewhat less prone to experience a greeting erection, but only because of their higher social status. When two hyenas engage in the greeting ceremony, it is the lower-ranking one that extends its phallus to the utmost.

Having observed this ceremony, Hans Kruuk suggested that the greeting rituals are crucial in maintaining social bonds and peacefully reestablishing social relationships. These functions are important because members of the clan often experience extended periods of separation while hunting. The greeting ceremonies are especially intense after a period of absence. But why the large erectable phallus in females? Kruuk proposed that the genitals of female hyenas are designed to mimic those of the males, an idea first developed by another ethologist, Wolfgang Wickler.[3] But why is genital mimicry required for effective greeting rituals? Most social animals, including lions and African wild dogs, have some form of greeting ritual but lack genital mimicry.

To understand the logic of Kruuk and Wickler, we need to briefly consider the ethological framework within which they were working. Konrad Lorenz, one of ethology's founding fathers, proposed that nonsexual social behavior involved two independent motivations: fear and aggression, which the participants manifest in varying degrees. Dominant animals express less fear and more aggression than subordinates. During the course of evolution, those behaviors expressing fear and aggression and, hence, dominance and subordination become ritualized in order to avoid physical conflict. It is particularly important that the subordinate individual is equipped with a behavior that can somehow turn off the aggression of the dominant one.[4] Exposing one's genitals may be a good way to turn off aggression in one's social superiors.

Wickler had long been fascinated by various sorts of mimicry in nature. Before his expansion of that concept, mimicry was conceived of as a relationship between two separate species, as in the paragon case involving the unpalatable monarch butterfly (the model) and the perfectly edible viceroy butterfly (the mimic).[5] In addition to documenting many new cases of this sort, Wickler proposed to extend the concept of mimicry in several ways. Of particular relevance here, Wickler hypothesized a number of cases of intraspecies mimicry, in which members of one sex evolved structures designed to look like a structure in the opposite sex.[6] For example, it was Wickler who first noted the function of the egg dummies on the anal fins of male *Haplochromis burtoni* in oral conception. Of more direct relevance here, however, was his proposal that the bright red butt of male hamadryas baboons—which has, for decades, titillated and repulsed schoolchildren on field trips to zoos—was designed by natural selection to mimic the estrous swelling of sexually receptive female baboons. Why? Because, Wickler hypothesized, the male swelling functioned in a greeting ritual known as "presentation," during which a subordinate male adopts the posture of a receptive female in the presence of a more dominant male. According to Wickler, this behavior turns off aggression in the dominant male by turning on—briefly—his sexual impulse. The swollen red butt of the subordinate males enhances the effect of the posture in this regard.

Kruuk envisioned a similar, aggression-inhibitor explanation for the female genitals in spotted hyenas. As Lorenz had documented in some detail, ritualized submissive behavior—as in, for example, dogs and wolves—often involves exposing a vital body part, the throat in the case of dogs.[7] And how could an animal possibly make itself more vulnerable than by exposing its genitals? The erect phallus is therefore a "flag of submission." This interpretation assumes that the greeting ceremony initially involved only males and that, in order for the females to participate along with the males, they needed to evolve male-like genitals, to turn off the aggression of more dominant hyenas and thereby avoid injury. Only having evolved this phallic flag of submission could the female hyenas, in essence, take from the males the reins of social control.

Though the mimicry (or flag of submission) hypothesis has some apparent plausibility, it seems far from meeting Williams's burden of proof standard

for adaptations; at best, Kruuk and Wickler have provided rather weak circumstantial evidence that natural selection has resulted in female spotted
hyenas with genitals that mimic those of males, certainly nothing that
would scare off the competition. Those who placed how-biology considerations in the foreground were certainly not impressed. One of the principal
ways in which the how-biology rejoinder differed from the adaptationist
explanation, and a way that is fairly characteristic in these disputes, was in
attempting to place the male-like female genitals in the context of other
spotted hyena traits, such as female aggression and social dominance. So,
before we proceed to the how-biology alternative explanations, it will be
helpful to provide a fuller picture of spotted hyenas. I will do so through a
hypothetical narrative analogous to the one for Fred and Barney. In stark
contrast to the Fred and Barney story, however, most of the following is
based on sound empirical evidence, not facile speculation.

The Story of Bea and Cleo

Six members of the spotted hyena clan set out for a hunt at dusk, toward a
mixed herd of zebras and wildebeests about a kilometer away. Seemingly in
no hurry at first, they saunter toward the herd and begin to fan out when
they are within 100 meters. As they quicken their pace, the zebras become
agitated and begin making their wheezing nasal warning cries. When the
hyenas get within 100 meters, the zebras and wildebeests begin to run, at
first in a uniform stampede, but soon the hyenas begin to splinter them off
and the herd breaks down into groups moving in all different directions.
Three of the hyenas single out a young male, not yet in his prime; they
pursue him with a relentless focus despite the seeming temptations offered
by other panicked zebras and wildebeests that occasionally streak across
their field of vision. Nipping at the zebra's heels and sometimes at other
parts of his body, they begin to wear him down. The exhausted zebra finally
stops and turns to confront his pursuers; he manages to keep the hyenas at
bay for a while, but then the other three members of the hunting party
appear and the victim is surrounded. They come at him from all sides, rip-
ping at his flesh; eventually he loses his balance and falls, still kicking and
screaming, until one of the hyenas finds the vulnerable underside of his
neck and clamps down with his powerful jaws, cutting off the flow of both
air to the lungs and blood to the brain. Another goes for the back of the
neck and the spinal cord; soon the zebra ceases his struggles. In the meantime, the other four hyenas have already begun dining on the soon-to-be
carcass. The esprit de corps seems to break down at this point; it is every
hyena for itself, as each tries to gobble down the most of the choicest morsels in a fierce competition with its companions.

Unfortunately for the hyenas, the commotion has attracted the attention of a pride of lions. Four lionesses approach the feasting hyenas with a purposeful but unhurried stride. At first, the hyenas redouble their efforts to consume the zebra quickly, but as the lionesses approach more closely, the hyenas rediscover their cooperative side and rush in unison at their adversaries. But now two large male lions join the fray, tipping the balance decisively against the hyenas. In a six-on-six situation, the hyenas stand no chance, so they begin to retreat. But just as the hyenas are about to concede defeat, they are reinforced with 10 other members of their clan. Now there is something like a temporary standoff, but the lions are less adept at these coordinated battles and they are outnumbered. Several hyenas have isolated a lioness and are harrying her mercilessly from all sides. She strikes back with her claws and teeth, but she gets no support from her pridemates, and the hyenas are crafty and patient. When several other hyenas join the fray, the lion pride retreats, leaving the lioness behind. Surrounded by a dozen hyenas, she soon meets her doom.

The hyenas now turn their attention back to the zebra carcass, which they begin to dispatch with amazing speed. Although the process looks chaotic, it is not. The alpha female, who was not part of the initial hunting party, is nonetheless the first to eat, once the lions have been dispatched. Alpha is soon joined by two of her adult daughters, Bea and Cleo, and they consume the choice internal organs. The other hyenas largely confine themselves to the periphery of the carcass at first. Only after Alpha and her two daughters are somewhat sated do they tolerate members of the Beta matriline, consisting of the progeny of Alpha's oldest sister. The males get whatever's left over.

Having eaten their fill, Alpha, Bea, and Cleo begin their journey back to the main clan headquarters, where all three have young pups awaiting them. But when they are about halfway home they are ambushed by the two male lions they had encountered earlier, still smarting from their defeat at the zebra kill. One of these males quickly runs down Alpha and dispatches her with a single neck bite. Once the lion is convinced she's dead, he leaves her corpse for the jackals and the vultures, for she has no food value for him; her death seems to be reward enough for his efforts.

Bea and Cleo escape the other lion, however, and make their way home. As the eldest daughter of Alpha, Bea is now the queen, as her sister recognizes, and the rest of the clan soon will. Even if the other clan members thought that Alpha was still alive, they would treat these two princesses with extreme deference upon their return, so both Bea and Cleo are greeted with enthusiastic displays of erect penises and clitorises by each of the rest of the clan members. By the time these greeting ceremonies have been completed, the rest of the hunting party, as well as their reinforcements, begin to straggle back in to the main camp, and a new round of phallic displays and inspections commences. Bea and Cleo decline, however, and go to their respective dens to check on their offspring.

When Bea arrives at her den, an underground chamber that extends several feet beneath the surface, she calls for her two female pups but only one emerges. This pup is bloody but seemingly unhurt. The blood, it turns out, is that of her sister, whom she killed while her mother was away hunting. Bea continues to call to the other pup for a while but soon lays down to suckle her surviving offspring, who is especially famished today, having expended so much energy in that mortal struggle with her sister. In only a single day, Bea has lost her mother and a daughter, but whatever the emotional toll she has incurred, she is now the queen and her surviving daughter is now the princess.

The story of Bea and Cleo reflects the recent knowledge gained from both field studies of spotted hyenas and research on captive individuals. I will first consider the field studies.

FIELD STUDIES AND THE WHY-BIOLOGY OF SPOTTED HYENAS

First consider in more detail some of the other spotted hyena traits mentioned in the story of Bea and Cleo, for they too will prove important in evaluating the why-biology explanations of the male-like genitals. This discussion is based largely on field studies, which have always provided the primary fuel for why-biology explanations.

Female Dominance

The first trait we will consider is female dominance and the related fact that female spotted hyenas are somewhat larger (heavier) than males, a sex difference that is not at all typical of mammals.[8] To appreciate the potential advantages of the spotted hyena condition, we need only consider their archenemies, the African lions.

As is true of spotted hyenas, adult lion females and their offspring constitute the core of lion society; males are fairly transient such that during the course of a female lion's lifetime she will associate with a number of male pride members. In contrast to spotted hyenas, however, male lions are substantially larger than female lions. This difference has important ramifications. At a kill, male lions eat first, though it is the females that usually do the hunting. Only when the males are sated do they allow the females to eat. When prey is relatively scarce, the cubs are vulnerable, both because their mothers may not be able to continue producing milk and because they are excluded from the carcass. In either case, starvation is a frequent result.

Lion males adversely affect the cubs in more direct ways as well. If the resident males are displaced by other males, the cubs are likely to be killed by the usurpers.[9] It has been suggested that infanticide benefits the new pride males because, by eliminating stepcubs, they cause the pride females to become sexually receptive much sooner than they otherwise would, increasing their own potential to sire cubs. Whatever the advantages of infanticide for male lions, it is obviously not a good thing from the females' perspective (not to mention that of her cubs). And it is an outcome that female spotted hyenas avoid by virtue of their greater size and power. Any male hyena foolish enough to attempt such an act puts his own life at serious risk. Moreover, because the males rank lower than females in the social hierarchy, they eat only after the females and their cubs have had their fill. As a result, the female spotted hyenas are more likely than lion females to sustain lactation during lean times, sparing the younger cubs, and the older cubs are more likely to derive enough sustenance to survive until things get better. The net result is that, on average, a female hyena wastes much less reproductive effort than a female lion.

This why-explanation for female dominance in spotted hyenas does seem to make sense of the condition by demonstrating its benefits; it does not, however, seem to generalize well. Why, given these advantages, have not lions or African wild dogs evolved a hyena-like social structure? Something more than ecology seems to be at work here.

Female Aggression

Female spotted hyenas not only dominate males; they also are more aggressive. According to one why-explanation, female aggression arose in the context of competition for food.[10] Spotted hyenas compete fiercely at a carcass. By comparison, lions are mild-mannered. Once hyenas have killed, say, a zebra, other hyenas soon appear, attracted by the commotion. Up to 30 hyenas may congregate at a kill, though only a relative few can feed at any one time. The competition is intense and nasty—a snarling, hissing, growling melee. Under these circumstances, docility is a grave defect; the meat goes mainly to the most aggressive, which therefore leave more offspring.

This explanation for female aggression, in contrast to the explanation for female dominance, does connect the trait more or less directly to the savanna environment and the large prey to be found there. One problem with this proposal, however, is that it is difficult to disentangle the advantages of aggression and social dominance in the context of food competition. The chaotic-seeming scramble for food is actually not all that chaotic. The consumption of the zebra is done in a fairly orderly and predictable manner. Initially, high-ranking females exclude all but their offspring from the kill. Only after they have removed the choicest morsels do lower-ranking females, and finally the males, gain access. But a hyena's social rank is not established through aggressive interactions; it is inherited in a conventional, castelike manner as we saw in chapter 6. So it is not clear that aggression per se is an important factor in determining how much food an individual will get. Nor does the putative advantage of aggression in females explain the lack of aggression in males.

Siblicide

Aggression in spotted hyenas begins at an early age. Hyena pups, which typically come in pairs, immediately set about to destroy each other once they hit the ground. Their remarkably advanced motor development and fully erupted teeth greatly facilitate this siblicide.[11] Most of the bites are directed toward the back and shoulders; once a cub sinks its teeth into a sibling, it commences to violently shake it. After a few days of this treat-ment, one cub emerges the victor. If the loser is not killed outright, it is doomed to a tenuous existence, because its access to its mother is restricted by the dominant sibling during the long period of dependence. As a result, many cubs die of starvation before weaning, and those that do survive grow much more slowly than the dominant sibling.

There have been claims, based, however, on very indirect evidence, that siblicide is more common among all-female pairs than mixed pairs or all-male pairs.[12] This suggests to the teleologically inclined an adaptation to the realities of hyena society. Same-sex siblings can reduce the fitness of a fe-male spotted hyena through social competition, so it is adaptive to elimi-nate them, or at the very least, dominate them.[13] Male siblings do not pose such a threat because they will emigrate once they are sexually mature and will be socially subordinate in the meantime. The fact that neonates have fully erupted teeth certainly seems to add some credence to an adaptive explanation of siblicide, though it does not add particular support to the view that siblicide is more common among female neonates.

LABORATORY STUDIES: THE HOW-BIOLOGY OF SPOTTED HYENAS

The why-explanations of female dominance, female aggression, and siblicide in spotted hyenas, although certainly more plausible than the why-explana-tions of the male-like genitals, are just as ad hoc. This is not a desirable state of affairs in any science. Indeed, as Aristotle, the father of why-biology, recognized long ago, it is a mistake to explain the traits of animals one at a time. Rather, he recognized that animals come in integrated packages:

> as one part of first-rate importance changes, the whole system of the animal differs greatly in form along with it. This may be seen in the case of eunuchs, who, though mutilated in one part alone, depart so much from their original

appearance and approximate closely to the female form. The reason for this is that some of the parts are principles, and when a principle is moved many of the parts that go along with it must change with it.[14]

On Eunuchs and Virile Females

This correlation of parts to which Aristotle refers can be understood only through how-biology considerations. His choice of eunuchs to illustrate this point may be particularly apropos here. In fact, female spotted hyenas could be the converse of eunuchs: they have unusually high levels of androgens (male, testosterone-like hormones) as high as those of males. Moreover, female spotted hyenas have much higher levels of androgens than female brown or striped hyenas, which lack female dominance and the male-like female genitals.[15] Gould and Vrba suggested, therefore, that the male-like female genitals are a byproduct of the females' high androgen levels. They then elaborated on this proposal to argue that the male-like female genitals were not an adaptation to anything, just an incidental byproduct of selection for female dominance, which required elevated androgen levels.[16]

Gould and Vrba's logic was essentially that of Aristotle on eunuchs. Because animals are highly integrated biological systems, a change in one part can have far-reaching consequences, which are not themselves adaptive, just byproducts. The condition known as congenital adrenal hyperplasia (CAH), which I discussed in chapter 9, lends credence to this view. In women with CAH, the fetal adrenal glands secrete abnormally large amounts of androgens.[17] CAH females are often indistinguishable at birth from male neonates, and until fairly recently their sex was often misdiagnosed.[18]

The reason that the female clitoris comes to resemble the male penis in human CAH females is that they are equivalent structures in an evolutionarily and developmentally deep sense known as homology. The same is true of the male scrotum and female labia. When exposed to testosterone or other androgens during development, female genitals will come to assume the form of typical male genitals. Conversely, male genitals will come to resemble those of the female in the absence of androgens. Under normal conditions, the genitals of a male fetus are exposed to testosterone from its own testes; those of the female fetus are not. Hence, the sex difference at birth. But when a female fetus or neonate is exposed to high androgen levels she will acquire male-like genitals.

As it happens, female spotted hyena genitals are not masculinized in the same way as those of CAH females. Rather, research on a captive colony at the University of California, Berkeley, revealed that high in utero androgen levels are generated through a novel mechanism in the female placenta.[19] But, as is true of CAH, this alteration from the normal mammalian state requires only a single mutation in a gene involved in the biosynthesis of testosterone and other gonadal steroids.

Clearly, CAH is not a case of adaptive mimicry. Yet the resemblance between the genitals of CAH females and typical human males can be no less striking than that between male and female spotted hyenas. At the very least, this indicates that the convergence in genital morphology between male and female spotted hyenas need not require sustained episodes of continuous selection on minor variants; it could arise instantaneously within a single generation.

What Gould and Vrba proposed was essentially a how-biology replacement for a why-biology hypothesis, and this replacement was widely accepted by biologists for the next two decades. Hard-core adaptationists were not deterred, however. They proceeded along several fronts. First, and most predictably, the hoary proximate/ultimate distinction was invoked by the sociobiologist Paul Sherman, in order to deflect the force of Gould and Vrba's argument.[20] Sherman claimed that the high in utero androgen levels explain only the "how" of male-like genitals; it constitutes only the proximate explanation. The flag of submission explains the "why" of it; it is the ultimate explanation. The two types of explanations—proximate and ultimate, how and why—are complementary, not mutually exclusive.

Sherman badly misses the point. Although it is true that the elevated-androgens explanation of the how-biology and the flag of submission explanation for the "why" are not mutually exclusive in principle, they are, in fact. Remember, our goal was to explain the resemblance between the female spotted hyenas' genitals and those of the males. An adaptationist begins with the assumption that this resemblance reflects design by natural selection and seeks to identify the environmental factor that makes the male-like genitals a good thing for a female spotted hyena to have. But our knowledge of the how-biology suggests a completely different explanation for the resemblance between clitoris and penis. Once the clitoris looked like a penis it could be used as a flag of submission, but that function does not explain why it looks like a penis, how it came to evolve a penis-like appearance. The proximate how-explanation also subverts any other ultimate why-explanation for this resemblance. If the how is all we need here, there is no need for a teleological adaptationist explanation of any kind.

A more interesting attempt to counter the byproduct explanation of male-like female genitals was prompted by research on captive hyenas that demonstrated the high cost of male-like female genitals.

A URETHRA RUNS THROUGH IT

Stephen Glickman, Lawrence Frank, and their associates have conducted long-term research in both the wild and on a captive colony of hyenas at the University of California, Berkeley. This research has made dramatically evident the disadvantages of male-like female genitals. For example, consider what it is like to give birth through a clitoris, even a big one. This is precisely what is required of female hyenas. The hypertrophied clitoris of spotted hyenas completely encloses the urogenital opening.[21] Hence, as in male mammals, the urogenital tract runs through the phallus. This arrangement poses no particular problem when it comes to voiding urine, but it certainly complicates the process of giving birth. And the first birth is particularly, well, laborious. If female humans were made to suffer during labor because of Eve's fateful bite from the forbidden fruit, one can only imagine what the serpent convinced hyena Eve to consume, because hyena labor is uniquely agonizing in ways that both human females and human males can empathize with.

Labor is particularly painful and protracted in females giving birth for the first time. The initial contractions are unremarkable, but things get worse when the fetus enters the clitoris. The female's agony is palpable and protracted. Once it enters the clitoris, the fetus moves very slowly because the clitoris lacks contractile muscles to help it on its way. In fact, were it not for the fact that the clitoris eventually rips open, the fetus would never get out. Even after a successful first birth, both mother and pup are exhausted and traumatized.[22]

In and of itself, this sort of suffering does not indicate bad design. Mother Nature (read natural selection), in maximizing fitness, does not minimize suffering. So, although we naturally empathize with the hyenas, we should not blame Mother Nature, or original sin, for their predicament. A good flag of submission can be worth a lot of pain in the grand evolutionary scheme. But consider some additional facts that should give even the most ardent adaptationist pause. A significant percentage of female hyenas die during their first parturition. In fact, aside from falling prey to lions, this may be the single greatest source of mortality for adult females.[23] Furthermore, a large portion of the first-born, as well as their twins, die as well. To understand why, consult figure 10.1. In a "normal" mammal the size of a spotted hyena, the fetus must traverse a distance of about 30 cm, in a straight line, from the cervix to the vagina; the hyena fetus, however, must travel about twice that distance and negotiate a hairpin turn in the process. Both the increased distance and the hairpin turn result directly from having to exit through a large clitoris pointing in the wrong direction. If the fetus gets stuck in that hairpin, both neonate and mother are doomed. In addition, though the cub must travel some 60 cm, the umbilical cord measures only 12 to 18 cm.[24] The umbilical cord is not elastic, so it breaks long before the neonate exits the clitoris. Now cut off from its oxygen supply, the cub must travel the rest of the distance before it asphyxiates, and once lodged in the clitoris, it cannot exit until the clitoris tears. Many of the first-born cubs die waiting for this to happen. The dead cub sometimes remains lodged in the birth canal for an extended period, endangering the life of the mother herself.

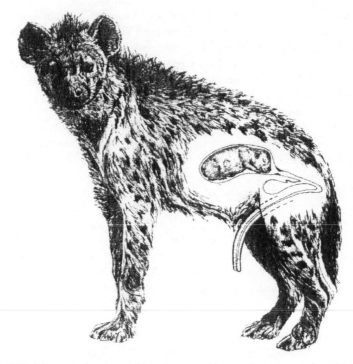

Figure 10.1 Urogentital system of female spotted hyena showing womb and birth canal
(Adapted from Frank 1997, p. 59)

Given the power of design typically accorded natural selection by neo-Paleyan teleologists, you might at least expect to find some compensatory mechanisms serving to ameliorate the birth process. Most obviously, a longer umbilical cord would seem to be in order. But the umbilical cord of the spotted hyena is no longer than that of the other two hyena species, which do not have to face the problem of giving birth through a clitoris.[25] And what about reducing the size of those neonates? Smaller pups would make for an easier passage through the birth canal. But natural selection seems to have ignored this option as well. In fact, spotted hyena neonates are larger and more precocious than newborn brown or striped hyenas.[26] Female spotted hyenas do benefit from the fact, that once ripped, the clitoris remains ripped, thereby greatly facilitating subsequent births.[27] But whether this failure to heal represents an evolved compensation or a generic response of this tissue remains to be established.

For any disinterested observer, the research on the Berkeley hyena colony raises questions as to whether the pseudo-penis reflects selection of any sort, much less selection for mimicry. But for a neo-Paleyan teleologist, the obvious costs of the male-like female genitals suggest that they must confer a huge selective advantage that more than compensates for the inconvenience of giving birth through a clitoris. Because the selective advantage of having a flag of submission seems inadequate to compensate for the problems imposed by the clitoris as a birth canal, adaptationists have attempted to find something else that male-like genitals might be good for and that would make male-like genitals a good thing to have. One novel hypothesis in this vein is that the male-like genitals also serve to prevent rape, thus providing female hyenas complete control over mate choice. But you would be hard pressed to find a mammal less in need of protection from rape than a female spotted hyena. Any male hyena foolish enough to attempt to rape a larger, more dominant and more aggressive female would probably find himself mauled and perhaps even castrated for his efforts.

The latest hypothesis that takes as its starting point the high costs of masculinized genitalia is a new version of the sexual mimicry hypothesis, called the "camouflage hypothesis."[28] Martin Muller and Richard Wrangham argue that females that possess male-like genitals are less likely to be attacked and killed by other females than are normally endowed females because of the protection they gain from appearing to be males. They claim that this advantage is particularly important for neonates, which, as we have seen, have a tendency to kill each other, although the extent and importance of siblicide are very much in dispute.[29] Their argument also rests heavily on the evidence that female neonates preferentially kill other females, evidence that can only be described as flimsy.[30] But rather than critique this new hypothesis in detail, I want to focus on two assumptions, based on teleological interpretations of the evidence, that motivated it, one obviously false and the other much less obviously so. Let's begin with the obvious falsehood.

Both the rape prevention and camouflage hypotheses were largely motivated by the research on the Berkeley colony demonstrating the high costs of male-like female genitalia. In a truly Paleyan move, Muller and Wrangham actually assert that these costs are evidence against the byproduct explanation, and for an adaptationist explanation, on the principle that the evidence of high costs are evidence ipso facto of even greater benefits. But this assertion cannot be right. It is true, that given the basic logic of natural selection, a trait will be selected for only when the benefits exceed the associated costs. But it does not follow that male-like genitals must therefore confer more benefits than costs. Rather, male-like genitals could also evolve, despite their obvious drawbacks, if they are correlated through their howbiology with a trait, such as social dominance, that confers benefits greater than these costs. The costs of male-like female genitalia are completely neutral with respect to the truth of the byproduct versus adaptation explanations for these genitals. All that the high costs of the male-like female genitals demonstrate is the miserable job that natural selection has done to meliorate them.

The second line of evidence that Muller and Wrangham cite against the byproduct explanation for hyena genitals is more compelling. The Berkeley group recently discovered that even when anti-androgens are administered early in development, female spotted hyenas nonetheless develop the male-like genitals.[31] This evidence—from the how-biology no less—is indeed problematic for the byproduct explanation, because it suggests that the male-like genitals can develop in the absence of androgens such as testosterone. If they can, the high circulating androgen levels in female spotted hyenas are seemingly irrelevant with respect to their male-like genitals. This irrelevance would seem to vindicate the approach of researchers such as Muller and Wrangham who claim that the male-like genitals represent a special adaptation of some sort rather than a byproduct of generic androgenic effects.

The evidence for androgen-independent development of spotted hyena genitals is less decisive than Muller and Wrangham claim, because spotted hyenas have a special mutation in their androgen receptor gene that is not found in other hyena species.[32] This mutation could render traditional anti-androgen treatments less effective in spotted hyenas than in other mammals. But let us assume for the moment that the evidence for androgen independence is beyond dispute. Would that situation decisively tilt the scales toward the sexual mimicry hypothesis? No. In order to understand why not, we need to consider three facts: (1) female spotted hyenas experience unusually high levels of androgens in utero; (2) such elevated androgen levels would cause the masculinization of genitals in all other mammals; but (3) genital development in spotted hyenas is seemingly not influenced by such high androgen levels. It is the last fact (3) that makes the sexual mimicry hypothesis or any other adaptationist explanation of spotted hyena genitals attractive. The first two facts are still a problem for this view, however. Why, given the fact that females experience elevated androgen levels, and the fact that such high androgen levels will produce, in other mammals, the genital phenotype we observe in spotted hyenas, would spotted hyenas need to evolve an androgen-independent mechanism to produce the same phenotype? It seems, then, that for both the byproduct explanation and the sexual mimicry explanation, we have some explaining to do with respect to the apparent androgen-independence of male-like genital development.

The sexual mimicry hypothesis presupposes female dominance, which was presumably achieved through androgen-mediated increases in their size and aggression. For some reason, however, the elevated androgen levels failed to influence genital development in the way they do in other mammals. So, secondarily, spotted hyenas must have evolved an androgen-independent means to masculinize the genitals in order to turn off aggression, especially in their female siblings. On this account, female dominance and masculinized female genitals evolved independently, and only female dominance was linked to an increase in androgen levels. At best, this is an inelegant explanation of spotted hyena genitals.

It is much more parsimonious to assume that, originally, the male-like genitals were a byproduct of the elevated androgen levels that resulted from selection for female dominance. The androgen-independence of genital development in this species would then have evolved secondarily in order to facilitate copulation and perhaps parturition.[33] But then how do we explain the continued male-like appearance of these genitals once androgen-independence evolved? Their male-like appearance must confer some mimicry-related advantage.

This explanation combines elements of the byproduct and adaptation views on spotted hyena genitals. The origin of male-like genitals was as a byproduct of the how-biology of selection for female dominance, but their male-like appearance is currently maintained by selection because of some benefit it confers in the context of spotted hyena social interactions.

Notice, however, that it still makes no sense to partition the explanatory work between proximate how-biology considerations and ultimate why-biology considerations. In fact, "ultimately," it will be the how-biological grunt work, not teleological speculations, that will prove decisive in explaining the evolution of spotted hyena genitals. In particular, much hinges on whether and to what degree genital development in spotted hyenas is androgen-independent. If it should turn out that genital development is not androgen-independent but rather follows the typical mammalian pattern, then we should consider the byproduct explanation proposed by Gould and Vrba an outright replacement, not only of the flag of submission hypothesis but of any other adaptive explanation as well. If, however, the recent evidence of androgen independence holds up, we will have another example of a how-biology modification of an adaptationist explanation. Either outcome points to the importance of how-biology in understanding the evolution of this peculiar trait and the inadequacy of teleological adaptationism.

11

Darwin's Temptress

In HIS classic study of religion among the Azande peoples of East Africa, Evans-Pritchard noted that most deaths were attributed to witchcraft, no matter what the proximate cause or how-explanation.[1] This observation is true of many tribal societies, where witchcraft is especially favored as an explanation for improbable causes of death such as lightning or snakebites. It's not that the Azande don't accept the how-explanation (how the victim accidentally stepped on the snake or how snake venom causes death); it's just that they are not impressed by these explanations, because they fail to answer the bigger why-questions: Why was the snake at that particular place and at the same time as the victim? Why this particular victim and not someone else? These are the ultimate questions that causal explanations, as conceived by scientists, cannot address. Whatever the how-explanation, the Azande will always go on to question why the events happened the way they did, and expect a teleological answer in return.

For the Azande, the particular sort of teleological explanation deemed acceptable is the malign intentions of someone who has mastered the power of witchcraft. Western religions promote the same attitude toward proximate how-explanations and ultimate (teleological) why-explanations but attribute the intentions to a designing God. Conceptions of how "hands-on" the deity is vary widely, but Azande-like reasoning remains prevalent in modern Westerners with religious worldviews, a fact that I came to appreciate first-hand during my summers as a firefighter for the National Forest Service.

A forest fire, it is often said, "has a life of its own"—at least, this was often said to me during my tenure as a firefighter. By "a life of its own," I think what was meant is that wildfires are erratic, often unpredictable, even seemingly capricious. But I was prepared to believe that without bestowing on wildfires the gift of life. Early on, I observed that after even the most severe conflagration, there are islands of unscathed vegetation in the midst of an otherwise incinerated landscape. This seemed to me a striking example of nature's randomness. I was similarly impressed by the seemingly random and arbitrary elements in fire-related injuries, whether caused by falls, falling rocks, falling trees, or falling helicopters. We tend to classify as "accidents" mishaps such as these, with a determinable cause, but for which the cause itself seems to have a random element. Getting struck by lightning is a paradigmatic accident in this sense. This conception of the accidental is not universally shared, however, as I discovered while on a fire crew. Though we all agreed about the chance element in the fire-related damage to the vegetation, some of my crewmates refused to acknowledge the role of chance in fire-related human injuries. This attitude was most dramatically evident after one incident in particular, which occurred on the steep slopes of the Santa Lucia Mountains above Big Sur, California.

I first spied the hurtling rock, about twice the size of a basketball, when it was about 50 meters up the slope from where I stood; from there, it touched the ground only twice before impact, and it was airborne for about the last 10 meters. The victim was struck flush in the middle of the back, and propelled through the air while draped around the rock like a towel, for a further 5 meters or so until he disappeared into some brush. His death seemed probable, at the very least, a severely damaged spinal cord virtually certain. Incredibly, however, he suffered only some whiplash and a few scratches and bruises. A few of my crewmates, notably those most outspoken in their Christian faith, viewed this event as an act of divine intervention— a miracle.

Others, however, settled on a more prosaic explanation. The brunt of the impact, it turned out, was born by a canteen and canister of extra fuel that the victim was carrying in a knapsack on his back at the time. Upon impact, they were flattened and ejected their contents, as a result of which much of the force from the hurtling rock was dissipated. On this much we all agreed. And for some, like myself, this was a sufficient explanation for our comrade's good fortune. For others though, the self-proclaimed "believers," this explanation, although unobjectionable, was only part of the story, and the insignificant part at that. The canteen and fuel canister were only the proximate source of his good fortune, on their view. The ultimate source, as they saw it, was the hand of God. They found it significant that the rock struck the only one among us who was carrying a canteen and fuel canister on his back. "Why did it not strike some other member of the crew?"

It was not clear to me "why" anyone *had* to be struck by this or any other rock. Moreover, the fact that the rock struck the only one wearing a knapsack with a canteen and fuel canister was not, it turns out, particularly noteworthy. The victim was struck primarily because he did not hear our warning shouts; he did not hear those shouts because he was "bucking" for the person operating the chainsaw, which is very noisy. Whoever had that duty always wore that knapsack.

But other "why" questions were admittedly less easy to answer: "Why did he have his back to the rock and not some other orientation?" And "why,"

given that his back was to the rock, did it strike him squarely on his knapsack and not, say, on the back of his head?"

It is the conviction that everything that happens is part of God's design
that compels the devout to seek teleological answers to these why questions
and to discount, much as the Azande do, answers of any other sort. Chance
is a particularly unattractive explanation for the devout. Only somewhat less
unacceptable are explanations that refer to well-understood causal principles
or processes, or to the material properties of canteens, water, and the human
body. For the true believers, these explanations are either beside the point
or just shallow. They are only proximate, not ultimate, explanations, much
as snake venom is only the proximate, not the ultimate, explanation of
death by snakebite.

Bishop Paley would certainly have concurred with my devout crewmates
as to how to go about explaining the incident at Big Sur, but he was much
more interested in linking broader patterns in nature, rather than singular
events, to God's design. He was particularly concerned to demonstrate God's
benevolence and power by way of the adaptedness of living things. What I
want to emphasize here is that Paley, in common with the Azande and
devout firefighters alike, had a two-tiered view of explanations that stems
ultimately from his teleology: proximate how-explanations and ultimate
why-explanations, the latter always trumping the former. Given this conception, your focus should be on the teleological why-explanations.

By the time Paley wrote his famous treatise, teleological natural theology
was largely a preoccupation of the English. A century earlier, however, these
teleological attitudes had been wider spread across Europe. And no one
embodied the systematizing teleological spirit more than the German philosopher Gottfried Wilhelm Leibniz. Most biologists first learned of the philosophy of Leibniz third-hand, by way of Gould and Lewontin's influential
(1979) critique of adaptationism, "The Spandrels of San Marco and the
Panglossian Paradigm," which alludes to Voltaire's influential critique of
Leibnizian philosophy, *Candide*.[2] Leibniz is perhaps most famous—or notorious—for his claim that this is "the best of all possible worlds." This pre-
cept, which flowed out of Leibniz's theological convictions, is the basis for
virtually all subsequent "natural theology," including that of Paley. It was
meant to solve a particular theological problem, the so-called problem of
evil that had come to bedevil enlightened believers once they in effect
abandoned the doctrine of original sin.

The enlightened solution to the problem of evil was that evil does not
really exist; it is only an illusion that results from our inability to grasp the
big picture. According to Leibniz, the afflictions we suffer contribute to our
greater good, so, as afflictions, they are only apparent. And "Imperfection in
the part may be required for greater perfection in the whole."[3] Leibniz's big-
picture apologetic did not appeal to all enlightened souls; it certainly did
not appeal to Voltaire, who ridiculed it in *Candide* (an examination of the

axiom "all is well"), which was written in the aftermath of the terrible earthquake that struck Lisbon in 1755. This horrible event shocked and disturbed the intelligentsia throughout Europe. Most troubling for many was the fact that the earthquake occurred in the most Catholic European city (30 churches were destroyed) during a holy festival (November 1, All Saints Day) at the precise time that the pious were celebating mass (9:40 A.M.).[5] The fact that the great Mosque of El-Mansur in Rabat was also destroyed seemed inadequate by way of compensation. Those whom the enlightenment had never touched were not particularly troubled by this disaster. The earthquake only increased John Wesley's zeal to sermonize about original sin. Many Protestants even celebrated the earthquake as divine retribution for Catholic misdeeds, though they were soon silenced by a severe earthquake that struck Protestant Boston less than three weeks later.

Voltaire was less outraged by this unenlightened response than by the "enlightened" big-picture apologists, whose views he traced to Leibniz. The story in his famous satire revolves around the bastard son of a baron, Candide, and his tutor, Professor Pangloss (*all-tongue*). Adhering to Leibniz's unchained teleology, Pangloss counseled Candide that:

> It is demonstrable . . . that things cannot be otherwise than they are; for, as all things have been created for some end, they must necessarily have been created for the best end. Observe, for instance, the nose is formed for spectacles, therefore we wear spectacles. The legs are visibly designed for stockings; accordingly, we wear stockings."[6]

Buoyed by this view, both Pangloss and Candide manage to remain optimistic despite an incredible series of travails, including the earthquake, which occurs on the day they arrive in Lisbon. But this is not a celebration of their indomitable spirit; it is a scathing indictment of their inability to entertain doubt about their premises. There is no Job-like tragedy in their suffering, only farce.

Pangloss may have been a caricature of Leibniz's philosophy, but certainly not of Paley's. Paley, as we have seen, also faced the problem of evil with Pangloss's big-picture optimism, even in the face of Voltaire's withering critique of this enterprise. Nor was Paley swayed by David Hume's anti-Leibnizian arguments in his *Dialogues Concerning Natural Religion*,[7] in which the Scottish philosopher skeptically addressed the assumptions of Natural Theology. Against the arguments of Leibniz and Paley for the perfection of nature, Hume, through his mouthpiece, Philo, asserted: "One would imagine that this grand production has not received the last hand of the maker, so little finished is every part, and so coarse are the strokes with which it is executed."[8]

And half a century before Paley rhapsodized about nature's booming buzzing profusion of happy creatures, Hume took a quite different view:

Look around this universe. What an immense profusion of beings, animated and organized, sensible and active! You admire this prodigious variety and fecundity. But inspect a little more narrowly these living existences. . . . How hostile and destructive to each other.[9]

And, contra Lebiniz and Paley, Hume eloquently expressed the view of nature "red in tooth and claw," long before Tennyson's post-Darwinian poem: "The whole presents nothing but the idea of a blind nature, impregnated by a great vivifying principle, and pouring forth from her lap, without discernment or parental care, her maimed and abortive children."[10]

You would think that Richard Dawkins, given his atheism and dark view of nature, would take Hume's side in this debate; in fact, he would seem to be Hume's disciple. But oddly, Dawkins defends Paley against Hume's skepticism.[11] For, though he shares Hume's religious skepticism, he does not share Hume's skepticism toward the argument for design. Consider, for instance, Hume's suggestion that the world is: "only the first rude essay of some infant deity, who afterwards abandoned it, ashamed of his lame performance."[12]

For Dawkins, as for Paley, Hume is off the mark here. Unlike Hume, Dawkins does not question Paley's evidence for design; his only quarrel with Paley concerns the source of this design. Because Dawkins's commitment to teleology supersedes his atheism, he can only view Hume's religious skepticism as premature and poorly motivated. It was not until Darwin found a suitable replacement for Paley's designing God that such religious skepticism of the sort advocated by Hume, and loudly championed by Dawkins himself, could be deemed intellectually respectable or even rational.[13]

Though I find Dawkins's conception of the relationship between Darwin, atheism, and rationality deeply flawed,[14] he is quite correct, and refreshingly candid, in emphasizing the debt that modern adaptationists owe to Paley. Darwin himself, as we have seen, was much taken by Paley, though much to the detriment, I would argue, to evolutionary biology as a science. But Darwin's Paleyan streak was meliorated by his recognition that all living things bear the imprint of their evolutionary history. It is this fact, that organisms are not novel creations but historical material entities, that ultimately invalidates Paley's teleological biology. But the trend among many self-described neo-Darwinians has been toward a godless variant of creationism in which evolutionary history is as insignificant in explaining organisms as it was for Paley. Each species is treated as a de facto novel creation, molded to its particular ecological niche by Mother Nature. This is the only way to make sense of that adaptationist claim by Alexander Graham Bell that "history is bunk."[15]

Exchanging God for natural selection is not, in and of itself, an improvement on Paley, from a naturalistic perspective. Not if the teleological mindset remains the same. Not if the conceit, common to the Azande, born-again firefighters, Paley and Leibniz alike, that teleological why-explanations trump all how-explanations remains. But this conceit is precisely what remains from Paley in Mayr, Dawkins, and many other neo-Darwinians. Unfortunately, as we have seen, it distorts our view of the evolutionary process and, ultimately, as in evolutionary psychology, our view of ourselves.

It is not that teleological thinking has no value in evolutionary biology; natural selection has quasi-teleological properties, but the ample disanalogies between natural selection and true design processes render teleological thinking of limited heuristic value.[16] Unfortunately, because of the kudzu-like properties of teleology, the limited heuristic view of teleology is unstable and quickly degenerates into a literalist fundamentalism. Daniel Dennett, though no evolutionary biologist, epitomizes this progression.

Let's Stance

Dennett has devised the most sophisticated and ingenious way to date to finesse the obvious disanalogies between natural selection and God-like design. It is called stancing. Dennett stances in order to explain nature teleologically without a commitment to its literally being so. Dennett, it should be emphasized, is first and foremost a philosopher of mind, and he first learned to stance in that context.[17] His stance "stance" represented something of a middle ground between those who would completely dismiss folk psychology categories, such as belief and desire, and those who would defend the unproblematic reality of those categories. By stancing, Dennett simultaneously acknowledged the defects in folk psychology and proposed to retain it because of its instrumental value as a means of predicting human behavior. He advised that we act "as if" our behavior is guided by our beliefs and desires, even though we know that, strictly speaking, it cannot be. This is what he refers to as the "intentional stance."[18]

Dennett's philosophy of mind informs his view of evolution. His self-proclaimed Panglossian evolutionary perspective is a way to justify his intentional stance psychology.[19] Though Dennett is well aware that natural selection is a bottom-up process, he invites us to conceive of Mother Nature as a programmer or engineer because, given the way our mind works, we cannot hope to understand nature otherwise. To which claim, the proper reply is: "Speak for yourself." Though Dennett finds it difficult to think about nature and evolution without a teleological mindset, many of us can do just fine without it. Moreover, as I have argued throughout this book, there are some distinct advantages to the nonteleological approach.

For example, we saw how the paranoiac mindset born of teleologically inspired why-questioning led us astray in understanding the clitoris of both spotted hyenas and human females. Once we opted to abandon that ap-

proach and seek the proximate causes—the how-biology—of the hypertrophied clitoris in spotted hyenas and the orgasmic responsiveness of the
clitoris in human females, we were able to devise a simpler and more principled explanation. Similarly, we saw that teleology took us nowhere in explaining the spectacular aptitude of mockingbirds to mimic other birds. The
ability of mockingbirds to mimic non-mockingbird sounds is simply something that cannot be helped given the plasticity of their song repertoire.
There is no adaptive rationale to mimic non-mockingbird sounds; it is just a
byproduct of the perceptual openness required of such versatile singers.
There is no ultimate explanation, only the prosaic proximate one.

Dennett's stancing about evolutionary teleology is made all the more effective because it is no longer clear whether Dennett is actually stancing or
not. Originally, his stancing was an ingenious attempt to use ironic distance
in order to legitimize a way of thinking that would otherwise have been
derisively dismissed outright—the verbal equivalent of a nod and a wink.
Dennett has, however become increasingly seduced by his own metaphors,
to the point that he has effectively stopped stancing. His irony has evapo-
rated, and he has become increasingly literalist in his attitude toward
design-talk and in his defense of Pangloss and Leibniz.[20] The "as-ifness" that
was so prominent in his approach to folk psychology has largely disappeared
in his biological ruminations such that he has come to consider unproblematic a view of biology that is much more folky than folk psychology ever
was. It seems that the only reason Dennett continues to pose as a stancer is
that it lets him claim allegiance to the materialist natural scientists, without
actually having to act like one.

Ultimately, whether Dennett is stancing or not does not matter much.
Not only are we not compelled to think about evolution as a design process;
we should not think about it in this way. The heuristic value of teleological
thinking has been vastly overrated by adaptationists, to the point that it has
become a dead hand encumbering further progress in understanding the
causal dynamics of evolution. Nowhere are the failures of the teleological
mindset more evident than in how it has been applied to one of the most
fundamental issues of evolutionary biology: sexual reproduction. As we have
seen, even the adaptationist Maynard-Smith would not claim that the problem of SEX has been solved, despite extensive use of the design stance. I
have further argued that it will never be solved if we do not move beyond a
teleological perspective, because historical factors, as embodied in genealogy
or phylogeny, are of paramount importance in determining which animals
reproduce sexually and which asexually. This fact should come as no surprise
to most evolutionists, but teleologists such as Bell prefer to ignore genealogical imperatives in principle. For them, the fact that virtually all vertebrates
reproduce sexually will continue to be a mystery. For that matter, the fact
that sex change among vertebrates is largely confined to teleost fishes will

also continue to perplex them, as will the virtual absence of conditional alternative mating tactics among mammals. Elephant seals, as we have seen, nicely represent the limitations of even the most principled teleological frameworks: first, in that they do not reproduce asexually because they are constrained by the how-biology of their vertebrateness to a sexual existence; second, that given their sexuality, they do not change sex—as teleost fishes with similar mating systems would—because of the how-biology of their mammalness; and third, that even given the fact that they must reproduce sexually without the benefits of sex change, they cannot even actualize the effective alternative mating tactics so common in fishes, also because of the how-biology of their mammalness. This single species is refutation enough of Dennett's Panglossian teleology. In a how-biology–informed perspective on evolution, there is a place for history but none for Pangloss, no matter how many nods and winks accompany him.

An emphasis on how-biology foregrounds not only evolutionary history but the material properties of organisms as well, which is why I sometimes refer to the how-biology perspective as robust materialism. In contrast, teleological Darwinians, in ignoring how-explanations, tend toward an anemic materialism or even an amaterialist perspective, in which the material properties of living things are ignored or at best paid lip service.[21] This is what I call "dry biology" or "virtual biology." One drawback of this dry biology is a tendency to misplace causes. For example, in ignoring one of the most fundamental material properties of the brain—its capacity for physical alteration as a result of experience—brain ecologists were prone to misconstrue cause and effect with respect to the association between hippocampal volume and food storing and other "spatial behaviors." More fundamentally, as we saw in the research on gonadotropin-releasing hormone neurons in *Haplochromis burtoni*, the same failure to appreciate the physical responsiveness of brains, this time at the level of the neuron, resulted in a mistaken assumption about cause and effect with respect to biological and social factors.

The problem only gets worse when we move to the level of the gene. Genes are biochemicals first and last, and as such they are physically alterable in the way of all biochemicals. Physical alterations in the base sequences are known as mutations. But genes experience other important sorts of physical alterations as well. For example, the chemical alterations that turn them on or off, that alter their expression. Among the causes of these latter alterations are neural and endocrine changes resulting from social interactions. Socially induced alterations in gene expression are actually commonplace but go unrecognized as such by teleologists because this basic aspect of the gene's materiality is ignored.

We now come to the final benefit of bringing how-explanations to the foreground: humility. Teleology-intoxicated evolutionists tend toward the most grandiose claims about the role of natural selection, not only in biology but for all manner of sociocultural phenomena as well. The dismissive attitude evolutionary psychologists display toward sociologists, social psychologists, and social anthropologists is unseemly but hardly unique. The human sociobiologists before them also exhibited this regrettable tendency. This hubris is not shared by nonteleological Darwinians, however. Once we removed the teleological blinders, we saw no reason to dismiss how-sociological explanations of putative human sex differences in spatial or mathematical cognition. In fact the how-sociology may be all we need to understand why men won't ask for directions. Some future version of evolutionary psychology, one worthy of the name, will embrace this fact.

NOTES

Notes for Chapter 1
Darwinian Paranoia

1. Paley 1825, pp. 317–18.
2. Ibid., p. 345.
3. LeMahieu 1976.
4. See especially Dawkins 1986.
5. Maynard-Smith 1969.
6. Brooke 1991, p. 198.
7. This quote comes from Dewey's *The Influence of Darwinism on Philosophy* (1910, p. 7). It was directed at Asa Gray, who attempted to reconcile Darwinism with a designing God.
8. Weiss 2002.
9. Williams 1966.
10. Mayr prefers not to use the term *teleology* owing to its past associations with natural theology and other unsavory (from a scientific perspective) ideas (for example, Mayr 1982, pp. 47–51) during his formative years. But what he advocates is no less teleological for his aversion to the term.
11. Mayr, who is given to wholesale reiteration, first promoted this dichotomy in 1961 (Mayr 1961), restated it in 1974 (Mayr 1974), and restated it again in his 1982 book (Mayr 1982, pp. 67–70), in which he characterizes the two biologies as "remarkably self-contained." This notion of the self-containment of "why" biology was also promoted by E. O. Wilson in his *Sociobiology* (1975) and has since then been most fervently advocated by sociobiologists and behavioral ecologists. In my critique (Francis 1992) of the proximate/ultimate cause distinction, I argued that, "ultimate causes" are neither ultimate nor causes. Mayr implicitly acknowledged that the term *ultimate cause* is unfortunate and has more recently taken to replacing it with *evolutionary cause*, though inconsistently. This terminological shift does not address the real issue, however, which concerns the role of teleological explanations in biology and the alleged autonomy that the term "ultimate cause" was meant to guarantee them. For what Mayr refers to as "ultimate explanations" or "evolutionary explanations" are just functional evolutionary explanations, not evolutionary explanations in the wide sense in which they are construed by those who are not ardent adaptationists. (When was the last time anyone ever referred to genetic drift as an "ultimate cause"?) Mayr's "ultimate" causes can be construed as "evolutionary causes" only if evolution is equated with natural selection, which is why this dichotomy is so attractive to adaptationists.
12. This sort of compatibilism is the position of even those ecumenical adaptationists who don't completely ignore proximate causes.
13. See, for example, Gilbert and Raff 1996; Raff 1996; Gerhart and Kirschner 1997; Arthur 2000; Wagner 2000; Gilbert and Bolker 2001; and Schuenk and Wagner 2001.

14. It is Jacob (1977) to whom we owe the natural selection-as-tinkerer metaphor.

15. The term *constraint*, however, does not fully capture the sort of interactions between selection and genealogy that I will highlight in this book. From my perspective, it is no less accurate to consider selection a constraint on genealogy as the reverse.

Notes for Chapter 2
An Orgasm of One's Own

1. Danielou 1994, p. 238.
2. Ibid., p. 171–78.
3. Lloyd 1993.
4. Kourada 1980.
5. Mori 1984.
6. Alcock 1987.
7. Hrdy 1988.
8. Hrdy 1981, 1988.
9. According to Hrdy (1977), when a new male takes over a group of females, he sets out to eliminate the dependent young sired by the erstwhile alpha male. As a result, females are brought into reproductive condition more quickly, and he does not waste resources on unrelated young. But see Sussman et al. (1995), who argue that the incidence of infanticide among primates in general has been exaggerated.
10. Lloyd 1993, p. 144.
11. Fox et al. 1970.
12. Aristotle, *Generation of Animals* 739a: 30–35.
13. Baker and Bellis 1993.
14. The authors do not present their data at all straightforwardly, but this much is clear: most female orgasms occurred before ejaculation, as would be expected of the sexually enlightened British population from which the data were drawn; but sperm retention was greatest in those orgasms that occurred after ejaculation. If the function of orgasms is to upsuck sperm, their timing should be better.
15. I added the bit about faking the orgasm, but it is certainly in the spirit of, and consistent with, the elaborate adaptive story told by Baker and Bellis.
16. Baker and Bellis 1993, p. 908.
17. Aristotle, *Generation of Animals* 739a: 20–25.
18. Ibid., 20–25.
19. Thornhill et al. 1995.
20. Kinsey et al. 1953.
21. Symons 1979.
22. See especially Lloyd 1993.
23. Symons 1979. It should be noted that Symons is generally enthusiastic about adaptationist explanations of human sexuality, and he has wholeheartedly embraced evolutionary psychology.
24. Gould (1987) notes that Charles Darwin's grandfather, Erasmus Darwin, himself pondered the function of male nipples and seriously considered the possibility that some males lactate. He also explored the ancient idea that, originally, humans

and other animals were hermaphrodites. On this view, nipples are atavisms without current utility, a radical move for a functionalist to make.

25. Alcock 1998. This article, entitled "Unpunctuated Equilibrium in the Natural History Essays of Stephen Jay Gould," was stimulated, according to Alcock himself, by the fact that graduate students in his own department had failed to identify what he considered egregious errors in another of Gould's articles entitled "The Diet of Worms and the Defenestration of Prague." Perhaps then, his venom partly reflected a sense of betrayal. The title of the diatribe makes mocking reference to Eldredge and Gould's idea of punctuated equilibrium, which, though it does not directly address the issue of adaptationism, is nonetheless a favorite target of adaptationists, because it appears to them to slight the role of natural selection in evolution. The fact that such an intemperate diatribe was published at all in a scientific journal is exceptional, but this is a case in which you really have to consider the source. Alcock's commentary appeared in the journal *Evolution and Human Behavior*, formerly, *Human Ethology and Sociobiology*, which has long been controlled by the truest of true believers, and the bar has been raised only since it became the "house" journal for evolutionary psychology.

26. Shakespeare, *A Midsummer Night's Dream* (V.i.18–20).

27. Glantz and Pearce 1989, p. 106.

NOTES FOR CHAPTER 3
SEX WITHOUT SEX

1. For example, the Mexican king snake (*Lampropeltis mexicana*), the rock rattlesnake (*Crotalus lapidus*), the Texas banded gecko (*Coleonyx brevis*), and the crevice spiny lizard (*Sceloporus poinsetti*).

2. Maynard-Smith 1976.

3. Even among professional biologists, the terms *recombination* and *mixis* have variable (narrower or broader) meanings (see Ghiselin 1988). Here, I will use the restrictive sense of recombination (i.e., crossing over), and I will use the term *mixis* in its broader sense. A distinction is often made between *automixis*, or selfing, in which one parent gives rise to genetically distinct offspring, and *amphimixis*, or outcrossing, in which two parents combine to produce different genetic combinations in their offspring.

4. This process, first posited by Gregor Mendel, is called independent assortment (or segregation). It ensures that paired maternal and paternal chromosomes have an equal probability of being represented in a given gamete. As such, each gamete will consist of an equal mixture of maternal and paternal chromosomes, though the probability that any two gametes will have the same mixture is extremely remote (2^n, where $n =$ the number of chromosomes in the genome).

5. There is also an asymmetric form of recombination called translocation, in which a piece of one chromosome is transferred to another without a reciprocal transfer. This can involve large pieces of a chromosome and hence a major genomic shakeup. Though usually deleterious, translocations are viewed by some biologists as an important evolutionary process (White 1973).

6. The volume edited by Michod and Levin (1988) provides a nice sample of the

various explanations of sexual reproduction. See also Stearns 1987; Tacel et al. 2001; Kondrashov 2001a,b; Otto and Leormand 2002.

7. Muller 1932.

8. I use Ghiselin's (1988) characterization of ecological theories.

9. Weismann 1891.

10. Williams (1975) presents models such as the "elm-oyster" in which fitness, though highly variable, has low (narrow-sense) heritability and is therefore lottery-like. The winners in any given generation win big, but because their particular felicitous combination of genes cannot be reliably inherited intact, and their fitness is so specific to microcontexts, there is selection to generate as many different genetic combinations as possible. His "Sisyphean genotypes" are the big winners that must be created anew each generation.

11. This is a fairly idiosyncratic view, which is a function of his more general belief that "natural populations can always evolve as fast as required of them and that failure to keep up with the demands of selection is never a cause of extinction" (Williams 1975, p. 157). For Williams, mutation rate itself is always under selection, and asexual populations can always generate as much variation by this means as is required. According to Williams, only the disappearance of a niche can cause extinction.

12. This is one of several topics in Ghiselin's (1974) *The Economy of Nature and the Evolution of Sex*, for which he invokes the division-of-labor concept of classical economics.

13. Bell (1982) adopts the term *tangled bank* (from Darwin's *Origin of Species*) in reference to hypotheses concerning the benefits of sexual reproduction with respect to spatial variation. Williams's (1975) elm-oyster and aphid-rotifer models would fall in the tangled bank category.

14. But see Williams 1975.

15. Van Valen (1973) was the first to use the term *Red Queen* in an evolutionary context. He argued that species go extinct at a constant rate because their predators, prey, competitors, and parasites are also evolving and in ways that cause the environment to deteriorate even as they adapt. Bell adopted this term with reference to the advantages of sexual reproduction by way of analogy with Van Valen's argument. Some Red Queen models of SEX include those of Levin (1975), Hamilton (1980, 1982), and Rice (1983).

16. W. D. Hamilton (for example, 1980, 1982) formulated the parasite version of the Red Queen idea for the advantages of sexual reproduction.

17. This is an example of a more general phenomenon known as frequency-dependent selection.

18. Explanations that advert to the long-term advantages of SEX generally require some form of group selection, which is anathema to orthodox Darwinians, though not to Hamilton.

19. For example, when fecundity is extremely high, heritability of fitness is low, and the environment is structured like a tangled bank on a very small scale. Although I agree with Williams's conclusion regarding the inadequacy of ecological models and the need to consider historical explanations, it is based on the faulty premise that when asexual populations do evolve in "higher animals," such as whiptails, they always out-compete and replace sexual populations. There is little if any

evidence for this claim. It seems to be motivated by his a priori reasoning that in low-fecundity organisms, sexuality is especially disadvantageous.

20. Though Williams, as we have seen, proposed that adaptation is an onerous concept that should be invoked only under special, fairly demanding, conditions, in practice he seems to recognize the onerousness of adaptive explanations only when the adaptation is attributed to groups of organisms. He has never been particularly critical of even obviously baroque adaptationist explanations as applied to individual organisms. During his tenure as editor of *American Naturalist* (1970–1975), that journal became the forum of choice for some of the most outré sociobiological speculation (see, for example, McKaye 1977, 1979 and the critique by Coyne and Sohn 1978). And Williams has recently become a totemic figure for evolutionary psychologists, compared with some of which, Bell is a shrinking violet.

21. Bell (1982, p. 90) states that "A more fundamental reason for objecting to the historical hypothesis is that it violates the axiom of perfection." (This axiom, although a hallmark of natural theology, is not obviously appropriate for biology.) For ardent adaptationists such as Bell, historical explanations that can make predictions about a trait's occurrence only on the basis of phylogenetic criteria are not nearly as satisfying as explanations that predict the phenomena on the basis of ecological factors.

22. Bell 1982, p. 90.

23. Buss 1987; Ruvinsky 1997.

24. Buss 1987, p. 125–26.

25. SEX qua mixis originated in prokaryotes (bacteria), for which it was not linked to reproduction (see, for example, Margulis and Sagan 1986).

26. Development, or ontogenesis, is a property of multicellular organisms, particularly those with differentiated cell types. Ruvinsky (1997) demonstrates that development is linked to meiosis in all three kingdoms of multicellular animals (animals, plants, fungi), though in a different manner in each. Buss (1987, p. 126) provides one hypothesis as to why this linkage is necessary.

27. I know this from having collected them there, much to the amusement of local inhabitants.

28. Carl L. Hubbs is widely considered the doyen of American ichthyology. He was especially knowledgeable about the family Poeciliidae and the closely related family Cyprinodontidae, more than a few members of which he was the first to discover and name. The discovery of Amazon mollys was reported by Hubbs and Hubbs (1932).

29. The common aquarium molly, *Poecilia latipinna*, has been selectively bred for various physical attributes. The melanistic form known as the black molly is particularly popular.

30. The sperm are required in order to stimulate embryogenesis, but none of the progeny of these matings evidence paternal genes. This mode of asexual reproduction is called gynogenesis.

31. Aeschelus referred to them as "The warring amazons, men-haters" (Hamilton 1940).

32. Unisexual populations of *Poeciliopsis* were discovered by Miller and Schultz (1959). The species *Poeciliopsis monicha* is the most common sexual ancestor for the unisexual populations. *Poeciliopsis lucida*, *P. occidentalis*, and *P. latidens* are the other

species that contribute to the generation of asexual populations (Wetherington et al. 1989). There are two categories of clones resulting from these hybridizations. Triploid clones reproduce gynogenetically like *P. formosa*, whereas diploid clones reproduce through hybridogenesis (Vrijenhoek 1993).

33. Poeciliids exhibit two distinct asexual (nonrecombinant) modes of reproduction. Amazon mollies (genus *Poecilia*) reproduce exclusively through a process known as gynogenesis, in which the sperm does not contribute genetically to the zygote; it merely stimulates embryogenesis. Triploid asexual populations within the genus *Poeciliopsis* also reproduce by means of gynogenesis, but diploid asexual populations of this genus reproduce by means of a process known as hybridogenesis in which the sperm is incorporated and the paternal genome is expressed in the zygote but only the maternal genome is transmitted to the gametes.

34. When unisexuals are common, the males should be more discriminating. Evidence for this assumption is largely indirect. For example, the percentage of fertilized eggs in unisexual females is negatively correlated with their abundance (Moore and McKay 1971). The direct evidence comes from Moore (1975, 1976), Keegan-Rogers and Schultz (1988), and McKay (1971), who found that subordinate males are more likely to inseminate asexual females than bisexual females. Unisexual females, in turn, should be more forward than bisexual females, in order to ensure that their eggs get fertilized. There is little evidence of such, however. There is some evidence that successful clones mimic females of sexual species (Lima et al. 1996).

35. Cole 1975; Vrijenhoek 1989.

36. The ancestry of *C. uniparens*, including the fact that two-thirds of its triploid genome is *C. inornatus*, was established through analysis of mitochondrial DNA (Densmore et al. 1989).

37. The term *parthenogenesis* is reserved for asexual organisms such as whiptails that do not require sperm in order to reproduce. Strictly speaking, there are no known parthenogenic fishes or amphibians.

38. The classic account of brain-hormone-behavior interactions during courtship comes from the work of Lehrman on ring doves, *Streptopelia risoria* (Lehrman 1965; see also Hutchison et al. 1996; Lea et al. 2001). Crews and his associates have done much to further our understanding in this area as well (for example, Crews and Silver 1985; Hartman and Crews 1996). The neuroendocrine mechanisms that regulate ovulation are particularly susceptible to modulation by courtship or copulatory behavior or both. Courtship modulation is more common among monogamous species, whereas copulatory modulation (including induced ovulation) is more common among polygynous species.

39. Crews et al. 1986; Crews 1987.

40. Individual whiptails engage in both male and female copulatory behavior, depending upon where they are in the reproductive cycle. Hence, a pair of individuals housed together will reciprocate. There is some evidence that dominant individuals engage in more of the receptive (female) behavior than subordinates and lay more eggs (Grassman and Crews 1987).

41. Lindzey and Crews (1986, 1988); Moore and Crews (1986). This type of ovarian cycle is typical of vertebrates that exhibit the "associated reproductive pattern," in which gonadal hormone levels are highly correlated with sexual behavior (Crews 1984, 1987). It is common but not universal among vertebrates.

42. The role of the ventromedial hypothalamus in receptive behavior appears to be highly conserved among tetrapod vertebrates (for example, Crews and Silver 1985; Pfaff and Schwartz-Giblin 1988; Sachs and Meisel 1988). Evidence for the importance of VMH in the expression of receptive behavior in whiptails is extensive (Crews et al. 1990; Wade and Crews 1991; Kendrick et al. 1995; Young et al. 1995; Godwin and Crews 1995).

43. For general references on sex differences in the POA see Gorski et al. 1978; Arnold and Gorski 1984; Commins and Yahr 1984; Kelley 1986; Yahr 1988; Adkins-Regan and Watson 1990; Crews et al. 1990. For information on POA sex differences in whiptails see Crews et al. 1990; Wade et al. 1993; Rand and Crews 1994.

44. Wade and Crews 1991; Wade et al. 1993; Crews et al., 1990 (brain size calculations were made after correction for body size differences between triploid *Cnemidophorus uniparens* and diploid *C. inornatus*).

45. Moore et al. 1985; Moore and Crews 1986.

46. Steroid receptors belong to a superfamily of transcription factors (Evans 1988).

47. Young et al. 1995a,b.

48. Young et al. 1995b.

49. The negative feedback influence of ER levels on estrogen is indirect. By increasing the sensitivity of certain neurons in the hypothalamus—which are connected to the median eminence of the pituitary—to estrogen, higher ER levels stimulate a reduction in the release of gonadotropin-releasing hormone (GnRH), which causes a decrease in gonadotropin secretion from the pituitary, which results in a reduction in estradiol release from the ovary.

50. Pfaff et al. 1994. Similar effects of progesterone and estradiol on female receptive behavior have been observed in anole lizards as well (Tokarz et al. 1981; Lindzey and Crews 1986).

51. As a result of the negative feedback interactions of steroid hormones and their receptors, we would expect an inverse relationship between steroid hormone levels and receptor abundance. Such is the case in primates. New World monkeys have much higher levels of both estradiol and progesterone, but much lower concentrations of their receptors, than Old World monkeys (Lipsett et al. 1985). Young et al. (1995) proposed that a similar compensation could explain the differences in both estradiol levels and sensitivity to estradiol in unisexual and bisexual whiptails. Whereas a typical diploid species such as the bisexual whiptail *Cnemidophorus inornatus* has two copies of each gene at a given locus, a triploid species such as the unisexual whiptail *C. uniparens* has three sets of chromosomes and hence three gene copies. For a given estrogen level, the parthenogen can potentially make 50 percent more ER than a sexual female. Because rising ER levels cause estrogen levels to drop through negative feedback, the higher ER levels in the parthenogens, relative to females of bisexual species, result in a greater reduction of estrogen levels in the parthenogens. The net effect is that the parthenogen gets more bang for a given picogram of estrogen than a sexual female, in the form of ER expression in the VMH, and, ultimately, sexual behavior. The lower estrogen levels in parthenogens relative to females from bisexual populations is simply a consequence of their higher ER levels, which itself is a consequence of their triploid genome. These generic interactions probably also explain why parthenogens are more sensitive to tes-

tosterone—of which they normally have none—than are the males of bisexual populations.

52. Pfaff et al. 1994. Griffo and Lee (1973) obtained similar results with gerbils.

53. This is what is commonly referred to as chemical castration (Bradford 1988; Lehne 1988).

54. Godwin et al. (1996) demonstrated these effects of progesterone on receptive behavior using gonadectomized whiptails that were also treated with estradiol benzoate. The reduction in female sexual behavior corresponded to a sharp decrease in both estrogen receptors and progesterone receptors in the ventromedial hypothalamus.

55. Grassman and Crews 1986; Moore et al. 1985.

56. Wade and Crews 1991; Wade et al. 1993.

57. Lindzey and Crews 1986, 1988, 1992. Lindzey and Crews (1988) demonstrated that it is progesterone, and not its metabolite RU486, that has this doughnut-stimulating effect. RU486 is an antiprogestin best known for its use in birth control (the morning-after pill marketed as mifepristone). This synthetic hormone binds to progesterone receptors with higher affinity than the progesterone itself but has no progesterone agonist properties. It is a pure progesterone antagonist in mammals (Witt et al. 1995).

58. In rats, progesterone augments the effects of testosterone in this regard, and its action is negated by the progesterone antagonist RU486. It should be emphasized that these results represent a feather in the cap for comparative endocrinology, because it is a case in which research on a nonmammalian species informed research on a mammalian species.

59. Progesterone is known to bind to androgen receptors, but with lower affinity than testosterone (Moguilewsky and Philibert 1985).

60. Lindzey and Crews 1993.

61. Maynard-Smith (1986, p. 110) actually cites the retention of meiosis in *Cnemidophorus* as an example of a sexual hang-up. I hope it is apparent by now that their sexual hang-ups extend well beyond the retention of meiosis.

Notes for Chapter 4
Transgendered

1. Anemones and corals (both hard and soft) belong to the class Anthozoa of the phylum Cnidaria (formerly Coelenterata). This phylum also contains the jellyfish (class Scyphozoa) and hydroids (Hydrozoa).

2. Nematocysts are capsules containing ejectible needle-like toxic elements, located in the tentacles. They are used to both capture prey and defend against predators. Nematocysts are found in all members of the phylum Cnidaria. The Latin word for nematocyst is *cnida*, for which the entire phylum is named.

3. During the course of acclimation to the anemone, the anemonefish acquires a protective mucus, but there is disagreement as to the source of this mucus. On one view, the protective coating comes from the anemone itself during the course of the acclimation process, as a result of which the anemone fails to identify the anemonefish as foreign. On this view, the acclimation process results in a form of chemical

mimicry (Schlicter 1972, 1976). On another view, anemonefishes have evolved a mucus that lacks the chemicals that normally stimulate nematocyst discharge (Lubbock 1980). If so, the elaborate acclimation ritual is either a relic or required to stimulate the fish's own mucus secretion.

4. The process of protandrous sex change in anemonefish was first described by Fricke and Fricke (1977).

5. Limbaugh (1961) was the first to bring attention to cleaning symbiosis; he conducted some early field experiments involving selective removal of cleaners of various species.

6. There has been increasing skepticism regarding the fitness benefits for the cleaned (for example, Losey 1972; Gorlick et al. 1978), but more recent evidence suggests that cleaners do provide fitness benefits to their clients (Poulin and Grutter 1996; Grutter 2001), although they also sometimes manipulate their clients through tactile stimulation alone (Bshary and Wurth 2001). A detailed account of the ethology of Labroides dimidiatus is provided by Potts (1973).

7. Sex change in the cleaner wrasse was first documented by Robertson (1972).

8. Ghiselin (1969) provides the basic outline of the size-advantage hypothesis for sex change. His framework was greatly elaborated by Warner et al. (1975). Robert Warner (of the University of California, Santa Barbara), in particular, has made substantial contributions to the development of the size-advantage hypothesis, with particular reference to sex change in fishes. I want to emphasize that this is a very good model indeed. Specifically, I completely concur with Warner (1988, 1989) that Shapiro's (1988a, 1989) criticisms of the size-advantage model are off the mark. My argument will be, rather, that more should be made of the fact that it cannot be applied to vertebrates other than teleost fishes.

9. There is increasing evidence that sperm is not as cheap as once thought, and sperm limitation is currently a hot topic in evolutionary ecology (for example, Shapiro et al. 1994). Among fishes, the amount of sperm an individual can produce is thought to put limits on fertilization success among group spawners but not among species (such as Amphiprion) that exhibit intimate courtship and lay a relatively small number of eggs.

10. Policansky (1982) surveyed the occurrence of sex change among plants and animals and noted the disparity between predicted and actual occurrence. Warner (1978) notes the lack of sex change among terrestrial vertebrates.

11. See Francis 1992.

12. Mammalian and avian sex chromosomes are highly heteromorphic, to a degree that is obvious under fairly low-power magnification. Not all sex chromosomes are heteromorphic, however. It is believed that low heteromorphism reflects the more recent evolution of genetic sex determination. Heteromorphism, primarily through the degeneration of one member of the pair, increases over evolutionary time. Few species of fish have sex chromosomes that can be distinguished karyotypically (Ebeling and Chen 1970; Gold 1979; Sola et al. 1981).

13. Indirect evidence for sex chromosomes is claimed from evidence of two kinds: (1) controlled breeding in species or populations with sex-linked color markers that are associated with single alleles (Kallman 1970) and (2) experiments in which an individual whose sex has been altered by hormonal manipulations (female → male) is mated with another (female) individual of the same "genetic sex" (for example,

Yamamoto 1961). Such matings sometimes result in all-female broods because, it is assumed, the parents are both genetic females. In some cases, the sex reversal is accomplished by removing the ovaries, after which testes sometimes develop in what was formerly a female (Becker et al. 1975; Lowe and Larkin 1975). (There are no reports of sex reversal following testis removal.) Neither procedure is effective in mammals. Steroid hormones do not effect gonadogenesis in mammals (Burns 1961; McCarrey and Abbott 1979), nor are there any reported cases of sex reversal following gonadectomy.

14. Reviewed in Yamamoto 1969, Yamazaki 1983.

15. By "developmental mechanism" I am referring to the actual process whereby gonadal fate is established. Aida (1921, 1936) established the XY mode of inheritance in medaka.

16. Yamamoto 1953, 1958.

17. Hunter and Donaldson 1983. Another problem with males, from a salmon farmer's point of view, is that, as discussed in chapter 5, a minority of them sexually mature at a small size, after which they cease growing and have less viability in saltwater (Hunter et al. 1982).

18. See for example Johnstone et al. 1978; Donaldson and Hunter 1982; Hunter et al. 1982.

19. Clemens and Inslee 1968; Guererro 1975; Tayamen and Shelton 1978; Nakamura and Iwahashi 1982; Yamazaki 1976. In some cases, all-female stocks are desirable (Bardach et al. 1972), and estrogen treatments have been successfully applied to this end (for example, Tayamen and Shelton 1978).

20. Burns 1961; McCarrey and Abbott 1979. Exogenous sex steroids seem to have more extensive effects on sexual differentiation in marsupial mammals than in placental mammals (Burns 1950). Perhaps the most dramatic effect of exogenous steroid hormones on sexual development in mammals is the so-called freemartin effect, first noted by Lillie (1916) in domestic cattle. Freemartins are sterile females that result from the influence of sex steroids excreted by their male twin in utero.

21. Cloning of zebrafish was first accomplished by means of a technique devised by George Streisinger at the University of Oregon (Streisinger et al. 1981). Diploid clones were generated from homozygous females. Eggs were first fertilized with ultraviolet-inactivated sperm. After the haploid set of maternal chromosomes had replicated, partitioning of two chromosomes sets into separate cells was prevented, by means of either heat shock or a combination of ether and high pressure. The result is a single diploid egg with two identical sets of chromosomes.

22. Bull (1983) reviews examples of temperature-dependent sex determination, which was first discovered by Charnier (1966) in a lizard of the family Agamidae. There is now a wealth of information on temperature-dependent sex determination, especially among turtles and crocodilians (see, for example, Pieau 1972; Bull 1980; Pieau et al. 1982; Crews et al. 1994; Pieau 1996; Shine 1999).

23. Pieau 1975; Bull et al. 1982.

24. For temperature-dependent sex determination in American alligators see Ferguson and Joanen 1982. Charnier (1966) reports this pattern in the lizard *Agama agama*.

25. According to Pieau (1996), this pattern of temperature-dependent sex determination is found in all crocodilians, including the American alligator. It has also been identified in some turtle species (for example, Yntema 1976, 1979; Bull et al. 1982).

26. For examples of exogenous hormone treatments overriding genetic sex determination in anurans see Witschi 1929, Richards and Nace 1978. Winge (1932, 1934) and Aida (1936) provided evidence for similar effects of hormonal treatments on the guppy *Poecilia reticulata*.

27. Rubin (1985) reported significant influences of pH on sex determination in several cichlids as well as in the poeciliid *Xiphophorus helleri*.

28. Francis 1983, 1984, 1987.

29. Francis 1984.

30. This impression was based on a paper by Becker et al. (1975), in which sex chromosomes were inferred from the sort of indirect evidence discussed in note 13.

31. In a two-factor (XX/XY or ZZ/ZW) system of sexual inheritance with Mendelian segregation, substantial variation from a sex ratio of 0.5 (= 1:1, male:female) is precluded (for example, Maynard-Smith 1978; Williams 1975; Charnov 1982).

32. This paper was by H. Schwier (1939).

33. The distinction between selection *for* and selection *of* is from Sober (1984, p. 97). In selecting for the propensity to win dyadic dominance encounters, I inadvertently caused the selection *of* maleness.

34. My selection procedure resulted in an increase in the likelihood that a given individual would become male under the environmental conditions that I maintained. One of the salient environmental features I identified was the degree to which the fish were crowded. The more fish maintained per unit volume of water, the greater the proportion that became females, regardless of dominance line. There were, however, significant differences in the sex ratios among dominance lines for any given density of fish. In essence, male development in the high-dominance line (HD) was buffered relative to that in the low-dominance (LD) fish, with respect to the crowding effects. Put another way, maleness in the HD line was canalized relative to that in the LD line.

35. Francis 1988b. The evidence that sex in Midas cichlids is determined by size-mediated social interactions among juveniles is presented in Francis and Barlow (1993).

36. Takahashi 1977.

37. Streisinger et al. (1981) mention that some clones are predominantly male. Bull (1983, p. 98) provides more specific information on interclone sex-ratio variation based on a personal communication from Streisinger. Obviously, I disagree with Bull's assertion that this variation can be attributed to "environmental noise." If social interactions determine sex in this species, we should consider it a deterministic process. It only looks like "noise" for lack of a convenient metric for the independent variable.

38. It was Bullough (1947) who first proposed that teleost sexual development is fundamentally protogynous. I expanded upon this notion in Francis 1992. The evidence for my thesis includes, for example, the fact that many species that do not change sex in nature exhibit a rudimentary form of hermaphroditism, as indicated by the presence of degenerate oocytes in adult males (Peters 1975; Naish and Ribbink 1990). This may reflect prematurational sex change of the sort observed in paradise fish and Midas cichlids. It is notable that the reverse form of rudimentary hermaphroditism (degenerate spermatic tissue in mature ovaries) has never been observed. Moreover, in those species that are known prematurational sex changers, all

individuals begin differentiating as females (Schwier 1939; Takahashi 1977; Davis and Takashima 1980; Takahashi and Shimizu 1983). True, or "simultaneous," hermaphrodites are also instructive in this regard. A number of sea basses (family Serranidae) are hermaphrodites (Smith 1965; Pressley 1981; Fischer 1984; Hastings and Petersen 1986) that may have themselves evolved from protogynous sex changers through the arrest of the sex-change process. Some of the sex-changing (female → male) members of this family concurrently possess functional ovaries and testes during this transition (Reinboth 1967; Hastings and Petersen 1986; see also Cole 1989 for a similar observation in a goby), and, in at least one species, there is a distinct hermaphrodite phase during which individuals produce both eggs and sperm, between the female and male stages (Hastings and Petersen 1986; Petersen 1987).

Hermaphroditic serranids require a partner to reproduce (they alternately exchange sperm and eggs), but *Rivulus marmoratus* is (internally) self-fertilizing (Harrington and Rivas 1958; Harrington 1961). These fish mature as hermaphrodites, but the testicular region of the gonads tends to increase with age; as a result, some older individuals become entirely male (Harrington 1971, 1975). But a few individuals appear to become males at smaller sizes and are referred to as primary males (Soto et al. 1992; Shakakura and Noakes 2000). Finally, the fact that among (postmaturational) sex-changing fishes, protogyny is so much more common than protandry, is also consistent with the fundamental protogyny of teleost sexual development.

39. Such species differences in the timing of critical developmental events have, since the publication of Gould's *Ontogeny and Phylogeny* (1977), increasingly interested those evolutionary biologists who have adopted a developmental perspective. The term *heterochrony* (= different times) is the general term used to describe such phenomena.

40. Female tissue is present from the earliest stages of sexual development in all the protandrous species examined to date. In both clownfish and porgies (family Sparidae), ovarian development both precedes and follows testicular development (Reinboth 1988; Shapiro 1992). In clownfishes, oocytes appear early during development, before the appearance of sperm tissue, and ovarian tissue is a prominent part of the gonad in both juveniles and males (Shapiro 1992; Godwin 1994).

41. Shapiro 1992.

42. The protogynous pattern of teleost sexual development may be related to a relatively unique embryological feature. In most vertebrates, the gonads exhibit two distinct germinal layers: the medulla and cortex (Lepori 1980; Van Tienhoven 1983; Adkins-Regan 1985, 1987). Testes differentiate from the medulla, whereas ovaries differentiate from the cortex. In teleosts, however, presumptive gonads consist entirely of the cortex homolog; there is apparently no medulla. It is noteworthy that the cyclostomes (hagfish and lampreys), the most primitive living vertebrates, share with teleosts the single gonadal primordium, and they are also fundamentally protogynous (Hardisty 1965a,b; Gorbman 1990). Cartilaginous fishes (sharks and rays), on the other hand, which have dual gonadal primordia as in tetrapods (Dodd 1960), are not at all sexually labile.

43. It needs to be emphasized that the teleost condition only superficially resembles that of mammals, in which female is also the default sex. Male mice and humans never experience even rudimentary ovarian development; instead, both testes and

ovaries arise directly from bipotential gonadal tissue. My claim is that, in fishes, the bipotential gonad stage and testicular development are separated by some degree of ovarian development.

44. Some of the best evidence for the lack of efficacy of sex steroid treatments compared with GnRH treatments in inducing sex change comes from studies of ricefield eels (*Monopterus albus*). Sex steroids have little effect on gonadal sex in this species (Tang et al. 1974; Chan et al. 1977), whereas both GnRH and gonadotropin treatments were quite effective in this regard (Tang et al. 1974b; Chan et al. 1975).

45. In studies on the bluehead wrasse (*Thalassoma bifasciatum*), it has also been established that both GnRH (actually a GnRH analog: Kramer et al. 1993) and gonadotropin (Koulish and Kramer 1989) are effective in inducing sex reversal.

46. Females of this protogynous species were ovariectomized before they became socially dominant (Godwin et al. 1996). When they became dominant, they underwent the same behavioral changes that occur during sex change, even inducing other females to spawn. These fishes did not undergo the typical color change that accompanies sex change in this species, indicating that gonadal hormones are required for that part of the transformation.

47. An alternative pathway for brain–gonad interactions during sex change was suggested by Crews (1993), on the basis of the observation of a direct neural connection between hypothalamus and gonad by Gresik (1973). Crews proposed that hypothalamic neurons might influence the gonads through such a direct neural connection. To date, there is no evidence for such a mechanism in sex-changing fishes.

48. There are two species of elephant seal. The vast majority of northern elephant seals (*Mirounga anguistirostris*) breed on Ana Nuevo Island and the adjacent mainland, north of Santa Cruz, California. The species, which was once reduced to 10 to 30 individuals (Bartholomew and Hubbs 1960; Hoelzel et al. 1993) as a result of overhunting, numbered around 127,000 in 1991 (Stewart et al. 1994). Southern elephant seals (*M. leonina*), which breed along the Patagonian coast, were also hunted extensively, but their numbers were never reduced to the extent of their northern counterparts. In 1990, their numbers were estimated at 664,000 (Laws 1994).

49. Le Boeuf 1974; Le Boeuf and Reiter 1988. Northern elephant seal beachmasters are somewhat less successful than their southern counterparts in impregnating their harem females and excluding other males, though they experience about the same success in securing copulations (Hoelzel et al. 1999). The difference between paternity and copulatory success among male northern elephant seals may reflect reduced fertility due to inbreeding depression as a result of their historical population bottleneck. It also seems to be true, however, that northern elephant seals are somewhat less polygynous than southern elephant seals, as indicated by the more extreme dimorphism in the latter (Haley et al. 1994).

50. Deutsch et al. 1994.

51. Great white sharks are the primary predator on elephant seals at Ana Nuevo, and the adjacent water is full of them during the breeding season. Despite occasional highly publicized attacks, true surfers are not deterred by such trifles, if, as at Ana Nuevo, the waves are consistently good.

52. One of the best discussions of the "constraints" on sex change in mammals and other tetrapods is that of Warner (1978). Because of his deeply functionalist perspective, Warner does not emphasize the moral I am drawing here, that mammals are precluded from changing sex by contingent historical factors.

NOTES FOR CHAPTER 5
ALTERNATIVE LIFESTYLES

1. The terms *alternative tactics* and *alternative strategies* have been used more or less interchangably. Gross (1996) has tried to differentiate the terms.

2. The sizes of the humps (i.e., the proportion of individuals within each) need not be the same, of course; that is, the humps could be, and often are, asymmetrical.

3. Bass (1996) provides a review of the differences between type I and type II males for a general audience. Behavioral differences in type I and type II males are the focus of Brantley and Bass (1994). Bass (1992) discusses these behavioral differences in the context of neurological differences. The dimorphism in sonic musculature is discussed by Brantley et al. (1993).

4. Ultrastructural (including mitochondrial) differences in the sonic musculature of type I and type II males were noted in Bass and Marchaterre (1989).

5. Bass and Marchaterre 1989; Bass 1992; Bass et al. 1994; Bass et al. 1996.

6. Perrill et al. 1982; Forester and Lykens 1986; Perrill and Magier 1988. There is much intra-individual variation in calling among tree frogs from hour to hour and day to day (for example, Friedl and Klump 2002).

7. Actually, some extreme adaptationists *will* claim that waylaying is a tactic, but we should dismiss this as a paranoiac delusion, given the evidence to date.

8. The mating system of bluegill sunfish and the alternative male reproductive types are described by Gross and Charnov (1980); Dominey (1980); and Gross (1991).

9. Gross 1982.

10. Early maturation was first identified in laboratory- and hatchery-reared Atlantic salmon (for example, Saunders and Henderson 1965; Thorpe 1975). Gross (1984, 1985) discusses early maturation in a Pacific (coho) salmon in the context of alternative reproductive tactics.

11. Because of the Atlantic salmon's commercial significance, with respect to both wild and farmed fish, there is an extensive literature on mature parr (see, for example, Taborsky 1994). Individuals that mature as parr are those that grow the most during the first year (Thorpe and Morgan 1978, 1980).

12. Gross 1984, 1996; Hutchings and Myers 1988; Fleming and Gross 1994; Thomaz et al. 1997.

13. There are three reproductive types in sex-changing species with alternative male tactics: females, type I males, and type II males. Because, in a sex changer, maleness and femaleness are not a given, these conditions should be considered tactics as well. Indeed, the most fundamental "decision" would seem to be whether to be a female or a male; if male, then secondarily, whether to be type I or type II. Male reproductive types would therefore result from a two-step decision process. In fact, however, as we shall see later, in fishes, type I males can result either from

females (via sex change) or from type II males (via growth), but none are born that way. Diandry was first noted by Warner and Robertson (1978).

14. The information on stoplight parrotfish discussed here comes from research at Glover's Reef, Belize, conducted by Cardwell and Lilely (1987; see also Cardwell and Lilely 1991a,b).

15. Wells (1977) used the term *satellite male* to refer to the "alternative tactics" of a number of frog species. Howard (1978) used the term *parasite* to refer to this tactic, which he distinguishes from a third tactic that he calls *opportunist*. His parasite and opportunist categories are combined here as satellite.

16. As I will discuss, this fact raises suspicion that the "satellite tactic" is not really a tactic at all in an evolutionarily relevant sense, but just the way smaller individuals make the best of a bad situation.

17. See Crews 1983 for an overview of his work on garter snakes to that point; Crews and Gartska (1982) give an account for a general audience.

18. Crews and Gartska 1982.

19. Female garter snakes emit through the skin sex pheromones that attract males (Gartska and Crews 1982).

20. Male mimicry of females is described by Mason and Crews (1985, 1986). Crews (1985) suggests that it may result from exposure to estrogen in utero, as a result of which she-males become feminized in several respects, including the transport of pheromones through the skin.

21. The she-male behavior may represent a case of evolutionary spite. It now appears that all male garter snakes pass through a she-male phase immediately upon waking from hibernation (Shine et al. 2000 a,b), while they are still weak and lethargic. They have no interest in courting while in this condition; their lethargy not only saves energy but causes other males to waste valuable energy in fruitless courtship. Although she-maleness may be adaptive (it could easily be an unselected byproduct), this behavior would certainly stretch the meaning of the term *alternative reproductive tactic*.

22. Hover 1985.

23. Thompson and Moore 1991.

24. Sinervo and Lively 1996.

25. Zamudio and Sinervo 2000; see also Sinervo et al. 2001.

26. The ruff mating system was described by van Rhijn (1973, 1991).

27. Lank et al. (1995) provide evidence that the alternative tactics are associated with a genetic polymorphism.

28. Hugie and Lank 1997, Widemo 1998.

29. See, for example, Clark and Galef 1995. Caro and Bateson (1986) discuss some putative cases among nonrodent mammals. See also Dunbar 1983 for a putative example of alternative reproductive tactics in a primate.

30. The intrauterine position effect was first discovered in house mice (*Mus musculus*) (vom Saal 1983, 1984, 1989).

31. For evidence of intrauterine position (IUP) effects on testosterone and aggression in mice, see vom Saal 1984, 1989; for similar effects in mongolian gerbils (*Meriones unguiculatus*), see Clark et al. 1986; Clark and Galef 1988; but see Simon and Cologer-Clifford 1991 for a skeptical treatment of IUP effects.

32. Vom Saal 1984, 1989.

33. For example, IUP effects on sex ratios have been reported in gerbils (Clark et al. 1993) and mice (Vandenbergh and Huggett 1994).

34. If intrauterine position is random, and we have no reason to expect otherwise.

35. Galdikas 1985; Mitani 1985.

36. Galdikas 1985; Delgado and Van Schaik 2000.

37. Maggioncalda et al. 1999; Maggioncalda and Sapolsky 2002.

38. Von Neumann and Morgenstern 1953.

39. See especially Maynard-Smith 1982.

40. Maynard-Smith and Price 1973; Maynard-Smith 1974; Parker 1974; Maynard-Smith and Parker 1976.

41. It remains a remote possibility that toadfish tactics are genetically determined.

42. Gross 1996, see also Gross and Repka 1998. The primary problem with Gross's classification of the ARTs is that he attempts to force these diverse phenomena into a narrow theoretic (game theory with genic strategists) framework that does not carve nature at its joints. See Brockmann 2001 for a much more useful classification of the ARTs.

43. You cannot effectively inseminate by spraying gametes if fertilization is internal, for obvious reasons. Therefore, sneaking tactics, which are so common in fishes, are not practicable in mammals.

44. For example, the guppy *Poecilia reticulata* (Houde 1997; Pilastro and Bisazza 1999); pygmy swordtails, *Xiphophorus nigrensis* (Zimmerer and Kallman 1989; Ryan and Causey 1989; Ryan et al. 1992); the platy, *X. maculatus* (Sohn and Crews 1977; McKenzie et al. 1983).

45. In his "Relative Plasticity Hypothesis," Moore (1991) proposed to use the standard dichotomy of steroid hormone actions into activational and organizational effects to explain the differences between "fixed" alternative tactics, in which an individual can adopt only one of the alternatives during the entire course of its life, and "plastic" alternative tactics, in which each individual adopts more than one tactic during its lifetime. According to the organizational/activational dichotomy, sex steroids act sequentially, at two different times in development. First, they act early in development (often prenatally), to organize the neural substrates that regulate both the behavioral and physiological differences between the sexes, or in this case, within-sex morphs. These initial hormonal effects are not sufficient to produce the full expression of dimorphic traits. The second type of hormonal effect occurs after maturation and is thought to "activate" those neural structures that were organized during early development. According to Moore, the category of fixed tactics includes those in which the differences in alternative male types can be traced to genetic factors, those in which the morph differences are regulated by hormonal factors acting early in development, and those in which morph differences are regulated by social interactions in juveniles. Whichever is the case, the male's tactic is determined before maturity, and there is no switching from one morph to another. In these cases, Moore posits that the morph differences are the result of organizing effects of hormones. Morph differences in species exhibiting plastic tactics, on the other hand, are the result of activational effects of hormones. Moore's relative plasticity hypothesis should be viewed as an attempt to identify the physiological basis for alternative male types, which is laudable. The evidence to date suggests that, although it applies well to his own study species, the tree lizard, *Urosaurus ornatus*

(for example, Hews et al. 1994; Knapp and Moore 1997), it does not generalize sufficiently to account for the alternative tactics found in fishes (Brantley et al. 1993). This shortcoming is perhaps not surprising, because in fishes and amphibians, as well as in those reptiles that lack sex chromosomes, the distinction between the organizational and activational effects of sex steroids is less apparent than in mammals (Crews 1993).

46. See, for example, Adams and Steiner 1988 for an overview on sexual maturation in mammals. The conservation of GnRH function across vertebrates is discussed in Dellovade et al. (1998).

47. Kallmann et al. 1944.

48. Mason et al. 1986a.

49. Mason et al. 1986b, 1987.

50. Kreiger et al. 1982; Gibson et al. 1984, 1997; Silverman et al. 1988; Silverman et al. 1992.

51. Klaus Kallman has been particularly active in this area (for example, Kallman 1975, 1989).

52. Kallman 1973.

53. Ibid.

54. The influence of the P locus on maturation was first linked to the maturation of pituitary gonadotropes by Kallman and Schreibman (1973); Halpern-Sebold et al. (1986) established that gonadotrope maturation was in turn, a function of GnRH expression. As in most fishes, there are several distinct populations of GnRH neurons in platyfish (Munz et al. 1981); GnRH expression first occurs in the terminal nerve in both early- and late-maturing males (Halpern-Sebold and Schreibman 1983).

55. Grober et al. 1994.

56. Grober et al. 1991.

57. For example, Schwanzel-Fukuda and Pfaff 1989; Wray et al. 1989a.

58. The discovery that preoptic area GnRH neurons originate in the olfactory placode was made simultaneously by two research groups (Schwanzel-Fukuda and Pfaff 1989; Wray et al. 1989a). Shortly thereafter, similar results were obtained in rhesus monkeys (Ronnekliev and Resko 1990, reviewed in Graham and Begbie 2000). Not all GnRH neurons originate from this placode, however. In particular, the midbrain (mesencephalic) population is thought to be of nonplacodal origin (see, for example, Muske 1993; Northcutt and Muske 1994).

59. The problem need not originate in the placode itself, however. For speculations about the etiological mechanism involved in Kallmann's syndrome, see Schwanzel-Fukuda and Pfaff 1989b; Schwanzel-Fukuda et al. 1992; Bouloux et al. 1992.

60. Wray et al. 1989b; Zheng et al. 1992.

NOTES FOR CHAPTER 6
SOCIAL INHIBITIONS

1. Fryer and Iles 1972; Greenwood 1991; Kornfield and Smith 2000. The term *adaptive radiation* is used to refer to the rapid proliferation of species within an (evolutionarily) short period of time. The term reflects the assumption that any such proliferation must be driven by adaptive imperatives in the process of filling unoc-

cupied niches. In the case of rift lake cichlids, the emphasis was on trophic, or feeding, adaptations (Liem 1980), which were thought to be enabled by their extra set of (pharyngeal) jaws (Liem 1973; see also Galis and Drucker 1996), in line with Vermeij's (1974) idea that rapid diversification is most likely when body parts have many independent elements (see Hori and Watanabe 2000 for an example of a particularly specific trophic adaptation). Much recent research, however, has focused on the role of female mate choice in driving the rapid speciation of rift lake cichlids (Dominey 1984; Turner and Burrows 1995; Seehausen et al. 1997). Sexual selection could potentially produce a rapid proliferation of species with much less niche partitioning (Genner et al. 1999; see also Goldschmidt and de Visser 1990) by creating numerous genetically distinct subunits within species (Van Oppen et al. 1998, Knight and Turner 1999). Though the niche-partitioning and sexual selection processes are not mutually exclusive (Galis and Metz 1998), it is more apt to label the rapid proliferation of species in these lakes a "species radiation" or "taxic explosion."

2. Nagl et al. 2000. Malawi is older, about 700,000 years (Meyer et al. 1990; Meyer 1993).

3. Some systematists have recently changed the scientific name from *Haplochromis burtoni* to *Astatotilapia burtoni*, but this change has not been widely accepted.

4. These egg dummies are actually somewhat bigger than real *Haplochromis burtoni* eggs and more brightly orange; so they may function as what ethologists refer to as supernormal releasers (Mrowka 1987).

5. Fernald has used the terms *macho* and *wimp* for territorial males and nonterritorial males, respectively, especially when promoting his research to the popular press. But those terms, though catchy, are quite misleading in implying permanent personality differences.

6. At least this was the story I heard when I first joined the Fernald lab, and I had no reason to doubt it, having not visited the field site myself. Subsequently, the explanation for the causes of territorial male turnover seems to have changed substantially. In Hofmann et al. 1999, Fernald claims (from personal observations) that hippos are the main cause of turnover as they stomp through the shallow breeding pools and mess up the territorial boundaries. Fernald never shared the hippo story with me, however. The hippo-caused turnover, much more so than the kingfisher-caused turnover, supports the conclusions that Hofmann et al. come to.

7. Fraley and Fernald 1982, Davis and Fernald 1990, Francis et al. 1993.

8. Hofmann et al. 1999. Based on the fact that under some conditions nonterritorial males grow faster than territorial males. It is often the case in fishes that reproduction inhibits growth to varying extents. Indeed, postreproductive females become larger than all males. (One such female, known as "Mongo," reached the size of a small trout.) What Hofmann et al. did not establish was whether the growth of nonterritorial males enhanced their success in contesting sites with established territorial males. The results of Francis et al. 1992 and personal observations suggest otherwise.

9. Slowtow et al. 2000.

10. Ibid.

11. Ibid.

12. The extended phallus is an aggressive display in baboons and many other primates. While visiting Samburu national park in northern Kenya, I was able to

directly experience this behavior. I spied a male baboon stealing a box of wafers from our unattended jeep, and I chased him, hoping he would drop them; he retreated into a large thorny bush, where he proceeded to eat the wafers with great deliberation, while sitting on his haunches, occasionally flashing his eyebrows and baring his fangs, his penis extended in my direction.

13. Hausfater 1975.

14. Frame et al. 1979; Packard et al. 1983; Creel and Creel 1991. The social suppression of female reproduction is also well documented in primates (Abbott 1984, 1987; French et al. 1989) and social mongoose species (Rood, 1980). Wasser and Barash (1983) provide a review of social suppression of reproduction among female mammals.

15. Frank et al. 1995.

16. Holekamp et al. 1996.

17. But there are some adaptationists who make what amounts to the converse claim, that those instances of incomplete or failed reproductive suppression of subordinates represents an adaptive "concession" on the part of dominant individuals, rather than a failure on their part. This is known as optimal skew theory (Vehrencamp 1983; Keller and Reeve 1994; Emlen 1995; Reeve et al. 1998; East et al. 2002). The basic idea is that it behooves dominant individuals to let their subordinates do some breeding in order to keep them from emigrating. This theory assumes, of course, that dominant individuals could potentially preclude all such breeding. It may be true that such concessions are practiced among social insects, but the evidence is much less impressive regarding social vertebrates (Clutton-Brock 1998). Rather, the more parsimonious explanation for such failures to suppress is that dominant individuals can exercise only limited control over reproduction in subordinates. The limited control hypothesis better explains the breeding of subordinates among meerkats (Clutton-Brock et al. 2001) and dwarf mongooses (Cant 2000) than does the concession hypothesis.

18. For an overview of social insects, see Wilson 1971.

19. The criteria for eusociality as formulated by Michener (1969) include, in addition to the reproductive division of labor, overlapping generations and cooperative care of the young.

20. Jarvis 1981. For a compilation of papers on reproduction in this species, see Sherman et al. 1991.

21. Lacey et al. 1991, Faulkes and Abbott 1993, Smith et al. 1997. Originally, it was thought that the reproductive suppression was mediated by pheromones, as in microtine rodents (for example, Vandenbergh 1988).

22. Clarke and Faulkes 1997.

23. Faulkes et al. 1991; Westlin et al. 1994.

24. Faulkes et al. 1990c.

25. Faulkes et al. 1991; Jarvis 1991.

26. Faulkes et al. 1994.

27. Faulkes et al. 1994; Faulkes and Abbott 1997.

28. Faulkes and Abbott 1991; Clark and Faulkes 1998. There is remarkably little aggressive behavior among males, even during succession (Lacey and Sherman 1991; Clark and Faulkes 1998).

29. The mole rat family (Bathyergidae; see Faulkes et al. 1997 for phylogenetic

information) includes, in addition to naked mole rats (*Heterocephalus glaber*), a number of other subterranean species of varying degrees of sociality (Bennett et al. 1999). The naked mole rat is the only member of the genus *Heterocephalus*, but there are several other eusocial mole rats in the genus *Cryptomys*, in addition to three genera of solitary mole rats (Honeycutt et al. 1991).

30. All members of the genus *Cryptomys* are cooperative breeders, in which reproduction is confined to a relative few individuals. The Damaraland mole rat (*Cryptomys damarensis*) is the most highly social member of the genus, achieving a degree of social complexity that approaches that of the naked mole rat (Bennett 1989; Bennett and Jarvis 1988; Jarvis and Bennett 1993).

31. Faulkes et al. 1994. In the much less social Mashona mole rat (*Cryptomys darlingi*), there is reproductive suppression in both sexes, but this does not involve the hypothalamus or pituitary in either sex (Bennett et al. 1997).

32. Clark et al. 2001. The same is true of other members of this genus (Burda 1995; Bennett et al. 2000; Spinks et al. 2000).

33. Or "spandrels," *sensu* Gould and Lewontin (1979).

34. Francis et al. 1992a.

35. Francis et al. 1992a; Soma et al. 1996.

36. This quote is from Gail Vines in the *New Scientist* (1992), in a report, for a popular audience, of the results from Francis et al. (1992b).

37. Francis et al. 1992b.

38. Francis et al. 1993.

39. Mark Davis had previously shown that social interactions regulate the size of GnRH neurons in juveniles (Davis and Fernald 1990).

40. The tactile nature of the signals was established in experiments conducted primarily by Mark Davis, with some contributions from myself, the results of which are unpublished. Davis demonstrated that the suppressive effects of social interactions occur only when the fish have tactile (not just chemical or visual) access to one another. Moreover, tactile contact alone (without visual interactions) is sufficient for the suppression of maturation.

41. The homolog of the adrenal gland in teleost fishes is the interrenal gland. I will nevertheless continue to refer to the stress axis in fishes as the HPA (rather than HPI) axis in order to avoid confusion.

42. Sapolsky 1982, 1983; Sapolsky et al. 1997. See Sapolsky 1987 for a review.

43. Sapolsky 1982, 1983. Virgin and Sapolsky (1997) provide a more nuanced and complex view of the relationship between social rank, testosterone levels, and cortisol levels, in which they distinguish between types of subordinate males on the basis of behavioral attributes that could be labeled "personality."

44. Rasmussen et al. 1983; Liu et al. 1997.

45. Sapolsky 1986; Rosenblum and Peter 1989.

46. In the African wild dog and the dwarf mongoose, for example, dominant individuals have higher basal glucocorticoid levels than subordinates (Creel et al. 1992; Creel et al. 1996; Creel et al. 1997), which indicates that, in these species, social suppression is achieved independently of the HPA axis. The same is also true of the ring-tail lemur (Cavigelli 1999). In other highly social species with reproductive suppression of subordinates, there is no association between rank and stress

hormones. This is true of, for example, the wolf (McLeod et al. 1996), two species of marmoset (Smith and French 1997; Saltzman et al. 1998), the cynomolgus monkey (Stavisky et al. 2001), and the cotton-top tamarin (Ziegler et al. 1995). Morever, in some female teleosts, social suppression of reproduction is not mediated by cortisol (Pankhurst et al. 1995). In all of these species, the reproductive suppression in subordinates is more likely to be adaptive for the subordinates than in species like *H. burtoni* and naked mole rats, in which suppression is a direct result of social stress.

47. Fox et al. 1997.

48. Chambers and Phoenix 1981; Orchinik et al. 1988; Borg et al. 1992. The most dramatic case in which the HPA axis stimulates the HPG axis occurs in the musk shrew (Rissman and Bronson 1987; Fortman et al. 1992; Schiml and Rissman 1999; Schiml et al. 2001), but this is about as asocial a mammal as can be conceived. And it is nevertheless the case that chronic stimulation of the stress axis in vertebrates depresses reproductive physiology (Christian 1964; Brann and Mahesh 1991; Blanchard et al. 1995; Berga 1995). In fishes, chronic social stress-induced reproductive suppression has been especially well characterized in salmonids (Winberg and Nilsson 1993; Pottinger et al. 1995; Elofsson et al. 2000).

49. In truth, some of my unpreparedness resulted from the fact that the results were reported prematurely—before their publication in a science journal—to the *New York Times* by Russell Fernald, who was not intimately involved in the experiments. The size differences in the GnRH neurons of territorial and nonterritorial males reported in the *Times* article were inaccurate—they were grossly inflated.

50. In practice, there is much waffling on the boundaries between intrinsic and nonintrinsic individual properties, such that proponents of this perspective often start out defending the explanatory primacy of intrinsic properties but end up justifying their position by referring simply to individual properties. This practice makes the position difficult to refute.

51. The classic work on chicken pecking orders was done by Schjelderup-Ebbe (1922).

52. See Collias 1943, for example. Moreover, theoretical work by Landau (1951) and Chase (1974) has demonstrated the a priori improbability that individual differences of any kind can produce linear hierarchies.

53. Evidence that levels of testosterone and other hormones are socially conditioned has been obtained in a wide variety of vertebrates, including *Haplochromis burtoni* (Hannes and Franck 1983) and other fishes (Hannes et al. 1984), as well as in primates and other mammals that I discuss later.

54. Chase et al. 2002.

55. Ibid.

56. Reviewed in Francis 1983. See also Ginsburg and Allee 1942; Kahn 1951; Beacham 1988; Beaugrand et al. 1991, 1996; Zucker and Murray 1996; Schuett 1997; Drummond and Canales 1998; Hsu and Wolf 1999, 2001; Johnstone and Dugatkin 2000.

57. Francis 1983.

58. There were actually three dominance lines; in the third line, the fish were selected for median dominance (MD) rank. In the dominance tests, the HD and LD

males were always matched with equal-sized opponents from the MD line. This approach controlled for inbreeding effects (Francis 1984).

59. This asymmetry in the effects of winning and losing has been found in a wide variety of organisms, but most of this research has been conducted on various fish species (reviewed in Francis 1983; see also Beacham 1988; Bakker et al. 1989; Chase et al. 1994; but see Hsu and Wolf 1999).

60. Francis 1987.

61. Chase 1982.

62. Frank 1986; Holekamp and Smale 1993; Smale et al. 1993; Jenks et al. 1995.

63. Frank 1986, 1997.

NOTES FOR CHAPTER 7
WHY DOES THE MOCKINGBIRD MOCK?

1. Kroodsma (1982) discusses the ins and outs of various ways of measuring repertoire size, or versatility. Jones et al. (2001) discuss problems with interobserver reliability in reading sonograms generally.

2. This method collapses the somewhat messy distinction between intrasong and intersong complexity, which, in any case, does not apply to continuous singers such as thrashers and mockingbirds (Read and Weary 1992).

3. Compared with mockingbird songs, the noises I endured while living above the Brooklyn-Queens Expressway were soporific.

4. Gilliard 1958, p. 339.

5. Ibid.

6. See reviews in Baylis 1982 and Hindmarsh 1986.

7. One of my "favorite" variants on this theme is Rechten's (1978) interspecific "Beau Geste" idea. According to the original Beau Geste hypothesis (Krebs 1977), individual male songbirds move around a lot and sing after each move, in order to fool potential rivals into believing that the area is already saturated with territorial males. The idea, and hence the name, was inspired by the legendary French Legionnaire who managed to single-handedly hold a fort against an army of North African foes by firing his rifle from various positions in the fortification. Rechten extends Krebs's idea to interspecific interactions in order to explain mockingbird vocal mimicry. He suggests that the function of mockingbird mimicry of other songbirds is an analogous form of territorial exclusion. Mockingbirds fool birds of other species with whom they might compete into "believing" that the area is already saturated with their conspecifics.

8. Kenyon 1972; Howard 1974; Dobkin 1979.

9. Striking body markings or colors that signal the ability to cause gross bodily harm are referred to as aposomatic coloration. In addition to skunks, other examples of aposomatic coloration include the poisonous coral snakes and noxious (to birds) monarch butterflies.

10. Adaptation, as conceived by Darwin, refers to the process whereby organisms are made more suitable with respect to an environment. On his view, following Lyell, the environment itself is first and foremost a suite of physical properties (for example, temperature, rainfall) but also includes such biological factors as predators and

prey. Because he did not consider other members of an organism's species to constitute part of its environment, he viewed sexual selection as nonadaptive.

11. See, for example, Lande 1981; Kirpatrick 1982. But adaptationists find runaway selection a much less satisfying explanation for mate choice because it does not invoke benefits for the choosy females.

12. Trivers 1972 was one of the first to formulate a "good-genes" hypothesis as an explanation for intersexual selection. It is the preferred family of explanations for ardent adaptationists (for example, Borgia 1979; West-Eberhard 1979; Thornhill 1980; Kodric-Brown and Brown 1984; Houtman 1992; Andersson 1994; Møller 1995, 1997; Petrie and Kempenaers 1998; Scheib et al. 1999). For good-genes proposals with particular reference to peacocks, see Petrie et al. 1991; Petrie 1992, 1994; Petrie and Kempenaers 1998; Møller and Petrie 2001.

13. See, for example Rohwer 1975; Dawkins and Krebs 1978.

14. Brown 1997.

15. Mitton and Grant 1984; Palmer and Strobeck 1986; Mitton 1993.

16. Allendorf and Leary 1986.

17. See, for example Weatherhead et al. 1999; Vollestad et al. 1999. But see also Landry et al. 2001.

18. Pomiankowski 1990.

19. Though, in fact, males could be ranked on a heterozygosity scale that could function much like an absolute genetic quality scale.

20. Waddington 1953; Van Valen 1962; Scharloo 1991; Palmer 1994.

21. Or so its advocates assume, two of the more outspoken of whom are, not surprisingly, Anders Møller and Randy Thornhill (Møller 1993; Watson and Thornhill 1994; Møller and Thornhill 1997; Møller and Thornhill 1998). Thornhill has been particularly eager to link fluctuating asymmetry to human sexual selection, including that which occurs via orgasms as discussed in chapter 2 (Thornhill et al. 1995; see also Thornhill and Gangestad 1993).

22. See, for example, Palmer 1996, 1999, 2000; Whitlock and Fowler 1997; Nachman and Heller 1999; Simmons et al. 1999; Bjorksten et al. 2000; Lens et al. 2002. Part of the skepticism concerns the heritability of fluctuating asymmetry. The meta-analysis performed by Møller and Thornhill (1997) demonstrating substantial heritability of fluctuating asymmetry has been roundly criticized on a number of grounds (Leamy 1997; Markow and Clarke 1997; Palmer and Strobeck 1997; Whitlock and Fowler 1997). In the only study to test the assumption that fluctuating asymmetry is heritable in a sexual trait, the results were not encouraging for the good-genes-as-symmetry crowd (Tomkins and Simmons 1999). But the heritability problem extends beyond fluctuating asymmetry in particular to all good-genes formulations. For example, Ohlsson et al. (2001) found that sexual ornaments in male pheasants reflect nutrition early in life and suggested that if nutrition were heritable the pheasant adornments would be indicators of good genes. But they offer no evidence or, for that matter, any reason at all to believe that nutrition attained early in life is heritable.

23. The handicap principle was formulated by Zahavi (1975; see also Zahavi 1977). It was initially severely criticized by evolutionary geneticists (for example, Arnold 1983; Kirpatrick 1986) even those such as Maynard-Smith (1976, 1978) who are otherwise sympathetic to functionalist explanations. Zahavi persisted none-

theless, and his handicap principle has undergone something of a revival of late, buoyed no doubt by the general recent ascendancy of extreme adaptationism within behavioral ecology, but also because it has been reinterpreted in a way that blurs the distinction between the handicap principle and Fisher's runaway selection (Iwasa and Pomiankowski 1994, 1999; Pomiankowsi and Iwasa 1998; Houle and Kondrashov 2001; but see also Grafen 1990 for a game theoretic argument for the handicap principle). Zahavi and Zahavi (1999) provide a book-length celebration of their perceived victory, as well as voluminous unbridled speculation as to how the handicap principle applies to all manner of social behavior (see Ryan 1998a for a useful review of the book). But even Zahavi should be impressed with the recent proposal that the autumn colors of many broadleaf trees is a handicap that warns insect predators of their vitality (Hamilton and Brown 2001).

24. Hamilton and Zuk 1982.

25. Read 1988; Hamilton and Poulin 1997; Poulin et al. 2000. Predictably, among the biggest boosters of the Hamilton-Zuk good-genes hypothesis is, once again, Anders Møller (Møller 1991, 1992; Saino and Møller 1994).

26. The immunocompetence handicap hypothess (ICHH) was originally formulated by Folstad and Karter (1992), based in part on known interconnections between the immune and reproductive systems (for example, Grossman 1984, 1985). In some ways, it represents a how-biology augmentation of the Hamilton-Zuk hypothesis, but the two ideas differ in other important ways as well. Whereas the Hamilton-Zuk hypothesis alleges female choice for "badges" that indicate genetic superiority in resistance to ever-changing pathogen populations, ICHH proposes mate choice for badges that indicate genetic superiority in maintaining reproductive effort in the face of the energy costs of mounting an immune defense to any pathogen (see Westneat and Birkhead 1998). For this reason, ICHH does not require, as does Hamilton-Zuk, frequency-dependent selection. ICHH does share with Hamilton-Zuk the need to demonstrate that female choice is based on heritable differences in pathogen resistance. Evidence of nonheritable sources of pathogen resistance such as were found by Griffith et al. (1999) is not welcomed by either of these good-genes hypotheses.

27. There is a lack of compelling evidence for ICHH despite intense efforts to find such (for example, Bortolotti et al. 1996; Kurtz and Sauer 1999; Poiani et al. 2000; Casto et al. 2001). For example, there is little if any evidence of a direct link between immunocompetence and fitness (Owens and Wilson 1999). At present, the best evidence for ICHH are a couple of correlational studies indicating that high testosterone levels may be immunosuppressive (Saino et al. 1995; Zuk et al. 1995). But Hasselquist et al. (1999) could find no evidence of testosterone-induced immunosuppression, whereas Peters (2000) showed that although testosterone-implanted superb fairy wrens evidenced some immunosuppression, in nature the males with the highest testosterone titers were more immunocompetent than were low-testosterone males. Results such as these have discouraged even Møller (Saino and Møller 1994; Møller et al. 1996; Saino et al. 1997) and have led him to suggest that cortocosteroids rather than testosterone are the main immunosupressants, an idea that is more in line with what endocrinologists already knew (see chapter 6; for specific criticisms of ICHH along these lines see, for example, Hews and Moore 1997; Hillgarth and Wingfield 1997a, b; Hillgarth et al. 1997). But if testosterone is not immunosuppressive, as well as the cause of the variation in the sexual trait that females are choosing, ICHH loses most of its force.

28. Kotiaho (2001) makes this point especially well.

29. Including the recent study of lion manes (West and Packer 2002), which has been much ballyhooed by good-genes advocates. With respect to ICHH, much rides on how expensive the immune system is to maintain. It is assumed that because of the costs of running an effective immune system, any diversion of resources to reproduction by testosterone must compromise it (this assumption is implied in Buchanan et al. 2001, for example). But, in fact, Raberg et al. (2002) found no evidence that the immune system is energetically costly to maintain, a finding that undercuts the ICHH.

30. Ryan et al. (1990; see also Ryan and Keddy-Hector 1992; Ryan 1998) present the case for Ryan's sensory exploitation hypothesis. Endler (1992) promotes a related idea called sensory drive, which he considers a broader category that includes sensory exploitation. Both ideas have been increasingly subsumed under the category "receiver bias" (for example, Ryan 1998b), which includes biases produced by more central nervous system processes and states. Some of the more interesting recent work on generalized receiver bias involves artificial neural networks (Phelps and Ryan 1998; Ryan and Getz 2000; Phelps et al. 2001; see also Enquist and Arak 1993, 1994).

31. James Gould broaches this idea in his (1982) book on ethology.

32. Those who point to sensory biases as explanations for the form of sexually-selected traits tend to view mate choice and species recognition as intimately related (Ryan and Rand 1993b). For instance, those attributes by which a female may recognize a male as a member of the right species may become exaggerated when coopted for the purposes of mate choice because of the fact that the receiver's sensory system has become tuned in a particular way in the context of species recognition. Biases in mate choice may also originate in the context of behavioral interactions that originally functioned to stimulate, or prime, the female reproductive system, an aspect of courtship completely neglected by adaptationists. In fact, the bias of female mockingbirds and other mimids for complex songs may have originated in this way, as male mockingbird song is known to restimulate the reproductive system of the mates once the brood has fledged (Logan et al. 1990). Finally, sensory biases may be quite general. For instance, to the extent that there is a female preference for symmetry, it may well reflect such a generalized esthetic bias (Enquist and Arak 1994; Jansson et al. 2002).

33. Originally, it was proposed by Endler (1980) that female guppys use the males' carotene-induced orange spots as an indicator of their "quality" (see also Houde 1987, 1997; Kodric-Brown 1989; Houde and Torio 1992). But Grether (2000; see also Grether et al. 2001) showed that, contrary to the predictions of this hypothesis, female preference for orange spots does not covary with geographic variation in the availability of carotenoids. Subsequently, Rodd et al. (2002) demonstrated that the female preference for orange spots is a pleiotropic effect of a sensory bias for the color orange that evolved in the context of feeding on orange-colored fruits (see Noor 2000 for a more general discussion of mating preferences as pleiotropic by-products of adaptive evolution in other contexts).

34. Ryan et al. 1990; Ryan and Rand 1995. See also Ryan and Rand 1999 for a review.

35. But not if the chucks come at the expense of his "whining." *Physalaemus coloradorum* males, unlike *P. pustulosus* males, emit a series of whines in rapid succession, and *P. coloradorum* females prefer the repeated whines over a single whine.

And, in a victory for symmetry, *P. pustulosus* females also prefer the repeated whines, even though *P. pustulosus* males cannot produce them (Ryan and Rand 1995). From this finding it was concluded that, ultimately, there is a permissive bias for complexity (Kime et al. 1998).

36. Basolo 1990, 1995a,b.

37. Female songbirds often sing as well, however, and in the tropics duetting species are common. Duetting, as the term implies, involves highly coordinated vocal exchanges between male and female mates. In some duetters, the male still sings more than the female, but in others, there is something like equality. Species in which females sing as much as males are exceptions, not the rule, however (Farabaugh 1982).

38. The idea that songs can be divided into two categories, those that function in male-male competition and those that function as sexual advertisements to females, was proposed by Byers and Kroodsma (1992) and received partial support from Beecher et al. (2000). For a broad review of the relationship between birdsong and sexual selection see Searcy and Andersson 1986. For reviews slanted toward good-genes interpretations of the evidence see Catchpole 1996, 2000.

39. For studies of varying success in linking repertoire size to one or another good-genes mate choice hypotheses, see, for example, Searcy 1984; Catchpole 1987; Lampe and Saetre 1995; Hasselquist et al. 1996; Szekely et al. 1996; Buchanan et al. 1999; Nowicki et al. 2000; Duffy and Ball 2002. Of these, I find Hasselquist et al. (1996) the most compelling. But these studies notwithstanding, a broadly comparative study found that there is little association between versatility and mate choice among songbirds (Read and Weary 1992).

40. Versatility should not be confused with vocal output or the total amount of singing. Møller (1991) and Saino et al. (1997) provide evidence for a relationship between parasite load and singing behavior only for output, not for versatility.

41. Not surprisingly, the most ardent functionalists, such as Møller, put the most positive spin on the available evidence (see Møller et al. 2000). Read and Weary (1990) found little evidence for a relationship between versatility and parasitism, and in Hamilton's own remarkably dispassionate review of the evidence for the Hamilton-Zuk hypothesis (Hamilton and Poulin 1997), he found little support from studies on birdsong.

42. For example, one combining the immunocompetence handicap idea with fluctuating asymmetry.

43. Searcy 1992.

44. For a general account of habituation and desensitization see Petrinovich 1984. Krebs (1976) discusses these processes specifically in the context of birdsong. For an adaptationist interpretation of the vocal limitations of grackles see Gray and Hagelin 1996.

45. This is not to say that mockingbirds, by virtue of their monogamy, are not subject to any intersexual selection, only that it must be less than that experienced by a polygynous species, if only because the difference in the variance of male and female reproductive success is necessarily less in monogamous than in polygynous species.

46. The sort of inertia to which I point here is often referred to as "phylogenetic inertia," which is simply the particular sort of resistance to change shared by members of a particular genealogical chunk.

47. Read and Weary 1992.

48. There is, of course, substantial variation in versatility within all three families.

49. Sibley and Ahlquist 1985.

50. Marler and Peters 1988.

51. It is also sometimes referred to as the "critical period" (for example, Marler and Peters 1988), a term that is a holdover from traditional ethology, with its assumption of precise genetic control of the offset and the onset. The critical period is, however, not as critical as ethologists would like and can be significantly extended by various means (Baptista and Petrinovich 1984; Eales 1987). For this reason, the term *sensitive period* is more appropriate.

52. Repertoire size data for white-crowned sparrows are from Baptista and King 1980. For northern juncos, see Williams and MacRoberts 1978, and for song sparrows see Mulligan 1966.

53. This view emerged in large part from some classic experiments by Konishi (1964, 1965) in which he demonstrated that when birds are deafened after laying down the "template" but before the onset of the sensorimotor phase, they cannot produce crystallized song, because of their inability to perform this trial-and-error matching. Birds deafened subsequent to song crystallization produce normal song (but see Brainard and Doupe 2000).

54. The most celebrated dialectic-endowed species is the white-crowned sparrow (Marler and Tamura 1962; Baptista 1977, 1985).

55. Morton 1982; Payne 1982; Rothstein and Fleischer 1987.

56. See, for example, King and West 1983; West and King 1985, 1988.

57. The study of learning of this sort, and in particular the social context of song development, has come remarkably late in the study of birdsong. But the study of birdsong has traditionally been confined to ethologists, and ethologists are natavists first and foremost, and like their learning constrained (see Johnston 1987, 1998, for a critique of ethological notions of innateness). The socially induced song alterations of cowbirds are simply a form of instrumental learning or operant conditioning. But ethologists, either from ignorance or antipathy to classical psychology, have touted this as a new form of learning called "action-based learning" (Marler and Nelson 1993). Marler (1997) has allowed as how "action-based learning is analogous to operant conditioning," but that is much too weak. Action-based learning is analogous to operant conditioning in the same way that H_2O is analogous to water. The wheel has been reinvented here, and rather late. Adret's (1993a,b) use of standard operant techniques to induce song imitation suggests that instrumental learning may play an important role in imitating tutors as well.

58. Brenowitz and Kroodsma 1996.

59. The neural changes associated with the initial phases of song learning are similar in age-limited learners such as zebra finches and open-ended learners such as canaries, though the timing can be substantially different. For information on the timing of neural development in the song nuclei of an age-limited learner see, for example, Bottjer et al. 1985, 1986; Konishi and Akutagawa 1985; Nordeen and Nordeen 1988; Bottjer and Sengelaub 1989; Nordeen et al. 1989; Sohrabji et al. 1990. For studies on the timing of neural development in the song nuclei of an open-ended learner see, for example, Nottebohm et al. 1986; Goldman and Nottebohm 1983; Alvarez-Buylla et al. 1988, 1998; Kirn et al. 1991. After initial song learning, however, open-ended learners such as canaries continue to maintain a high degree of

neural plasticity (Alvarez-Buylla et al. 1988; Kirn et al. 1991), which is assumed to be related to their ability to learn new songs into adulthood.

60. The focus in neuroethology is on only one of Tinbergen's four projects, the one concerned with causation. (For a discussion of Tinbergen's four questions, see chapter 8.) Neuroethologists further confine themselves to investigating the neural mechanisms that cause particular sorts of behavior. Neuroethologists have been particularly successful in identifying the neural substrates for birdsong (see Nottebohm 1989; Konishi 1989, 1994; Brenowitz et al. 1997; Brenowitz and Kroodsma 1996).

61. Immelmann 1969; Eales 1985, 1987; Bohner 1990.

62. Arnold 1975.

63. Nottebohm et al. 1986.

64. Nottebohm (1981) originally emphasized how the seasonal changes in HVC volume seems to prepare canaries to acquire new songs and lose old ones at the appropriate times in the breeding cycle. Subsequent studies, however, suggest a less straightforward adaptationist interpretation. For instance, though Baker et al. (1984) found a lack of seasonal variation in HVC volume in the white-crowned sparrow, which is what you would expect in an age-limited learner, Tramontin and Brenowitz (1999) did find seasonal variation in the incorporation of new neurons into the song nuclei of another age-limited learner, the song sparrow (see also Tramontin et al. 2001). Why would age-limited learners that do not acquire new songs as adults show the same seasonal variation as canaries? Nottebohm's thesis certainly does not help us understand this similarity in the neurobiology of open-ended and age-limited learners. Moreover, even in canaries the linkage between the addition of new neurons to HVC and the acquisition of new vocal memories does not look to be as tight as Nottebohm suggested (Leitner et al. 2001).

65. Nottebohm et al. 1981.

66. Aside from Nottebohm's original study, others that identified a significant relationship between HVC and repertoire size were Canady et al. 1984; Ward et al. 1998; and Airey and DeVoogd 2000. On the other hand, Kirn et al. (1989), Brenowitz et al. (1991), Bernard et al. (1996), and MacDougall-Shackleton et al. (1998) all failed to identify an association between HVC size and repertoire size.

67. Canady et al. 1984; Kroodsma and Canady 1985.

68. Canady et al. 1984.

69. DeVoogd et al. 1993. DeVoogd was subsequently involved in another study of this sort, this time restricted to European warblers of the family Sylviidae (Szekely et al. 1996). The eight species investigated exhibited a wide range of repertoire sizes (measured as syllable numbers: Catchpole 1980, 1986; Catchpole et al. 1984). Members of the genus *Acrocephalus* are renowned songsters, whereas members of the genus *Locustella* sing some of the simplest songs in birddom. As we would expect, based on the results of the earlier study, *Acrocephalus* warblers had larger HVCs than *Locustella* warblers. None of the other song nuclei (RA, lMAN, and Area X) exhibited similar sorts of size variation.

70. Farabaugh 1982, Levin 1996.

71. Farabaugh 1982.

72. In addition, HVC neurons do not form synaptic connections with RA in zebra finch females (Nottebohm and Arnold 1976; Konishi and Akutagawa 1985).

73. Brenowitz et al. 1985; Arnold et al. 1986; Brenowitz and Arnold 1986.

74. Nealen and Perkel 2000.

75. It is also standard practice, given a brain-behavior correlation of this sort, to assume that the causal arrow points from brain to behavior. Fortunately, in recent years alternative causal scenario have been recognized, and in the case of the marsh wrens, tested. Brenowitz et al. (1995) distinguished three possibilities: (1) the standard view, that HVC size determines song repertoire size; (2) repertoire size determines the size of the HVC; or (3) the size of each is determined by an independent factor (say, estrogen level in juveniles). They tested scenario 2 by exposing lab-reared eastern marsh wrens to small or large tape-recorded marsh wren repertoires, which I will refer to as impoverished versus enriched conditions, respectively. They found no difference between the HVC sizes of the wrens reared under enriched conditions and those reared under impoverished conditions, though the song repertoires developed by the two groups matched their experimental conditions. And they reasonably concluded, with palpable relief, that it could not be the case that repertoire size determines HVC size. Scenario 2 had been falsified. Interestingly, however, there was a much greater proliferation of dendritic processes (and presumably synapse number) in the enriched wrens, though this apparently did not affect the gross volume of the HVC. This result raises a question that will become increasingly central: neurobiologically, what is the significance of these gross size measures of song nuclei, or of any other brain region for that matter? The significance of dendritic growth is well established in neurobiology. The meaning of gross size differences is much less clear. This question notwithstanding, these researchers should be commended for moving beyond the assumption that in brain-behavior associations, the brain is always the cause and behavior the effect.

76. Though there are no sex differences in the amount of singing or versatility in this duetting species, surprisingly, both the HVC and the RA are twice as large in males as in females. Furthermore, there are many more neurons in the nuclei of the males than in those of the females; and the soma size of the neurons in the HVC is much larger in males (Gahr et al. 1998). Female bush shrikes are able to produce vocal behavior identical to males, with much less brain space and fewer and smaller neurons in the HVC. Bush shrikes have a brain dimorphism waiting for a behavioral dimorphism.

Are the bush shrikes really as idiosyncratic as they seem at first glance? Gahr argued that they are not. He suggested that if those species that do not sing at all are excluded, there is little correlation between the magnitude of the sex difference in singing and sex differences in the size of the HVC. That would not be good news for those who want a simple deterministic relation between brain and behavior as it is related to sex. This analysis, however, leaves something to be desired, in that it did not take into account the confounding effects of taxonomic relationships, but rather treated each species as an independent data point, whether they were members of different families, the same families, or congeneric with other species included in the data set. The comparative analysis of sex differences by MacDougall-Shackleton and Ball (1999) is a significant improvement in this regard, and indeed the first analysis of this kind to explicitly incorporate the comparative techniques (independent contrasts) that have become standard in evolutionary biology (see, e.g., Harvey and Pagel 1991; Martins 2000). It is therefore noteworthy that MacDougall-Shackleton and Ball did find significant covariation between the sexual dimorphism in HVC volume and the dimorphism in vocal repertoire, though HVC volume accounted for less than half of the variance in the dimorphism in versatility.

77. A study by Airey et al. (2000) is one of the first attempts to directly link HVC size to mate choice in a songbird (the European sedge warbler), but the results are somewhat enigmatic. Although females preferred males with complex songs, and complex songs were correlated with larger HVCs, successful (at breeding) males did not have any larger HVCs than unsuccessful males.

78. Much of this research, as you might expect, concerns the effects of gonadal hormones on song nuclei at the cellular (for example, Kirn et al. 1991) and subcellular (for example, DeVoogd and Nottebohm 1981; Gurney and Konishi 1980) levels (see DeVoogd 1991 for a review). One of the most interesting findings from this research is that the song nuclei appear to have evolved hormonal controls on sex differentiation that differ from the mammalian-style hormonal controls found elsewhere in the body of songbirds (see Schlinger and Arnold 1991; Arnold and Schlinger 1993; Schlinger 1994, reviewed in Balthazart and Ball 1995). Even more interesting, much of the sexual differentiation of the song nuclei seems to be under genetic rather than hormonal control (Arnold et al. 1996; Arnold 1996, 1997; Schlinger 1998; Wade et al. 1999).

79. Read and Weary 1990, 1992.

80. In the discussion that follows I benefited greatly from a paper by Rebecca E. Irwin (1988). Irwin applies a developmental framework (in the spirit of Gould 1977) in explaining variation in versatility among songbirds. In particular, she suggests that differences between species such as sparrows that have small repertoires and those such as mockingbirds with large ones may reflect differences in the timing of developmental events related to song acquisition. The continuous singing and wide variety of song elements characteristic of plastic song in sparrows greatly resembles the adult song of mockingbirds. Irwin proposed that the progression from subsong to crystallized song is accelerated in sparrows, relative to wrens, and that this acceleration might account for the differences in repertoire size. She further suggested that repertoire size may be largely a function of the amount of time spent in plastic song. If it is, the time between the onset of plastic song and song crystallization should be shorter in sparrows than in thrushes or mockingbirds. And it is. Mockingbirds, it seems, never progress beyond plastic song to crystallized song.

81. Unfortunately, the state of songbird systematics is such that we cannot say with confidence whether age-limited learning is the primitive (baseline) condition or the evolutionarily derived and more recent condition. Using rather suspect logic, ethologists have tended to assume that age-limited learning is primitive, whereas mockingbird-like virtuosity requires a special selectionist explanation. It has been widely assumed that primitive songbirds evolved from ancestors that sang "innate" songs. On this view, songbirds have been progressively freed from genetic constraints on song learning to become more versatile singers, primarily as a result of sexual selection. Mockingbirds should then represent the apotheosis of songbird evolution. Research on the eastern phoebe (*Sayornis phoebe*) seems to support this view. Phoebes are not true songbirds (oscine); they lack the highly developed syrinx, for example. Phoebes are, however, closely related to songbirds. It is considered significant, therefore, that phoebes acquire their songs without any auditory feedback, and hence, without any learning. Deafened juvenile phoebes grow up to sing just as well

as their normal counterparts (Kroodsma and Konishi 1991). The "innate" song of the phoebes is to be expected on the standard view of songbird evolution. But other close relatives of the songbirds are some of the most versatile singers and open-ended learners on the earth. The lyrebird family (Menuridae) is composed of only two species, both confined to Australia. Lyrebirds are highly polygynous, and males are a combination of peacock and mockingbird. Like peacocks, they have extremely impressive feathery finery—albeit of a subdued palette by peacock standards—which are shown to maximum advantage in their characteristic courtship pose. And like mockingbirds, they sing with the versatility of a Las Vegas lounge-lizard. All of this to attract females, which, however, can expect nothing but sperm from their ardent wooers. Lyrebirds rival even the spectacular New Guinea birds-of-paradise in the intensity of intersexual selection. And they have the kind of vocal repertoire that would be expected if vocal repertoires are shaped by sexual selection. If lyrebirds, not phoebes, represent the baseline for songbird evolution, mockingbird versatility constitutes the ancestral condition. This is not to say that mockingbirds should then be considered more primitive than sparrows, only that complex song may be an ancestral trait. If mockingbird-like versatility is ancestral, there is little reason to invoke sexual selection to explain the current versatility of mockingbirds. Moreover, age-limited learning, not open-ended learning, would require the most explanatory work. Obviously, the phylogenetic story, whatever it turns out to be, should frame our explanations of repertoire sizes. Moreover, even given our current knowledge of phylogenetic relationships, it behooves us to attend to taxonomic patterns in our explanations of this attribute.

82. Bertram's (1970) monograph is on the vocal behavior of the Indian Hill mynah. Hindmarsh (1984, 1986) studied repertoire development, including mimetic elements, in the starling. The scientific name of the common starling (*Sturnus vulgrais*) indicates the low esteem in which it is held in its native Europe. Here, in North America, it is regarded even more dimly. Introduced in the in the 1890s, starlings have flourished in the New World (and in Australia as well) to the detriment of native hole-nesting species such as the bluebird. These villains thrive because, like their fellow Old World transplant, house sparrows (*Passer domesticus*), they tolerate well the company of humans. Unlike house sparrows, however, they take to relatively pristine habitats as well, which is why they pose such a threat to native birds.

83. Hindmarsh 1984.

84. Laskey 1944; Wildenthal 1965.

85. Adaptationists prefer to believe that evolution will inevitably eliminate such mistakes, but we would expect such mistakes to be eliminated by selection only if they impose a fitness-reducing cost. It is especially likely that such costs are minimal in birds with complex repertoires, in which case the details of the song will be much less important than they would be for, say, a sparrow.

86. For example, the very versatile marsh warbler (*Acrocephalus palustris*), mentioned earlier, incorporates mimetic bits into its songs (Dowsett-Lemaire 1979; see also Chisholm 1932 for examples of mimicry among lyrebirds and other Australian birds).

87. Mimicry is common in the versatile North American wrens, especially among the younger birds (Kroodsma and Byers 1998).

88. Marler and Peters 1982. Vernon (1973) also noted that juvenile mimicry was

common in many African songbirds that do not mimic as adults (see also Thorpe 1958; Eberhardt and Baptista 1977).

1. Rothstein and Robinson 1994; Trine et al. 1998.

2. Sherry et al. 1993; Reboreda et al. 1996. A similar sex difference was also observed in another species of cowbird (Astie et al. 1998) with similar parasitic habits.

3. A paper by Krebs et al. (1996) inspired this term. The participants in this program tend to refer to it as cognitive ecology (Healy and Braithwaite 2000). More recently, the term *neuroecology* has been promulgated (Bolhuis and Macphail 2001).

4. You could make a case that brain ecology constitutes a subdiscipline within evolutionary neurobiology, as I have done (Francis 1995). Since 1995, I have come to think of these endeavors as much less complementary than opposed.

5. Hendriks-Jansen (1996) provides a particularly cogent appreciation of ethology from the perspective of someone working in situated robotics.

6. The four questions were posed in Tinbergen 1963.

7. There has long been a minority of ethologists who emphasize development (for example, Bateson 1981a,b; ten Cate 1989). The question of phylogeny, however, was almost completely ignored after Lorenz's (1941, 1954) early efforts. Recently, the study of behavioral phylogeny has been revived (for example, Brooks and McLennan 1991; Martins 2000) as part of a broader reemphasis of historical factors in evolution.

8. It is now customary for behavioral ecologists to at least pay lip service to Mayr's "proximate causes" (for example, Krebs and Davies 1997). But even when they do so, the proximate explanations are always subordinated to the "ultimate" (teleological) explanations because of their adaptationist predilections.

9. Two of the earliest forays into optimal foraging were Krebs et al. 1977, 1978. There is now a vast literature on optimal foraging; the subject has become something of an exemplar (*sensu* Kuhn 1961) for behavioral ecology in general.

10. Gray's critique was all the more effective for being presented in a "low-key" manner. It was later published as Gray 1987.

11. Reviewed in VanderWall 1990.

12. Tomback 1980; VanderWall and Balda 1981; Sherry 1984; Stevens and Krebs 1984; Balda and Kamil 1989.

13. Tomback 1980.

14. VanderWall 1982; Balda and Turek 1984. See Balda and Kamil 1992 for a demonstration of the impressive memory of nutcrackers under controlled laboratory conditions.

15. The test was an operant task referred to as spatial non-matching-to-sample (SNMTS: Hampton and Shettleworth 1996b).

16. Clayton and Krebs 1994; Healy and Krebs 1995.

17. A number of studies comparing storers and nonstorers in the laboratory have failed to demonstrate the superiority of storers at spatial tasks (for example, Hilton

and Krebs 1990; Healy 1995; Hampton and Shettleworth 1999). Just as I was completing this manuscript, I came across an excellent review and detailed critique of the behavioral ecologists' approach to cognition by Macphail and Bolhuis (2001), in which they document numerous other failures to demonstrate that storers outperform nonstorers on spatial tasks. Although I agree with much of their analysis, I do not endorse their general process view of learning, which I consider the opposite extreme to the view of learning promoted by brain ecologists (as well as the evolutionary psychologists, whom I discuss in the next chapter).

18. The attempts to liken the avian hippocampus to the HVC include those of Krebs et al. (1989), Healy and Krebs (1993), and Clayton and Soha (1999).

19. Hampton et al. 1995; Healy and Krebs 1996. These comparisons are for hippocampal volume relative to telencephalon volume and body size. It should be noted that the bird homolog of the mammalian hippocampus (for example, Bingman et al. 1989; Erichsen et al. 1991) differs from it in important ways (Butler and Hodos 1996) and is usually referred to as the "hippocampal formation," rather than, simply, hippocampus. Nevertheless, in what follows I will refer to the avian homolog of the mammalian hippocampus, as hippocampus, rather than hippocampal formation, because for my purposes here, the difference does not make a difference.

20. Krebs et al. 1989; Sherry et al. 1989; Healy and Krebs 1992c; Basil et al. 1996.

21. Scoville and Milner 1957; Milner 1966.

22. For some idea of the diversity of opinion on hippocampal function, see Scoville and Milner 1957; O'Keefe and Nadel 1978; Rawlins 1985; Squire and Zola 1998; Squire 1992; Eichenbaum 1996; Wallenstein et al. 1998; Eichenbaum et al. 1999; Eldridge et al. 2000; Fortin et al. 2002.

23. This view was motivated in large part by the discovery of "place cells" in the rat hippocampus that fire when an animal is in a particular location (O'Keefe and Dostrovsky 1971). It was then suggested that these place cells constituted a Cartesian representation of the environment, each cell signaling a particular coordinate locus. Thus, the hippocampus could be construed as a cognitive map of the sort that Tolman (1948) envisioned (O'Keefe and Nadel 1978; see also O'Keefe and Burgess 1996; Nadel 1991). On the behavioral side, there is increasing evidence that simpler mechanisms, such as vector integration, can explain the results formerly attributed to cognitive mapping even in such mental athletes as nutcrackers (Gibson and Kamil 2001) and humans (Gibson et al. 2001).

24. See, for example, Cohen and Eichenbaum 1991; Eichenbaum et al. 1994; Zola-Morgan et al. 1994; Eichenbaum 2000. The message from these publications is that the hippocampus functions in a much broader category of mnemonic aptitudes than remembering spatial cues. The notion that the hippocampus functions exclusively as a cognitive map has been called into question even in rats. The discovery that place cells fire only when the animal is moving in a particular direction (Muller et al. 1994; see also Froehler and Duffy 2002) indicates that place cells are not simply place indicators. Moreover, it was subsequently discovered that place cell firing corresponds to salient aspects of space related to goal cues rather than to a conserved suite of spatial cues that define the environment (Gothard et al. 1996). Even later, it was discovered that place cells are influenced by the subject's recent experience (Frank et al. 2000). Finally, Hampson et al. (1999) identified particular

areas of the hippocampus that process nonspatial information. These studies all suggest that the hippocampus as cognitive map view is simplistic.

25. Jerison 1973. See Iwaniuk et al. (1999) for a dramatic counterexample to the principle of proper mass.

26. Gibson (2001) gives a nice account about how the bigger-is-better principle applies to the primate cerebral cortex.

27. See, for example, Jacobs et al. 1990; Sherry et al. 1992.

28. The influence of genome size on the optic tectum in amphibians is particularly instructive here. The optic tectum is an obvious feature of the brains of all vertebrates. Output from the tectum controls visually mediated movements directed toward specific locations in space. For example, projections from the tectum to the spinal cord are involved in orienting the head and body to important stimuli. Among amphibians (frogs and salamanders) the tectum plays a particularly important role in the capture of insect prey by means of precisely directed projections of the tongue. There is marked variation in the development of these lingual skills among amphibians; the most acrobatic tongue thrusters belong to a family of lungless salamanders (Plethodontidae) and, in particular, a speciose Neotropical subgroup known as the bollitoglossines (Roth 1987; Wake 1987). Roth and his collaborators have shown that tectal morphology is largely a function of a species' genome size, not functional demands on the tectum (Roth et al. 1988; Roth et al. 1990 a,b; Roth et al. 1994). The larger the genome, the larger the cells in the tectum (and elsewhere in the body).

29. The same lack of impressive brain volumes is true of other men of eminence as well. Franz Joseph Gall, one of the founders of phrenology, had a particularly meager brain, and neither the phrenologist Paul Broca (who was particularly concerned to demonstrate the correlation between brain size and intelligence) nor the mathematician Karl Friedrich Gauss weigh in with particularly impressive results (Gould 1981). More recently, Schoenemann et al. (2000) found no correlation between brain size and intelligence within families.

30. In fact, Hampson et al. (1999) found anatomical specializations within the hippocampus for nonspatial representations, and Zhao et al. (2001) observed distinct transcriptional boundaries, indicating functional differentiation, within the hippocampus (though see Kim and Baxter 2001 for evidence of nonmodular organization in this structure). The best studies on the relation between spatial memory and hippocampal function take this complexity into account (for example, Schwegler et al. 1988, 1991). There has been some effort among brain ecologists to move beyond gross volume measurements (for example, Montagnese et al. 1993; Lavenex et al. 2000), but in general the brain ecology investigations of the hippocampus lag far behind, for example, the evolutionary neurobiology investigations of the optic tectum by Roth and his colleagues, with respect to their neurobiological sophistication.

31. In this study, hippocampal volume varied more with genealogy than with ecology (Balda and Kamil 1998). The study included four jay species three of which the casual observer might find difficult to tell apart: the pinyon jay (*Gymnorhinus cyanocephalus*), the scrub jay (*Aphelocoma coerulescens*), and the Mexican jay (*Aphelocoma ultramarina*). These three species are members of a New World lineage of corvids. The fourth species, Clark's nutcracker, is the only member of a Eurasian lineage to make it to the New World. All four species depend on stored pinyon pine

seeds to some degree. Pinyon jays and Clark's nutcrackers are the most dependent on stored seeds for their survival. They also perform much better than the Mexican and scrub jays at recovering hidden food in the laboratory. A brain ecologist would expect therefore that pinyon jays and Clark's nutcrackers would have a larger hippocampus than that of the other two jays. As predicted, the Clark's nutcracker has a relatively large hippocampus; but the hippocampus of pinyon jays is the same size as those of the Mexican jay and the scrub jay, to which it is more closely related. A subsequent comparative study proved even messier from the perspective of brain ecologists. Gould-Beierle (2000) found that pinyon jays and scrub jays (from the New World lineage) outperformed nutcrackers and jackdaws (from the Old World lineage) on a laboratory spatial memory task (radial maze). This outcome makes no sense from a perspective of brain ecology because jackdaws, unlike the other three corvids, do not cache food at all. These results are consistent with the phylogenetic relationships of these species, however.

32. See Healy and Krebs (1992b) for a comparison of marsh tits and blue tits, and Healy and Krebs 1992a for a comparison of marsh tits and great tits. In an attempt to put the best face on these results, the authors suggested that hippocampus size is more closely related to the amount of stored "information" than to the duration of memories.

33. Healy and Suhonen 1996.

34. Again, the authors exerted much effort to put the best spin on these results in their discussion section (Healy and Suhonen 1996, pp. 77–78).

35. See Moore et al. (1992) for a wonderful experiment demonstrating how maternal licking stimulates motor neuron development in the spinal cord of rats. A couple of examples closer to home may also help make this point. Paulescu et al. (2000) demonstrated that different parts of the brain are activated in English speakers than in Italian speakers when listening to their own language. This is a clear case of a culturally induced physical alteration in the brain. More generally, the highly specific response capacities of particular subsets of neurons (that some would call modules) are not genetically determined but rather the result of experiences that derive from interactions with the environment (Johnson et al. 2000).

36. For example, Woodruff-Pak et al. (2000) demonstrated that classical conditioning of the eye-blink reflex results in increased volume of the cerebellum of humans. With respect to the hippocampus in particular, Maguire et al. (2000) reported that taxi drivers experience enlargement of the hippocampus as a result of their job requirements. Moreover, Clayton herself (1998) reported increasing evidence of experiential effects on the hippocampal volume in food-storing birds.

37. This distinction was first formulated by Greenough et al. (1987).

38. Many other examples exist in birds of particular organs or tissues that grow in advance of their need (Piersma and Lindstrom 1997). For example, flight muscles hypertrophy (Gaunt et al. 1990) and fat stores increase (Weber and Piersma 1996) before migration.

39. See, for example Clayton and Krebs 1994a,b; Healy and Clayton 1994; Clayton 1995, 1996. Similar results were obtained in a study on corvids (jackdaws and European magpies: Healy and Krebs 1993).

40. Clayton 1995a,b.

41. Clayton and Krebs 1994; Clayton 1996.

42. Clayton and Krebs 1994a; Clayton 1995b.

43. Clayton 1996. Clayton subsequently backed off somewhat from her endorsement of the hippocampus-as-gonad view (Clayton and Soha 1999), arguing instead that the hippocampus has some gonad-like characteristics and some muscle-like characteristics. In essence, she now argues, the pattern of hippocampal growth indicates that food storers are primed to learn spatial cues.

44. Clayton's (1996, p. 157) argument for the experience-expectant (hippocampus-as-gonad) view largely rests on the fact that there is no statistically significant difference in the hippocampal volumes of experienced birds sacrificed after 3, 5, or 8 trials. That is not much of a spread, however, especially given the small sample sizes. What happens to the hippocampus after 50 trials? She also notes that the increase in hippocampal volume occurs before day 44, when there is a dramatic upsurge of storing in these birds. The implication is that the hippocampal growth anticipates the memory demands. But, at best, this is extremely weak evidence, even for her highly hedged conclusions.

45. Again, it is instructive to use the HVC as a standard in evaluating a putative adaptation in the avian hippocampus, this time for experience-expectance. Seasonal changes in HVC actually precede experience-expectant seasonal changes in the gonads (Tramontin et al. 2001), though the functional significance of this seasonal variation is somewhat murky, as I discussed in chapter 7. Clayton does not come close to demonstrating that the hippocampus is HVC-like in this regard. Interestingly, there are purported to be seasonal changes in the hippocampus of songbirds as well (see, for example, Barnea and Nottebohm 1994; Smulders et al. 1995; Smulders and Dhondt 1997), including cowbirds (Clayton et al. 1997). Much has been made about the fact that, among parids, food storing reaches its peak in late autumn, which also happens to be the time of year when the hippocampus is largest in black-capped chickadees. It should be noted, however, that for all we know, the same occurs in non-foodstoring species. Moreover, Lavenex et al. (2000) showed that the seasonal changes observed in these studies were actually age-related developmental differences, not evidence of adult plasticity under the influence of day length. This conclusion accords with the results of Krebs et al. (1995), who observed an effect of photoperiod on food storing but not on hippocampal volume. So again there is little in the way of experience-expectance in the hippocampus.

46. For example, Clayton (1996, p. 157) states that "*one interpretation* of these results is that the one or two seeds stored before day 44 *may* have been sufficient to stimulate the growth of the hippocampus" [emphases added]. Yes, that's one interpretation, and the most congenial to brain ecologists, but certainly not the most obvious. See also Clayton (1995a, p. 2806), where she concludes that, "these results *might* be taken to *suggest* that plasticity in hippocampal volume is an anatomical specialization of the food storer's brain." Yes, and those same results *might* be taken to *suggest* any one of a number of other conclusions as well. What these results do not suggest is a compelling reason to mention this particular one of the many things they "might be taken to suggest."

47. Clayton 1995a.

48. Evidence that the salient adaptation in food storers is related more to the motivation to store food than to innate differences in the hippocampus comes from the brain ecologists themselves. For example, Clayton and Dickinson (1999) found

that the food-storing motivation is independent of hunger in scrub jays. Also of note is the study of Pravosudov and Clayton (2000), in which they showed that mountain chickadees reared under food-deprived conditions, as a result of which they were more motivated to store food, showed enhanced spatial ability relative to mountain chickadees fed ad lib.

49. This is where you would typically get the ecumenical argument about a lack of conflict between proximate (learning to attend to spatial cues) and ultimate (selection for increased hippocampal size) causes. But this argument misses the essential point that if the differences in hippocampal volume result from generic processes found in, say, all birds, the differences observed in the species under consideration are unselected byproducts of behavioral differences; and if the differences in hippocampal volume are unselected byproducts of behavioral differences in the propensity to store food, the brain ecologists, as opposed to the behavioral ecologists, have no story to tell.

50. Gaulin and Fitzgerald 1986, 1988, 1989.

51. Gaulin and Fitzgerald 1986.

52. There is another obvious prediction that, unfortunately, has not been tested: male meadow voles should do better in these maze learning trials than male pine voles.

53. Jacobs et al. 1990.

54. For a fascinating account of kangaroo rat kidneys, see Schmidt-Nielsen 1964.

55. Merriam's kangaroo rats also tend to shift burrow locations and use trails less frequently than bannertail kangaroo rats.

56. Jacobs and Spencer 1994.

57. It should be noted, however, that the largest hippocampus among bannertail kangaroo rats belonged to a female, an indication, certainly, that these sex differences are not terribly robust, a finding that should have provoked more consternation than the authors were ready to acknowledge.

58. Reboreda et al. 1996.

59. For a genetic—as opposed to developmentalist—version of intergender hitch-hiking, see Lande 1980. Lande sees genetic correlations between the sexes as a constraint on sexual selection's ability to cause adaptive intersex divergence. Reeve and Fairbairn (2001), on the other hand, argue that "genetic correlations" little constrain sexual selection, at least with respect to body size.

60. This isn't the whole story. With respect to body size, Rhen (2000) showed that, theoretically, selection on the larger sex can increase the frequency of sex-limited alleles and hence increase size dimorphisms, under some conditions. But he provided no evidence that such conditions are common.

61. If you browse through any field guide for birds, you will notice all manner of gradations in the degree of sexual differences in coloration (sexual dichromatism), even among species with highly colorful males. These differences in the sex differences largely reflect the degree to which the coloration affects female fitness.

62. Red-winged blackbirds provide an excellent example of intergender hitch-hiking. Male redwings use their colorful epaulets in territorial displays. But female redwings have epaulets too, though they don't use them in territorial displays. In fact, female epaulets have no function at all; they exist solely because of genetic correlations between males and females and because there was no selection against

redwing females with epaulets (Muma and Weatherhead 1989). Kimball and Ligon (1999) argue that, among birds, bright colors in both sexes should be considered the ancestral condition and that sex-limited (usually to females) cryptic coloration subsequently evolved in some species. This scenario is consistent with my thesis that unless what's good for the goose is bad for the gander, the sexes will tend to resemble each other even if only one sex benefits.

63. Antlers and horns as products of sexual selection are discussed by Henshaw (1971) and Lincoln (1994).

64. Females could also evolve antlers if they came to serve another function. For example, in caribou (reindeer), they function in female-female competition for food (Lincoln 1994).

65. Caribou (called reindeer in Europe) are the only members of the deer family in which females possess antlers. On the other hand, horned females are the rule among cattle, goats, and sheep. And even among antelopes, there are more species in which females have horns than species in which females lack horns. For example, according to my tally of African antelopes (from the Collins Field Guide to the Mammals of Africa, Haltenorth and Diller 1988) there are 39 species in which females have horns and 29 species in which females lack horns.

66. Potential female benefits from horns need to be considered too. Females of several species of oryx, for example, use their rapier-like horns to ward off predators such as spotted hyenas and lions, and their horns are nearly as large as those of their male counterparts. The point remains, however, that whatever those benefits, they would never have evolved without sexual selection on males.

67. It is not at all apparent to me that nest parasitism should result in an increased capacity for spatial learning. The argument here is much less compelling than that used by the brain ecologists who study voles or food-storing birds.

68. Gaulin (1995; see also Jacobs 1996) is typical in his offhand comment that the cost of increased hippocampal volume must be energetic. This argument is more explicit in those attempts to suggest selection on hippocampal volume by demonstrating seasonal variation in its size along the lines of the HVC in songbirds (Smulders and Dhondt 1997). Neural tissue is energetically expensive (Aiello et al. 2001), but no more so than digestive tissue, heart tissue, or liver tissue; and there is no correlation between the amount of neural tissue and basal metabolic rate (BMR) among mammals (MacNab and Eisenberg 1989; Martin 1996). Even evidence that the extremely large human brain requires extra energy relative to other mammals is extremely hard to come by (Aiello and Wheeler 1995). Perhaps birds are more constrained than mammals with respect to additions of neural tissue because of flight. But in lieu of any actual experimental data, talk is cheap when it comes to the energy costs of neural tissue, especially with respect to specific brain parts.

69. Clayton, who actually uses the phrase "use it or lose it" in the title of one of her research reports (Clayton 1995b), does not seem to appreciate this implication.

70. See, for example, Goodfellow and Lovell-Badge 1993.

71. For an overview of androgen receptors see Lamb et al. 2001.

72. McEwen et al. 1977. For an overview of aromatase, see Simpson et al. 2002.

73. Lamb et al. 2001.

74. Gaulin (for example, 1992, 1995) is something of an exception among brain

ecologists in recognizing that intergender hitchhiking must be overcome if sexual dimorphisms are to evolve.

75. Among brain ecologists, Sara Shettleworth has shown the most interest in grounding the enterprise in how-considerations, though in her case it is how-psychology rather than how-biology (see Shettleworth 2001). Alan Kamil, in some moods, is also more inclined than the typical brain ecologist toward at least an inclusive perspective with respect to animal learning (see Kamil 1988). Both Shettleworth and Kamil, however, remain far too committed to teleological adaptationism from my perspective.

Notes for Chapter 9
Why Men Won't Ask for Directions

1. See, for example, Lee and Devore 1968; Lovejoy 1981; Tooby and Devore 1987; Silverman et al. 2000. The "man the hunter" part of the hunter-gatherer story was originally stimulated by the discovery that australopithicines consumed African game (Dart 1953). But there are certainly dissenters with respect to the importance of hunting in early hominids. For example, Blumenschine and Cavallo (1992) argue that scavenging, not hunting, was the most important source of protein. There is even reason to question whether the savanna was in fact where early hominids evolved. A long derided alternative hypothesis, first proposed by Hardy (1960) and elaborated upon by Morgan (1982), that hominids evolved in an aquatic or aquaboreal habitat, resembling current mangrove environments, and subsisted primarily on shellfish, has recently been revived by Verhaegen et al. (2002) on the basis of fossil evidence. If true, this would definitely undermine the entire hunter-gatherer edifice upon which so much adaptationist speculation about human behavior is premised.

2. For a sample of what impressive fabrications adaptationists have woven from the story of Fred and Barney, see Buss (1994) on the origins of human lust; Thornhill and Thornhill 1983, 1992 on two different versions on rape as an adaptive strategy; Daly and Wilson 1988 on our propensity to kill each other; Pinker 1994 on the evolution of language; Miller 2001 on the evolutionary rationale for art; and Gallup 1995 on the adaptiveness of homophobia. But my personal favorite is a book (Quinnett 1996) that explains, on the basis of adaptationist logic, the allure of fishing—and it has nothing to do with aquatic ancestry.

3. See, for example Tiger and Fox (1971) and Eibl-Eibesfeldt (1971) for examples of human ethology.

4. Perhaps as a result of some oedipal dynamic, the evolutionary psychologists adopted a surprisingly (given their common adaptationist ideology) competitive attitude toward their sociobiological forebears (for example, Symons 1989). The sociobiologists took exception to this impudence (for example, MacDonald 1991). The tensions between the two camps erupted into a spirited exchange, to which one entire volume of the journal *Ethology and Sociobiology* was devoted. The contributors from the evolutionary psychology camp included Symons (1990), Barkow (1990), and Tooby and Cosmides (1990). (John Tooby and Leda Cosmides played particularly large roles

in the development of the evolutionary psychology framework.) In addition to Richard Alexander (1990), the sociobiology camp was represented by Turke (1990), among others of his followers. (See also Betzig 1989 and Smuts 1991.) More recently, followers of Alexander, Smith et al. (2001a), have sought to differentiate the Alexander version of sociobiology from that of the evolutionary psychologists.

5. For a good introduction to evolutionary psychology, see the collection of essays in Barkow et al. 1992. Other representative work includes Cosmides 1989; Tooby and Cosmides 1989, 1990, 1992; Cosmides and Tooby 1992. Buss's works (1995, 1999) are good introductions to the evolutionary psychology of psychological sex differences.

6. Geary 1996. See also Geary (1999b) and Geary et al. (2000). Geary (1998) also discussed sex differences in math, along with a host of other psychological sex differences. See also Joseph (2000) for an attempt to relate sex differences in spatial reasoning to the story of Fred and Barney. The idea that spatial aptitudes are directly related to mathematical performance is controversial (for example, Manger and Eik-keland 1998), but for the sake of argument, I will assume that Geary is right here.

7. The idea that male homosexuality is adaptive was first broached by Wilson (1975); he suggested that male homosexuality was promulgated through kin selection because homosexual males help raise their relatives' offspring.

8. It has become increasingly important to anyone who would biologize cognitive sex differences to demonstrate differences between homosexuals and heterosexuals as well. To this end, Hall and Kimura (1995) reported that, on a spatial-motor targeting task that favored men, homosexual men performed no better than women. Moreover, Wegesin (1998b) reported that homosexual men performed no better than women in a mental rotation test. Gladue and Bailey (1995), however, could find no difference between homosexual and heterosexual men in visual/spatial reasoning tasks. Hall and Kimura (1995) and Kimura (1996) suggest that homosexual men may show a male-typical or female-typical cognitive profile depending on the nature of the cognitive task. For reasons that will become clear later in this chapter, I do not put much stock in the conclusions of these studies.

9. This is true of mammals but not of fish, as I discussed in chapter 6.

10. DiPietro 1981; Eaton and Enns 1986; Maccoby 1988. There are also sex differences in the choice of play objects and activities (for example, Berenbaum and Hines 1992). There is considerable overlap in choice, however, and the range of differences in each sex generally exceeds the mean between-sex differences.

11. I include humans and bonobos in the category of notable exceptions.

12. Gorski et al. 1978.

13. A similar dimorphism has been observed in other rodents (Commins and Yahr 1984; Hines et al. 1985) and ferrets (Tobet et al. 1986).

14. Lesions of the SDN produce little in the way of behavioral changes (Arendash and Gorski 1983; DeJonge et al. 1989).

15. The most extensive research in this regard has been conducted on humans (Swaab and Fliers 1985; Swaab and Hofman 1988; Allen et al. 1989), in which the SDN has been subdivided into four areas referred to as the interstitial nuclei of the anterior hypothalamus (INAH-1–4). The most reliable sex difference among these four nuclei is in INAH-3, which is larger in human males than females (Byne et al. 2001).

16. Breedlove and Arnold 1981; Forger and Breedlove 1986; Forger et al. 1992.

17. Forger and Breedlove 1986, 1987.

18. This point is made quite effectively by Breedlove (1994).

19. Swaab et al. 1985; Swaab et al. 1992; Swaab and Hofman 1995; Wegesin 1998; Gur et al. 1999; Duff and Hampson 2001; Swaab et al. 2001.

20. See, for example, Aguirre et al. 1996. There are exceptions: O'Keefe has been keen to demonstrate hippocampal involvement in human navigation and more generally the similarities between humans and nonhuman mammals with respect to the neural substrates of navigation (for example, O'Keefe et al. 1998; Maguire et al. 1999). Others have implicated the hippocampus as one of several components of the neural substrate for navigation in humans (Sandstrom et al. 1998; Gron et al. 2000).

21. Falk (2001) takes the prize for the most globalizing view of neural sex differences as they are related to any one of a number of psychological sex differences, including spatial cognition. Falk builds on the work of Gur et al. (1999) on sex differences in the proportion of gray and white matter, among other tidbits, to reach the conclusion that male and female brains are wired completely differently. This is the neurobiological equivalent of the women-are-from-Venus-men-are-from-Mars psychology.

22. Brain lateralization studies have become something of a subdiscipline within cognitive neurobiology since the celebrated split-brain solution to severe epilepsy, a significant improvement on the temporal lobotomy experienced by H. M. (see chapter 8). Subsequent studies on split-brained patients demonstrated that language was processed primarily in the left hemisphere and spatial information in the right, at least among right-handers (Sperry and Gazzaniga 1967). Subsequently, there has been an explosion of much less reputable research, publicized in popular journals such as *Psychology Today*, that expanded the lateralization thesis to include virtually every psychological category, including emotions, musicality, and others.

23. In the inflationary phase of lateralization research, it was inevitable that some would look for sex differences in lateralization. It had been observed that in stroke victims it is easier to predict the neuropsychological deficit in men than in women, if you know whether the lesion is on the right or left side (reviewed in Levy and Heller 1992). This finding suggested to some that a sex difference in lateralization might somehow explain the putative sex differences in spatial reasoning favoring males and in verbal reasoning favoring females. The resulting research has produced massively contradictory results (see McGlone 1980 for a review). For example, whereas Potter and Graves (1988) reported better interhemispheric transfer in women and left-handed men, and Kulynych et al. (1994) found more asymmetry in language lateralization in men, and Georgopoulos et al. (2001) reported sex differences in lateralization of spatial object construction, Frost et al. (1999) found no sex difference in language lateralization, and Gur et al. (2000) reported that women are more, not less, lateralized for spatial tasks. More fundamentally, given the amount of research effort devoted to this endeavor, the evidence for sex differences in any form of psychological lateralization is remarkably thin. The confusing results from the study of Voyer and Bryden (1990) epitomize this research area. But this dearth of convincing or clarifying evidence has not deterred others from pressing on. For example, Crucian and Berenbaum (1998) looked for but could not find a negative correlation

between spatial reasoning and emotionality—an outcome they had expected on the basis of their understanding of sex differences in brain lateralization. Russo et al. (1999) found only very weak evidence that women are less lateralized than men, and Beaton (1997) concluded from their study that the evidence was inconclusive. Worse, Eviatar et al. (1997) could find no evidence at all for a sex difference in asymmetry, nor could Ueker and Obrzut (1993; see also Hoffman and Polich 1999). These results prompted Hamberg (2000) to conclude that the hypothesis of sex differences in lateralization has been refuted.

24. Much of this research was motivated by the grandiose theory of brain development of Geschwind and Galaburda (1987), in which sex differences in brain lateralization played a central theoretical role, and testosterone levels functioned as the primary independent variable. It was recognized early on that, among its other defects, this grand unified theory of neurobiology was completely untestable, unlike its counterparts from physics (see McManus and Bryden 1991, for an unsuccessful attempt to rectify this problem).

25. The largest bundle of fibers connecting the left and right hemispheres is the corpus callosum (CC), so the report by de Lacoste-Utamsing and Holloway (1982; see also Clarke et al. 1989; Allen et al. 1991; Oka et al. 1999) of a sex difference in this structure was greeted with considerable excitement. They reported that the width of the posterior portion of the CC, known as the splenium, is wider in women than in men. But research on sex differences in the CC, which has burgeoned with the advent of new imaging techniques such as functional magnetic resonance imaging (fMRI) and positron emission tomography (PET), is rife with negative results (for example, Bell and Variend 1985; Byne et al. 1988; de Courten-Myers 1999). Such conflicting results indicate that any sex difference in this structure is subtle at best. Another smaller fiber bundle, known as the anterior commissure (AC), connects the two hemispheres and is reported to be larger, relative to body weight, in women than in men (Allen and Gorski 1986, 1991, 1992). But, as with the CC, it is unclear what the etiology of this sex difference is and what its functional significance is. With respect to etiology, an early report that a sex difference in the CC was present at birth (de Lacoste et al. 1986) has never been replicated (Clarke et al. 1989; Allen et al. 1991; see also Bell and Variend 1985). The preponderance of the evidence indicates that this sex difference is not congenital (Breedlove 1994), nor has it been connected to hormonal factors. And what about function? Researchers who see the size differences in these two bundles as functionally significant assume that a larger CC or AC implies more axons communicating between the two hemispheres, which results in less functional specialization on each side. However sex differences in one part of the CC do not necessarily indicate a difference in the number of axons (Juraska and Kopcik 1988; Kopcik et al. 1992). Moreover, it may well be the case that hemispheric specializations would require more sharing of information and hence more axons (Breedlove 1994).

26. Ellis (1894).

27. Woolley 1910, p. 340.

28. See Shields 1975 for a critique of functionalist approaches to explaining psychological sex differences in general.

29. See, for example, Eagly (1995) and Richardson (1997).

30. Hyde and Plant 1995.

31. Maccoby and Jacklin (1974).

32. See Halpern 1997, for a recent example.

33. Silverman and Eals 1992, p. 533.

34. See, for example, Maccoby and Jacklin 1974; Caplan et al. 1985; Jacobs and Eccles 1985; Linn and Petersen 1985; Halpern 1986; Sanders et al. 1986; Hyde 1990; Hyde et al. 1990; Halpern 1992; Alyman and Peters 1993; Masters and Sanders 1993; Voyer et al. 1995; Caplan and Caplan 1997.

35. Which is an indication that the category "spatial ability" is nebulous and ill-defined (Caplan et al. 1985; Voyer et al. 1995). See Linn and Petersen (1985) for an unsuccessful attempt to classify and partition spatial abilities.

36. For example, the Porteus maze test was the basis for most of the earliest claims for sex differences in spatial cognition (Porteus 1965; Harris 1978); these studies have now been thoroughly discredited on the grounds of poor statistical analyses (Caplan et al. 1985). Similarly, robust sex differences on two other standard tests, the Rod and Frame test and the Piaget water level test, have also been discredited (for example, Hyde and McKinley 1997; Heller et al. 1999). More common has been the dramatic down-sizing of claims for sex differences in the face of subsequent contradictory results. This has led some researchers to conduct meta-analyses on previous studies, which themselves have led to contradictory claims (Linn and Petersen 1985; Hyde 1990; Voyer et al. 1995; Hyde and McKinley 1997).

37. Shepard and Metzler 1971; Vandenberg and Kuse 1978. There are disputes about whether these tests indicate sex differences in "innate" spatial abilities or sex differences in the response to the test situation and/or performance factors, such as whether the test is timed (Goldstein et al. 1990; Voyer 1997; Masters 1998).

38. Linn and Petersen 1985.

39. Clutton-Brock et al. 1977; Cheverud et al. 1985; Fleagle et al. 1980; McHenry 1992.

40. Actually, the situation may be more analogous to a bird species in which a subset of the female population does not differ from the males at all. For in some studies, it was reported that, although males did better than most females on mental rotation tests, there is a subset—right-handed women with non–right handed relatives—that "scores" just as well as the males (Casey and Brabeck 1989, 1990; Pezaris and Casey 1991; Casey et al. 1992). This result is a real puzzler from an adaptationist perspective.

41. Kolakowski and Malina 1974; Jardine and Martin 1983.

42. Kolakowski and Malina 1974.

43. Hall and Kimura 1995.

44. What is perhaps more surprising than this amalgamation is the dearth of studies on the relationship between mental rotations and navigation or wayfinding, which is critical to Geary's thesis. Two studies that do report such a correlation are those of Silverman et al. (2000) and Saucier et al. (2002). But Malinowski (2001), found that women navigated as well as men. Other researchers have looked for and found sex differences in navigation, without relating them to sex differences in mental rotations, through a variety of virtual maze tests (Astur et al. 1998; Moffat et al. 1998). Such sex differences are often explained in terms of a male superiority in processing Euclidean information, the relation of which to mental rotations has yet to be established. The idea that males are superior navigators given only Euclidean

cues is an old one, as is the idea that women primarily use landmarks to navigate whereas males use geometry (see O'Keefe and Nadel 1978, who hypothesize sex differences in the amount of G, for geometry, and L, for landmark, cells in the hippocampus). Sandstrom et al. (1998) found evidence that males and females do indeed use different cues, but Galea and Kimura (1993) reported that although males performed better than females given only Euclidean cues, there was no sex difference with respect to the use of landmarks versus Euclidean cues. On the other hand, Gillner and Mallot (1998) found no evidence that either sex constructed cognitive maps of any sort, whether based on landmarks or on Euclidean cues. They proposed rather that humans use what is known as "vector integration" based on local views and movements (see also Gibson 2001).

45. Silverman and Eals 1992.

46. Eals and Silverman 1994. See also McBurney et al. (1997) for an analogous test. Postma et al. (1998), however, reported a male advantage in a similar test using object arrays on a computer screen.

47. There is also evidence of a difference in the 2D:4D ratio among homosexual and heterosexual women (Williams et al. 2000), suggesting a role for early hormonal milieu. Another nonfunctional sex difference that may also be related to the prenatal hormonal environment is in dermatoglyphic characters (whorls and ridges) of the sort used in fingerprinting (Bener 1979).

48. Ghiselin (1996) and Frost (1998) both point to this confusion by Geary in his (1996) target article. (See Temeles et al. 2000, for a good example of a sexual dimorphism unrelated to sexual selection.) Geary uses a particularly poor example of a sex difference due to sexual selection in his (1999a) attempt to educate psychologists on the matter. In reference to an illustration of a male oryx (an African antelope), he quite mistakenly asserts that female oryx lack horns. In fact, however, the horns of the female oryx are actually somewhat longer, albeit thinner, than those of the males—an example of the general rule described in chapter 8, that, among horned beasts, both sexes tend to have these adornments, in contrast to the antlered members of the deer family. This is an example of why psychologists should not learn their evolution from evolutionary psychologists.

49. This logic reveals just how shallow are evolutionary psychology's roots in mainstream evolutionary biology. Every species on the earth, not just the human one, has experienced a unique selective milieu. But no evolutionary biologist would conclude from this fact that there is no value in comparative analyses. In their radical separation of humans from the rest of nature, evolutionary psychologists have converged on the most extreme social science critics of sociobiology.

50. This term comes from Gould and Lewontin's (1979) influential critique of adaptationism.

51. Benbow and Benbow 1984; Benbow 1988. But see Tomizuka and Tobias (1982) for a different interpretation.

52. Hyde et al. (1990, p. 149) obtained an overall effect size (averaged over all samples) of 0.15. Benbow and Stanley's (1980) widely cited report of somewhat larger gender differences on the SAT (0.39 in 1979) was from samples of mathematically gifted individuals (see also Benbow and Stanley 1983, and Benbow 1988) that cannot be generalized to the population as a whole. Their male advantage at the high end is consistent with a female advantage overall (Becker and Hedges

1984). Moreover, substantial sampling problems exist in the Benbow and Stanley studies, given the way their subject pools were assembled (Hyde and McKinley 1997). Unfortunately, the results of that study have had adverse consequences for potential female mathematicians, not least because their own parents tended to rate their aptitudes lower as a result of the publicity surrounding it (Jacobs and Eccles 1985).

53. Kimball 1989; Wainer and Steinberg 1992.

54. This is indeed the conclusion of Wainer and Steinberg (1992).

55. Gallagher et al. (2000) make a similar argument about the reason for the disparity between standardized test scores and classroom achievement in females.

56. Stanley and Stumpf 1996.

57. Actually, Geary (1999a) did take a stab at the sex differences in scores on standardized tests of political science knowledge (p. 276). He calls the male advantage in the political sciences "intriguing," and consistent with his hunter-gatherer perspective. He argues that "males are more interested in political power than females" across all cultures, and therefore the greater male interest in political science and history "probably reflects the male concern for dominance and power."

58. According to one survey (Ethington 1990), there was no area of mathematics evidencing a clear male superiority among 13-year-olds across countries. In another study of 13-year-olds (Hanna 1989), sex differences in mathematical achievement were reported to vary widely from country to country; in most countries, there was no male advantage at all. Even in Germany, the sex difference among high achievers is smaller than in the United States (Wagner and Zimmerman 1986).

59. Kimball 1989; Hyde et al. 1990.

60. Among African Americans and Hispanics, girls outperform boys on standardized tests (Schratz 1978). In a meta-analysis that excluded SAT data, Hyde et al. (1990) found no sex differences in scores for African Americans, Hispanics, or Asians. In a Hawaiian study, girls outperformed boys in all ethnic groups, but the female advantage was lowest among Caucasians (Brandon et al. 1987).

61. Feingold 1988; Hyde et al. 1990. The same trend has also been observed for mental rotations (Stumpf and Klieme 1989; Richardson 1994), but see Masters and Sanders (1993) for contrary results.

62. Before the current push to link sex differences in cognition to the organizing effects of gonadal steroids, the emphasis was on genetic differences. Stafford (1961) was the first to explicitly propose that sex differences in cognition are X-linked—that is, the result of dosage differences between the sexes due to genes on the X chromosome, for which females have two copies and males only one (see also Yen 1975). This is the explanation for sex differences in the most common form of color-blindness, for example, but it makes no sense in this context and the idea is widely considered to have been debunked (for example, Boles 1980; but see Stavnezer et al. 2000).

63. Money and Dalery 1977.

64. Ehrhardt et al. 1968a,b; Berenbaum and Hines 1992; Meyer-Bahlburg et al. 1996; Berenbaum 1999.

65. Resnick et al. 1986; Nass and Baker 1991; Hampson et al. 1998.

66. Helleday et al. 1994. They found no difference in verbal ability.

67. Perlman 1973; Baker and Ehrhardt 1974; McGuire et al. 1975.

68. Baker and Ehrhardt 1974; McGuire et al. 1975; Sinforiani et al. 1994.

69. Kester et al. 1980; Ehrhardt et al. 1985; Meyer-Bahlburg et al. 1995.

70. Hines and Shipley 1984; Hines and Sandberg 1996.

71. Masica et al. 1971. AIS males reared as males, however, show a typical male sexual identity (Money and Ogunro 1974). These results indicate that, at the very least, there is a complex interplay between biological and social factors in typical sexual development.

72. Imperato-McGinley et al. (1991) provide one of the few studies to report an effect of AIS on spatial cognition, but most of the tests they used do not generate reliable sex differences and are therefore difficult to interpret.

73. Perlman 1973.

74. Young boys engage in a more rough and tumble style of play than girls (Beatty 1984; Williams et al. 1990).

75. Williams et al. 1990, p. 97.

76. Collear and Hines 1995; see also Hines 2000. Grimshaw et al. (1995) did report sex differences in the mental-rotation skills of 7-year-olds, which they claimed were evidence of organizing hormonal effects. But these results, which run counter to those of previous studies, at best provide weak evidence for such organizing effects.

77. Geary and other evolutionary psychologists also look for evidence of the activating effects of sex hormones on cognition to support their functionalist speculations. In this research, the goal is to relate circulating hormonal levels to cognitive performance. An inordinate amount of the research claiming such activating effects on spatial cognition originates at the University of Western Ontario, where Doreen Kimura is something of a doyenne (for example, Hampson 1990; Kimura and Hampson 1994; Kimura 1996; Moffat and Hampson 1996). Little of Kimura's own research has been published in peer-reviewed journals; rather, her favored venues are abstracts from scientific meetings and review articles (see Fausto-Sterling 1985 for an excellent discussion of the peculiar way in which Kimura disseminates her results). In any case, her results make little of sense from an adaptationist evolutionary perspective. For example, according to Kimura, among males it is those with low, not high, testosterone levels who perform best on mental rotations and other spatial tests (Kimura 1992), and men perform best during spring, when their testosterone levels are at their seasonal lows and during the afternoons when they reach their daily lows. The message here is that Johnny should never take a geometry test on an October morning.

Outside of the University of Western Ontario, the association between testosterone and spatial cognition is much less apparent. Although Neave et al. (1999) reported a correlation between testosterone and spatial cognition, Liben et al. (2002) could find no evidence that testosterone has an activating effect on spatial abilities, and O'Connor et al. (2001) found that testosterone actually inhibited spatial abilities, while enhancing verbal skills; to further complicate matters, Wolf et al. (2000) reported no effect of testosterone on either spatial or verbal abilities. Finally, the evolutionary psychologists Silverman et al. (1999) reported that, although testosterone was correlated with spatial ability, changes in testosterone had no effect on spatial ability. The obvious interpretation of these results, though resisted by the authors, is that, whatever the correlation between testosterone and spatial cognition, there is no indication of a causal role for testosterone vis-à-vis spatial skills.

Activating hormonal effects are no less murky for women. In one attention-grabbing study, Hampson (1990) reported that women's spatial ability varied inversely with estradiol levels; women perform best during menses, when estradiol levels are lowest, and worst during the follicular phase, when estradiol levels are high. Manual dexterity test results exhibited the opposite pattern. Similar results were obtained by Hausmann et al. (2000) and McCormick and Teillon (2001). The latter reported that during the menstrual phase, women score as high as men on spatial tests. Perhaps then, when Wilma and Betty were menstruating, they left their basket weaving aside and joined the boys in the hunt. Alas, Postma et al. (1999) found that women performed best on spatial tests during the nonmenstrual phase, the very opposite of Hampson's results. (They also found no correlation between spatial skills and circulating testosterone levels among these women.)

78. By the second grade, both males and females classify math as a "male activity" (Klein 1989). It is in this context that we should view the important role of self-concept (Nash 1975; Spence 1985; Signorella et al. 1988). Girls who view themselves as masculine do better than those who view themselves as feminine; the same is true of boys (Signorella and Jamison 1985). That this situation cannot be explained by within-sex hormonal differences is evident from the study by Bronson (1998), in which he showed that girls perform less well on a test if it is described as a test of spatial ability than they do when it is not described in that way. Moreover, Hammer and Dusek (1996) cite a study by Sadker and Sadker (1994) that showed that teachers view girls as less adept at math even when they get better grades than boys. This may be one reason girls take fewer math courses than boys, once it becomes elective (Linn and Hyde 1989)—another important experiential factor influencing scores on standardized tests.

79. Hyde et al. 1990. A similar pattern has been reported for a variety of spatial tests. For example, Barnfield (1999) could find no sex differences in children aged 4 to 16 for the Silverman and Eals (1992) test that favors females; they did, however observe the pattern reported by Silverman and Eals in adults. As for mental rotations, Hall et al. (1999) reported finding no sex differences in children from grades 5 to 8, and Karadi et al. (1999) found a sex difference in adults but not in 9-year-olds (see also Roberts and Bell 2000b). Levine et al. (1999), however, reported sex differences in children as young as 4.5 years, yet another example of the wildly conflicting results so typical of this research area. One obvious explanation for the late emergence of these sex differences is the increasingly differentiated experiences of boys and girls during the course of their education, including is a sex difference in the propensity to take math courses when they become elective. Benbow and Stanley (1980), while minimizing the influence of differential course selection, nonetheless found that more of the mathematically gifted girls than boys cease taking math (or science) classes in college. Mathematically gifted girls report parental resistance to enrolling in advanced algebra courses (Fox et al. 1983). In addition, on average, boys also have more access to mathematical problem-solving experience outside of the classroom (Kimball 1989), even among the mathematically gifted (Benbow and Stanley 1980). For example, girls report problems convincing their parents to buy them Legos (Casserly 1980). Lately, even Benbow (in Lubinski et al. 2001) has placed more emphasis on such sociocultural factors in explaining the performance of the mathematically gifted than she did in the studies for which she is most widely known.

80. And Lunneborg et al. (1982) reported that women who participated in sports had higher math self-confidence. Sexually differentiated experiences extend well beyond sports, however. Recently, it was shown that the high participation of boys in computer games involving joysticks explains their superiority on computerized mental rotations tests. Similarly, Roberts and Bell (2000a) reported that sex differences on 2D mental-rotations disappear after a period of computer familiarization. More generally, practice with rotated figures increases scores on mental rotations tests in both sexes (Schaefer and Thomas 1998). Similar effects of practice were also observed in other spatial tests (Kaplan and Weisberg 1987).

81. Evolutionary psychologists, who are inordinately fond of acronyms, refer to the sociocultural opposition as SSSM (standard social science model) (Tooby and Cosmides 1992).

82. Wilson (1975) was the first to Darwinize along sociobiological, as opposed to ethological, lines in the last chapter of that book. He later expanded his human sociobiology in his 1978 book *On Human Nature*. But it was Richard Alexander (see especially Alexander 1979) who initiated the form of sociobiology that most influenced social scientists and especially anthropologists. He created what we might call the (University of) Michigan School of Sociobiology, which differed in important respects from Wilson's version, not least in avoiding the latter's inclination toward genetic determinism. You might consider Alexander's the vanilla version of human sociobiology to Wilson's rocky-road version. Some representative work of the Michigan school includes Chagnon and Irons 1979; Borgerhoff-Mulder 1991; Smith and Winterhalder 1992; Hill and Hurtado 1996; Hawkes et al. 1997; Sherman and Reeve 1997; Alexander et al. 1998; Smith 1998; Voland 1998. Evolutionary psychologists tend to refer to this work as Darwinian Social Science, or DSS.

83. This contrast is drawn most forcefully by Cosmides and Tooby (1987).

84. Fodor (1974, reprinted in Fodor 1981) provides one of the most sophisticated arguments for the specialness of psychology.

85. Skinner 1957.

86. Chomsky 1959.

87. According to Bruner (1990, p. 36) the term *folk psychology* was "coined in derision by the new cognitive [read computationalist] scientists." Bruner is himself an unapologetic proponent of folk psychology.

88. This emphasis is evident in his (1978) collection of essays entitled *Brainstorms* (which I still consider Dennett's best work). See especially his essay entitled "Why the Law of Effect Will Not Go Away."

89. Dennett (1995, p. 395) refers to Skinner as a "greedy reductionist" in a passage that, along with an anecdote about Bertrand Russell, is the high point of an otherwise regrettable book.

90. This attitude was already becoming evident in Fodor's (1968) work.

91. Chomsky makes this analogy between mind parts and body parts especially clearly in his 1980 article "Rules and Representations."

92. Chomsky 1965, 1968.

93. Fodor 1983, 1985.

94. PDP first began to reach a broad audience as a result of the publication of studies by Rumelhart and McClelland 1986 and McClelland and Rumelhart 1986. Rumelhart's (1989) work is a good general account of how connectionists deal with

some of the central issues in cognitive psychology. For a connectionist-informed critique of Fodor-style modularism, see Karmiloff-Smith (1992).

95. Symons (1992, p. 142) writes, "There is no such thing as a 'general problem solver' because there is no such thing as a general problem." This is an undiluted nonsequitur. The fact, if it is a fact, that there are no general problems in no way discredits the efficacy of a general problem solver under many conditions.

96. Young 1970.

97. I heard David Buller give a very nice talk on the Freudian aspects of evolutionary psychology at a meeting in Seattle; this was later published (Buller 1999) in Hardcastle 1999.

98. Randy Thornhill is a good example of someone who has taken advantage of evolutionary psychology's empirical opacity. His original adaptive theory of human rape (Thornhill 1980; Thornhill and Thornhill 1983), constructed within the Darwinian social science tradition, made some specific predictions that were proved false. His subsequent conversion to the evolutionary psychology framework allowed him to proceed with his adaptationist speculation unimpeded by this previous failure, because it is now couched in terms of deep psychology rather than behavior (Thornhill and Thornhill 1992). See Sterelny (1995) and Buller (1999) for good general discussions of this problematic aspect of evolutionary psychology.

99. See, for example, Barton and Harvey 2000; Clark et al. 2001; de Winter and Oxnard 2001; and Redies and Puelles 2001; for emphasis on the non-modular aspects of brains, see Farah 1994; Duncan and Owen 2000; and Kaskan and Finlay 2000.

100. See Duchaine et al. (2001) for a particularly egregious example of this hubris, in which they chide neuroscientists for a lack of theory, and propose to remedy the situation with evolutionary psychology.

101. Chomsky 1988; Fodor 2000.

102. See Foley (1996) for a particularly trumped-up treatment of the theoretical value of the notion of the environment of evolutionary adaptedness (EEA).

103. Several philosophers of science have written broader critiques of the evolutionary psychology program. I particularly recommend Sterelny (1995), Buller (1999), Davies (1999), Lloyd (1999), Grantham and Nichols (1999), and Downs (2002). A good biological critique of evolutionary psychology written by an adaptationist sympathizer is Wilson (1994). Davies et al. (1995) provide an excellent critique of the modularity thesis (see also Mameli 2001) in general, and of Cosmides' research on the Wason card test in particular.

104. Tooby and Cosmides (1992) evidence the most disdain for the social sciences ever motivated by an adaptationist evolutionary perspective.

NOTES FOR CHAPTER 10
A TEXTBOOK CASE OF PENIS ENVY?

1. Aristotle, for example, refers to the alleged hermaphroditism in hyenas but correctly refutes the claim (*History of Animals* 579b: 16–30).

2. Kruuk conducted an extensive field study on spotted hyenas living in the Serengeti from 1964 to 1968, which was published in a 1972 book. His research did much to dispel the myth that spotted hyenas rely primarily on scavenging for sustenance.

3. Wickler mentioned the idea in several publications in German (for example, Wickler 1965).

4. Within the ethological framework, behavior is assumed to be controlled by fundamental motivations such as hunger, sex, and aggression, which often conflict with each other for behavioral expression. In social interactions, the best ways to reduce aggressive behavior is either to stimulate a "motivational center" for an incompatible behavior or remove the stimulus for aggression.

5. There are actually at least two commonly recognized forms of mimicry. That of the viceroy and monarch butterflies is referred to as Batesian mimicry, after a famous nineteenth-century naturalist and lepidopterist, Henry W. Bates, who first described the phenomenon on the basis of his collections of Brazilian butterflies (1862). In Batesian mimicry, a species without defenses mimics a species with defenses.

6. Wickler's category "automimicry" includes a wide variety of animal features that have a signaling function. Most of the examples he cites involve sexual behavior. In automimicry, both the model and the mimic exist within a given species; usually one sex exhibits a morphological feature that serves as the (evolutionary) model for members of the other sex. In the case of the homosexual pseudo-copulatory behavior of male hamadryas baboons, the males are alleged to possess a red rump because it resembles the estrous swelling of sexually receptive females. This feature is supposed to be adaptive in that it inhibits aggression in more dominant males by turning on their sexual behavior. Presumably, subordinate males that lacked a red rump would experience more aggression and hence would be likely to get injured. This and other cases of automimicry are discussed in Wickler's (1968) book. The hamadryas baboon example is discussed in pages 230–34.

7. This is Lorenz's interpretation of submissive behavior in dogs, in which the subordinate awkwardly twists its neck, exposing its throat. Sometimes the subordinate ends up twisting its whole body, rolling over on its back, and wimpering in the process. If it really is fearful, it also urinates; fear is also why "house-broken" dogs sometimes have lapses when disciplined or in the presence of strangers. (See, for example, Lorenz 1950.)

8. Evidence that adult female spotted hyenas tend to be heavier than males is provided by Hamilton et al. (1986). This is not at all typical of mammals (Ralls 1976).

9. Bertram 1975, 1976.

10. This is referred to as "the competition-aggression hypothesis," which has been advocated by the Berkeley researchers, particularly Laurence Frank (see especially, Frank 1986, 1997). There are two primary problems with this hypothesis. First, as I discussed at length in chapter 6, the link between androgens and aggression is not at all straightforward (Francis et al. 1992a), and still less so is the link between androgens and social dominance (Francis 1988a; Francis et al. 1992a). Second, the relative lack of aggression in males exposed to those same androgens is mysterious. Nor is it clear why early exposure to androgens should result in a greater increase in female than in male body size. The sibling aggression hypothesis suffers from the same assumption of a close connection between androgens and aggression, as well as the puzzle as to why prenatal androgen exposure would affect the females more than the males.

11. Frank et al. 1991.

12. Frank et al. 1991.

13. The adaptive siblicide hypothesis has been strongly advocated by East et al. (1993) and East and Hofer (1997). On this view, high androgen levels in utero represent an adaptation primarily for siblicide, which is itself adaptive because singletons grow faster than twins. The male-like phallus of females was then, initially at least, a byproduct of this selection.

14. Aristotle, *Generation of Animals* 766a: 23–28.

15. Racey and Skinner 1979.

16. Gould and Vrba 1982. Gould also addressed the issue of spotted hyena genitals in one of his columns for *Natural History*, entitled "Hyena Myths and Realities" (1981).

17. CAH is a syndrome with several distinct etiologies, the most common of which involves a deficiency in the enzyme 21-hydroxylase, which results in a deficiency in cortisol, which is manufactured in the adrenal glands. The feedback pathways normally involved in cortisol synthesis compensate for the cortisol deficiency by supplying more of its (metabolic) precursors. But, because the cortisol pathway cannot use them, these precursors become available for the adrenals' metabolic pathway for androgen synthesis, resulting in a surfeit of androgens (and progesterones). The abnormal exposure to androgens begins prenatally, and genital masculinization is present at birth.

18. The degree of masculinization in CAH females is variable; in extreme cases, female neonates have been misidentified as males (for example, Money and Dalery 1977).

19. The proposed mechanisms for the masculinization of female spotted hyenas focus on the elevated androgen levels to which they are exposed in utero. Neither of the two current hypotheses proposes a CAH-like involvement of the adrenal glands. The first of these two hypotheses is the oldest: Matthews (1939) suggested that it is the ovaries of the female fetuses themselves that produce the extra androgens. More recently, Lindeque and Skinner (1982) further developed this hypothesis (see also Lindeque et al. 1986; van Jaarsveld and Skinner 1991). There is evidence, however, that the ovaries begin to synthesize steroids only after the rise in androgen levels (Licht et al. 1998).

According to the second hypothesis, the placenta is the source of the extra androgens, particularly androstenedione (Yalcinkaya et al. 1993). This hormone is manufactured in the placentas of all mammals, but, in spotted hyenas, unusually high levels are present there (Glickman et al. 1987). Normally, androstenedione is converted into estradiol by the enzyme aromatase, but spotted hyenas have unusually low levels of this enzyme. Under these conditions, androstenedione is converted into testosterone by the enzyme 17-B-hydroxysteroid dehydrogenase. Hence, the elevated testosterone levels ultimately result from a deficiency in the enzyme aromatase. Similar aromatase deficiencies have been reported in some human females with masculinized genitalia (Shozu et al. 1991; Ito et al. 1993; Conte et al. 1994).

The specific cause of the aromatase deficiency is unknown, but Yalcinkaya et al. (1993) suggest that a nongenetic (or epigenetic) mechanism may play an important role in transmitting the masculinizing syndrome from one generation to the next. It is known that when testosterone is administered to young rats, they experience a reduction in the ovarian follicles and granulosa cells where androstenedione is normally converted into estradiol, but the (thecal) cells that produce androstenedione are not affected. Hence, exposure to high levels of testosterone in utero, may result not only in masculinized genitals but also in an ovary that produces less estrogen than it otherwise would, and which, in turn, masculinizes the next generation. This process would augment the effect of a mutation in any gene involved in the regulation of aromatase levels.

20. Sherman 1988. It is more appropriate to consider this trait in the context of Williams's contrast between adaptations and byproducts. In assessing whether special features of the spotted hyena, such as the enlarged clitoris, should count as an adaptation or a mere byproduct of the evolution of a correlated character, it is not sufficient to identify a putative function, or something for which it is good. The hypertrophied clitoris may very well prove useful in greeting ceremonies, without, however, having been selected (qua hypertrophied clitoris) for that end. In determining the extent to which hyena genitals have been shaped specifically for this or any other function, it is especially useful to determine the extent to which it constitutes a specific evolved response to some environmental factor, as opposed to a generic feature that is unrelated to external conditions. The more generic the developmental process to which hyena traits such as the enlarged clitoris can be attributed, and the more global the effects of alterations in this process, the less likely that the trait constitutes a specific adaptation to local conditions and the more likely that it is merely a correlated byproduct of alterations in other hyena parts. Androgen sensitivity of genital development is a very generic property of mammals. Moreover, alterations in the steroid hormone milieu during early development are known to have diverse effects in mammals, ranging from those on brain and behavior to those on growth and coloration, to name a few. In that light, we should be careful about attributing the exceptional female spotted hyena phallus to selection. The best evidence for selection will be signs that specific phallic features reflect alterations in the generic mammalian developmental process that do not influence other hyena parts.

21. Matthews 1939.

22. Glickman et al. 1993; Frank and Glickman 1994.

23. Frank et al. 1995; Frank 1997.

24. Frank 1997.

25. Ibid.

26. This size difference is largely related to the fact that the gestation period of spotted hyenas (110 days) is considerably longer than that of the brown and striped hyenas (about 90 days) (Frank 1997).

27. Subsequent births are less traumatic and, more important from an evolutionary perspective, less dangerous for both mothers and neonates. Nonetheless, parous spotted hyenas in zoos experience higher rates of birth complications than other carnivores (Frank and Glickman 1994).

28. Muller and Wrangham 2002.

29. Smale et al. (1999) argue that siblicide is far less common than proposed by Frank et al. (1991) and Hofer and East (1997). They argue that fighting among neonates commonly occurs in the context of establishing dominance. The loser in this battle is permanently subordinate, so there is no need to kill it. They also attribute the high rate of singletons emerging from dens to singleton births rather than siblicidal reduction.

30. The evidence for siblicide has been indirectly inferred from the number and sex of infants that emerge from the natal dens. Moreover, Smale et al. (1999) argue that the low number of female-female twins may result from skewed sex ratios at birth.

31. Drea et al. 1998. See also Glickman et al. (1998) for further evidence of the androgen-independence of genital development (but see Licht et al. [1998] for evidence of androgen dependence).

32. Avila et al. (1996) identified a possible mutation in an androgen receptor gene in spotted hyenas. They noted that the androgen receptors of spotted hyenas differ from those of striped hyenas by a single amino acid in the steroid-binding domain. They suggested that this amino acid substitution may change the sensitivity of the androgen receptors to various androgens, perhaps rendering them more promiscuous.

33. For there are important differences in the genitalia of male and female spotted hyenas despite their superficial similarity (Neaves et al. 1980). Whereas the urethra of males is surrounded by corpus spongiosum, that of females is suspended beneath loose connective tissue, which allows for the dilation of the canal and hence entry of the penis during copulation. More important, female, but not male, hyenas have a substantial retractor muscle above the urethra, which, when activated, considerably shortens the clitoris, thereby facilitating entry of the penis. The loosely packed connective tissue surrounding the urethra of the clitoris provides great elasticity relative to the male penis and is taxed to the utmost during parturition.

NOTES FOR CHAPTER 11
DARWIN'S TEMPTRESS

1. Evans-Pritchard 1937.

2. Gould and Lewontin 1979.

3. *Candide* was published in 1759 and almost immediately was ordered to be burned by the Great Council in Geneva. Not surprisingly, Voltaire denied that he was the author, but everyone in France knew otherwise.

4. This quotation is from Leibniz's *Theodicy*, an attempt, in part, to reconcile Catholics and Protestants, published after his death in 1710. A similar sentiment was later expressed by Alexander Pope in his *Essay on Man*: "All partial evil is universal good."

5. Durant and Durant 1965, vol. 9, pp. 720–21.

6. Voltaire, *Candide*, p. 20.

7. The *Dialogues* were first published in 1779, three years after Hume's death.

8. Hume, *Dialogues*, p. 120.

9. Ibid., p. 121.

10. Ibid., p. 121.

11. Dawkins 1986.

12. Hume, *Dialogues*.

13. Dawkins 1986. With characteristic aplomb, Richard Dawkins asserts that Darwin made the academy, if not the world, safe for atheists such as himself. After Darwin, and only after Darwin, was it reasonable to be an atheist, according to

Dawkins. (Dawkins also believes that, after Darwin, it was unreasonable to be other than an atheist, though Darwin himself was agnostic).

14. Aside from playing into the hands of creationists, this equating of Darwin, atheism, and skepticism about the argument from design is problematic on a number of grounds, but primarily in being extremely and narrowly provincial. Dawkins's claim is provincial in three distinct ways. First, it reflects his preoccupation with atheism at the expense of other forms of skepticism. Second, Dawkins ascribes to science, and particularly biology, too much of the credit for advancing this skepticism. And third, Dawkins focuses too narrowly on a distinctly English tradition in framing the issue.

15. Bell 1982.

16. In his metaphor of the blind watchmaker, Dawkins's mistake, and it is fundamental, is to accept Paley's premise that a watch and an organism reflect deeply analogous processes—he refers to organisms as living watches. To agree, that is, to that they both came to exist through some sort of design process and that, for purposes of understanding their complex organization, the fact that organisms are products of biological evolution, whereas watches are not, is irrelevant. But consider some obvious disanalogies between watches and organisms. Watches, in virtue of being truly teleological artifacts, are constituted according to an external ideal, in the form of a schematic model or prototype that exists before the manifest watch. Organisms, on the other hand, cannot be so constituted, once we give up the idea of a designing God or Platonic form. Whatever order and adaptedness they have, they do not have by virtue of their conformity to some preexisting external ideal, without which it makes no sense to say they were designed.

The lack of design in even the most celebrated of animal parts, the vertebrate eye, is apparent at the most basic level, that of their structure with respect to the transduction of visual signals into electrical ones. The sort of inefficiency in this process that we observe in vertebrate eyes would not be expected if they had been designed. And to the extent that eyes reflect functional excellence, it is solely the result of the contribution they made to their bearers in their competition with other eye bearers. The imperative here is not to conform to an ideal, it is just to better the competition. And there is no reason, except theological inertia, to expect those imperatives to converge. Design is a top-down process. You begin with a representation of the final product, then you seek ways to realize it through manufacture of its parts. Natural selection, however, is straightforwardly bottom-up. Variants occur vis-à-vis parts of existing organisms, which are retained or not according to how they affect the functioning of that organism relative to other variants. Natural selection is simply the ex post facto winnowing of these random variants according to environmental contingencies. Dawkins's attempt to naturalize Paley is only the culmination of a trend in modern Darwinism to explain bottom-up processes in a top-down manner. But to the extent that there is a meaningful analogy to be made between watches and organisms, it does not result from treating organisms as products of top-down design, but rather in treating watches as products of bottom-up processes. That is, the extent to which watch properties reflect their contingent history rather than a priori design.

17. Dennett is but one of a number of philosophers who migrated out of the philosophy of mind into evolutionary biology in the process of finding ways to natu-

ralize psychological categories (for example, Kim Sterelny, Peter Godfrey-Smith, and Paul Griffiths). Dennett is the most extreme adaptationist among them, and the least well-grounded in biology, but he has had the most influence in shaping popular (as opposed to professional) thinking about evolution.

18. Dennett's intentional stance originally resembled Kant's general opinion of teleological thinking as something, that, though obviously defective, cannot be helped.

19. Dennett originally, however, held out more hope that insights gained from the intentional stance could ultimately be cashed out in a more naturalistically respectable manner. The intentional stance would then act as a bridge between folk psychology and a truly scientific, or naturalized, psychology. But with respect to central problems such as purposive behavior, Dennett suspected that neuroscience would have little of relevance to say, at least in the foreseeable future. Instead, Dennett has always leaned more heavily on the computationalist view of the mind, derived from Turing-machine functionalism. He finds the computer program analogies especially useful in explaining purposive behavior. A computer program accounts for purposive behavior as resulting from internal states in the form of information structures that represent goal states. The intentionality of these representations, however, is not intrinsic, but rather can be traced to the intentions of the programmer. But from whence does animal intentionality derive if animals have no equivalent of a programmer? There are two options here. Either claim for animals' intrinsic intentionality or identify an extrinsic source of intentionality analogous to the computer programmer. Dennett chose the latter. This is the move he calls "passing the buck to biology" (not neurobiology as favored by the Churchlands but evolutionary biology). Dennett suggests that we view Mother Nature as the programmer, the ultimate (and omnipotent) engineer.

20. Dennett was clearly stancing in earnest with regard to folk psychological categories. He was still stancing, but much less self-consciously, when he looked to natural selection to ground his approach to explaining minds. By Darwin's Dangerous Idea, however, his stance stance had become disingenuous.

21. This trend is in large part due to the increasing popularity of the information idiom, which I would argue has helped legitimize not only a dematerialized biology but a form of crypto-vitalism as well. That, however, is a subject for another book.

BIBLIOGRAPHY

Abbott D. 1984. Behavioral and physiological suppression of fertility in subordinate marmoset monkeys. *American Journal of Primatology* 6: 169–86.

——. 1987. Behaviorally mediated suppression of reproduction in female primates. *Journal of Zoology* 213: 455–70.

Adams LA and Steiner RA. 1988. Puberty. *Oxford Review of Reproductive Biology* 10: 1–52.

Adkins-Regan E. 1985. Non-mammalian psychosexual differentiation. In: *Handbook of Behavioral Neurobiology: Neurobiology of Reproduction*. Adler N (ed.). Plenum, pp. 43–76.

——. 1987. Hormones and sexual differentiation. In: *Hormones and Reproduction in Fishes, Amphibians and Reptiles*. Norris DO and Jones RE (eds.). Plenum, pp. 1–29.

Adret P. 1993a. Operant conditioning, song learning and imprinting to taped song in the zebra finch. *Animal Behaviour* 46: 149–59.

——. 1993b. Vocal learning induced with operant techniques: an overview. *Netherlands Journal of Zoology* 43: 125–42.

Agrawal A. 2001. Sexual reproduction and the maintenance of sexual reproduction. *Nature* 411: 692–95.

Aguirre G, Detre J, Alsop D, et al. 1996. The parahippocampus subserves topographical learning in man. *Cerebral Cortex* 6: 823–29.

Aida T. 1921. On the inheritance of color in a fresh-water fish, *Aplocheilus latipes* Temminck and Schlegel, with a special reference to the sex-linked inheritance. *Genetics* 6: 554–73.

——. 1936. Sex reversal in *Aplocheilus latipes* and a new explanation of sex differentiation. *Genetics* 21: 136–53.

Aiello L and Wheeler P. 1995. The expensive tissue hypothesis. *Current Anthropology* 36: 199–211.

Airey D and DeVoogd T. 2000. Greater song complexity is associated with augmented song system anatomy in zebra finches. *Neuroreport* 11: 2339–43.

Airey D, Buchanan K, Szekely T, et al. 2000. Song, sexual selection, and a song control nucleus (HVC) in the brains of European sedge warblers. *Journal of Neurobiology* 44: 1–6.

Alcock J. 1987. Ardent adaptationism. *Natural History* 96: 4.

——. 1998. Unpunctuated Equilibrium in the Natural History Essays of Stephan Jay Gould. *Evolution and Human Behavior* 19: 321–36.

Alexander GM, Swerdloff RS, Wang C, et al. 1998. Androgen-behavior correlations in hypogonadal men and eugonadal men. II. Cognitive abilities. *Hormones and Behavior* 33: 85–94.

Alexander R. 1979. *Darwinism and Human Affairs*. University of Washington Press.

——. 1990. Epigenetic rules and darwinian algorithms: the adaptive study of learning and development. *Ethology and Sociobiology* 11: 241–303.

Allen L and Gorski R. 1986. Sexual dimorphism of the human anterior commissure. *Anatomical Records* 214: 3.

——. 1990. Sex differences in the bed nucleus of the stria terminalis of the human brain. *Journal of Comparative Neurology* 302: 697–706.

——. 1991. Sexual dimorphism of the anterior commissure and massa intermedia of the human brain. *Journal of Comparative Neurology* 312: 97–104.

——. 1992. Sexual orientation and the size of the anterior commissure in the human brain. *Proceedings of the National Academy of Sciences, USA* 89: 7199–202.

Allen LS, Hines M, Shryne JE, et al. 1989. Two sexually dimorphic cell groups in the human brain. *Journal of Neuroscience* 9: 497–506.

Allen LS, Richey MF, Chai YM, et al. 1991.

Sex differences in the corpus callosum of the living human being. *Journal of Neuroscience* 11: 933–42.

Allendorf F and Leary R. 1986. Heterozygosity and fitness in natural populations of animals. In: *Conservation biology: the science of scarcity and diversity*. Soule M (ed.). Sinauer: 57–76.

Alvarez-Buylla A and Nottebohm F. 1988. Migration of young neurons in adult avian brain. *Nature* 335: 353–54.

Alvarez-Buylla A, Theelen M and Nottebohm F. 1988. Birth of projection neurons in the higher vocal center of the canary forebrain before, during, and after song learning. *Proceedings of the National Academy of Sciences, USA* 85: 8722–26.

Alvarez-Buylla A, Kirn J, and Nottebohm F. 1990. Birth of projection neurons in adult avian brain may be related to perceptual or motor learning. *Science* 224: 901–3.

Alvarez-Buylla A, Garcia-Verdugo JM, Mateo AS, et al. 1998. Primary neural precursors and intermitotic nuclear migration in the ventricular zone of adult canaries. *Journal of Neuroscience* 18: 1020–37.

Alyman C and Peters M. 1993. Performance of male and female children, adolescents and adults on spatial tasks that involve everyday objects and settings. *Canadian Journal of Experimental Psychology* 47: 730–47.

Amundsen T. 2000. Why are female birds ornamented? *Trends in Ecology and Evolution* 15: 149–55.

Andersson M. 1994. *Sexual selection*. Princeton University Press.

Arendash G and Gorski R. 1983. Effects of discrete lesions of the sexually dimorphic nucleus of the preoptic area or other medial preoptic regions on the sexual behavior of male rates. *Brain Research Bulletin* 10: 147–50.

Aristotle. 1984. *The Complete Works of Aristotle: The Revised Oxford Translation*. Barnes J (ed.). Princeton University Press, pp. 1111–218.

Arnal C, Cote I, and Morand S. 2001. Why clean and be cleaned? The importance of client ectoparasites and mucus in a marine cleaning symbiosis. *Behavioral Ecology and Sociobiology* 51: 1–7.

Arnold A and Gorski R. 1984. Gonadal steroid induction of structural sex differences in the nervous system. *Annual Review of Neurosciences* 7: 413–42.

Arnold AP. 1975. The effects of castration and androgen replacement on song, courtship, and aggression in zebra finches (*Poephila guttata*). *Journal of Experimental Zoology* 191: 309–26.

——. 1996. Genetically triggered sexual differentiation of brain and behavior. *Hormones and Behavior* 30: 495–505.

——. 1997. Sexual differentiation of the zebra finch song system: positive evidence, negative evidence, null hypotheses, and a paradigm shift. *Journal of Neurobiology* 33: 572–84.

Arnold AP and Schlinger BA. 1993. The puzzle of sexual differentiation of the brain and behavior of zebra finches. *Poultry Science Reviews* 5: 3–13.

Arnold AP, Bottjer SW, Brenowitz EA, et al. 1986. Sexual dimorphisms in the neural vocal control system in song birds: ontogeny and phylogeny. *Brain, Behavior and Evolution* 28: 22–31.

Arnold AP, Wade J, Grisham W, et al. 1996. Sexual differentiation of the brain in songbirds. *Developmental Neuroscience* 18: 124–36.

Arnold SJ. 1983. Sexual selection: the interface of theory and empiricism. In: *Mate choice*. Bateseon P (ed.). Cambridge University Press, pp. 67–107.

Arnold W and Dittami J. 1997. Reproductive suppression in male alpine marmots. *Animal Behaviour* 53: 53–66.

Arthur W. 2000. The concept of developmental reprogramming and the quest for an inclusive theory of evolutionary mechanisms. *Evolution and Development* 2: 49–57.

Asoh K and Kasuya M. 2002. Gonadal development and mode of sexuality in a coral-reef fish. *Journal of Zoology* 256: 301–9.

Astheimer L, Buttemer W, and Wingfield JC. 2000. Corticosterone treatment has no effect on reproductive hormones or aggressive behavior in free-living male tree sparrows, *Spizell arborea*. *Hormones and Behavior* 37: 31–39.

Astie A, Kacelnik A, and Reboreda J. 1998. Sexual differences in memory in shiny cowbirds. *Animal Cognition* 1: 77–82.

Astur R, Ortiz M, and Sutherland R. 1998. A characterization of performance by men and women in a virtual Morris water task: a large reliable sex difference. *Behavioral Brain Research* 93: 185–90.

Avila D, Wilson J, Glickman S, et al. 1996. Sequence analysis of the gene encoding the androgen receptor (AR) of the spotted hyena. *Program of the International Congress of Endocrinology* 10: 188.

Avitsur R, Stark J, and Sheridan J. 2001. Social stress induces glucocorticoid resistance in subordinate animals. *Hormones and Behavior* 39: 247–57.

Baker MC, Bottjer SW, and Arnold AP. 1984. Sexual dimorphism and lack of seasonal changes in vocal control regions of the white-crowned sparrow brain. *Brain Research* 295: 85–89.

Baker RR and Bellis MA. 1993. Human sperm competition: ejaculate manipulation by females and a function for the female orgasm. *Animal Behaviour* 46: 887–909.

Baker S and Ehrhardt A. 1974. Prenatal androgen, intelligence and cognitive sex differences. In: *Sex Differences in Behavior*. Friedman R, Richart R, and Vande Wiele R (eds.). Wiley, pp. 53–76.

Bakker T, Bruijn E, and Sevenster P. 1989. Asymmetrical effects of prior winning and losing on dominance in sticklebacks (*Gasterosteus aculeatus*). *Ethology* 82: 224–29.

Balda R and Kamil A. 1989. A comparative study of cache recovery by three corvid species. *Animal Behaviour* 38: 486–95.

———. 1992. Long-term spatial memory in Clark's nutcrackers, *Nucifraga columbiana*. *Animal Behaviour* 44: 761–69.

———. 1998. The ecology and evolution of spatial memory in corvids of the southwestern USA: the perplexing pinyon jay. In: *Animal Cognition in Nature*. Balda R, Pepperberg I, and Kamil A (eds.). Academic Press, pp. 29–64.

Balda R and Turek R. 1984. The cache-recovery system as an example of memory capabilities in Clark's nutcracker. *Animal Cognition*. Roitblatt H, Bever T, and Terrace H (eds.). Erlbaum, pp. 513–32.

Balthazart J and Ball GF. 1995. Sexual differentiation of brain and behavior in birds. *Trends in Endocrinology and Metabolism* 6: 21–29.

Baptista L. 1977. Geographical variation in song and song dialects of the migratory white-crowned sparrow, *Zonotrichia leucophrys pugetensis*. *Condor* 79: 356–70.

———. 1985. The functional significance of song sharing in the white-crowned sparrow. *Canadian Journal of Zoology* 63: 1741–52.

Baptista L and King J. 1980. Geographical variation in song and song dialects of montane white-crowned sparrows. *Condor* 81: 267–84.

Baptista LF and Petrinovich L. 1984. Social interaction, sensitive phase, and the song template hypothesis in the white-crowned sparrow. *Animal Behaviour* 32: 172–81.

Bardach JE, Ryther JH, and McLarney WO. 1972. *Aquaculture*. Wiley-Interscience.

Barkow JH. 1990. Beyond the DP/DSS controversy. *Ethology and Sociobiology* 11: 341–51.

Barkow J, Cosmides L, and Tooby J (eds.). 1992. *The Adapted Mind: Evolutionary Psychology and the Generation of Culture*. Oxford University Press.

Barnard C, Behnke J, and Sewell J. 1996. Social status and resistance to disease in house mice (*Mus musculus*): status-related modulation of hormonal responses in relation to immunity costs in different social and physical environments. *Ethology* 102: 63–84.

Barnea A and Nottebohm F. 1994. Seasonal recruitment of hippocampal neurons in adult free-ranging black-capped chickadees. *Proceedings of the National Academy of Sciences* 91: 11217–21.

Barnfield A. 1999. Development of sex differences in spatial memory. *Perceptual and Motor Skills* 89: 339–50.

Barrette C. 1977. Muntjacs. *Evolution* 31: 169–76.

Bartholomew GA and Hubbs CL. 1960. Population growth and seasonal movements of the northern elephant seal, *Mirounga angustrirostris*. *Mammalia* 24: 313–24.

Barton R and Harvey P. 2000. Mosaic evolution of brain structure in mammals. *Nature* 405: 1055–58.

Basil J, Kamil A, Balda R, et al. 1996. Differences in hippocampal volume among food storing corvids. *Brain, Behavior and Evolution* 47: 156–64.

Basolo A. 1990. Female preference predates the evolution of the sword in swordtail fish. *Science* 250: 808–10.

———. 1995a. A further examination of a pre-existing bias favouring a sword in the genus *Xiphophorus*. *Animal Behaviour* 50: 365–75.

———. 1995b. Phylogenetic evidence for the role of a pre-existing bias in sexual selection. *Proceedings of the Royal Society of London, B* 259: 307–11.

Bass A. 1992. Dimorphic male brains and alternative reproductive tactics in a vocalizing fish. *Trends in Neuroscience* 15: 139–45.

———. 1996. Shaping brain sexuality. *American Scientist* 84: 352–63.

Bass A and Grober M. 2001. Social and neural modulation of sexual plasticity in teleost fish. *Brain, Behavior and Evolution* 57: 293–300.

Bass A, Hovarth B and Brothers E. 1996. Nonsequential developmental trajectories lead to dimorphic vocal circuitry for males with alternative reproductive tactics. *Journal of Neurobiology* 30: 493–504.

Bass A, Marchaterre M, and Baker R. 1994. Vocal-acoustic pathways in a teleost fish. *Journal of Neuroscience* 14: 4025–39.

Bass A and Marchaterre M. 1989. Sound-generating (sonic) motor system in a teleost fish (*Porichthys notatus*): sexual polymorphism in the ultrastructure of myofibrils. *Journal of Comparative Neurology* 286: 141–53.

Bates H. 1862. Contributions to an insect fauna of the Amazon Valley. *Transactions of the Linnean Society, London*.

Bateson P. 1981a. Discontinuities in development and changes in the organization of play in cats. In: *Behavioral Development*. Immelman K, Barlow G, Petrinovich L, and Main M (eds.). Cambridge University Press, pp. 281–95.

———. 1981b. Ontogeny of behaviour. *British Medical Bulletin* 37: 159–64.

Bateson P (ed.). 1991. *The Development and Integration of Behavior: Essays in Honour of Robert Hinde*. Cambridge University Press.

Batty K, Herbert J, Keverne E, et al. 1986. Differences in blood levels of androgens in female talapoin monkeys related to their social status. *Neuroendocrinology* 44: 347–54.

Baughman E and Dahlstom W. 1968. *Negro and White Children*. Academic Press.

Baylis JR. 1982. Avian vocal mimicry: its function and evolution. *Acoustic Communication in Birds* 2: 51–83.

Beacham J. 1988. The relative importance of body size and aggressive experiences as determinants of dominance in pumpkinseed sunfish, *Lepomis gibbosus*. *Animal Behaviour* 36: 621–23.

Beaton A. 1997. The relation of planum temporale asymmetry and morphology of the corpus callosum to handedness, gender, and dyslexia: a review of the evidence. *Brain and Language* 60: 255–322.

Beatty WW. 1984. Hormonal organization of sex differences in play fighting and spatial behavior. *Progress in Brain Research* 61: 315–27.

Beaugrand J, Goulet C, and Payette D. 1991. Outcome of dyadic conflict in male green swordfish, *Xiphophorus helleri*: effects of body size and prior experience. *Animal Behaviour* 41: 417–24.

Beaugrand J, Payette D, and Goulet C. 1996. Conflict outcome in male green swordtail fish dyads (*Xiphophorus helleri*): interaction of body size, prior dominance/subordination experience, and prior residency. *Behaviour* 133: 303–19.

Becker B and Hedges L. 1984. Meta-analysis of cognitive gender differences: a comment on an analysis by Rosenthal and Rubin. *Journal of Educational Psychology* 76: 583–87.

Becker P, Roland H, and Reinboth R. 1975. An unusual approach to experimental sex inversion in the teleost fish, *Betta* and *Macropodus*. In: *Intersexuality in the Animal Kingdom*. Reinboth R (ed.). Springer-Verlag, pp. 236–42.

Bee M. 2002. Territorial male bullfrogs (*Rana catesbiana*) do not assess fighting ability based on size-related variation in acoustic signals. *Behavioral Ecology* 13: 109–24.

Beecher M, Campbell S, and Nordby J. 2000. Territory tenure in song sparrows is related to song sharing with neighbours, but not to repertoire size. *Animal Behaviour* 59: 29–37.

Begin J, Beaugrand JP, and Zayan R. 1996. Selecting dominants and subordinates at conflict outcome can confound the effects of prior dominance or subordination experience. *Behavioural Processes* 36: 219–28.

Bell G. 1982. *The Masterpiece of Nature: The Evolution and Genetics of Sexuality.* University of California Press.

Bell A and Variend S. 1985. Failure to demonstrate sexual dimorphism of the corpus callosum in childhood. *Journal of Anatomy* 143: 143–47.

Benbow C. 1988. Sex differences in mathematical reasoning ability in intellectually talented preadolescents: their nature, effects, and possible causes. *Behavioral and Brain Sciences* 11: 169–83.

Benbow CP and Benbow RM. 1984. Biological correlates of high mathematical reasoning ability. *Progress in Brain Research* 61: 469–89.

Benbow C and Stanley J. 1980. Sex differences in mathematical ability: fact or artifact? *Science* 210: 1262–64.

———. 1983. Sex differences in mathematical reasoning ability: more facts. *Science* 222: 1029–31.

Bener A. 1979. Sex differences and bilateral asymmetry in dermatoglyphic pattern elements on the fingertips. *Annals of Human Genetics* 42: 333–42.

Bennett N. 1989. The social structure and reproductive biology of the common mole-rat, *Cryptomys h. hottentotus* and remarks on the trends in reproduction and sociality in the family Bathygeridae. *Journal of Zoology London* 219: 45–59.

Bennett N and Jarvis J. 1988. The social structure and reproductive biology of colonies of the mole rat *Cryptomys damarensis* (Rodentia: Bathyergidae). *Journal of Mammalogy* 69: 293–302.

Bennett NC, Jarvis J, Faulkes CG, et al. 1993. LH responses to single doses of exogenous GnRH by freshly captured Dam-araland mole rats, *Cryptomys damarensis. Journal of Reproduction and Fertility* 99: 81–86.

Bennett NC, Faulkes CG, and Spinks AC. 1997. LH responses to single doses of exogenous GnRH by social mashona mole-rats: a continuum of socially induced infertility in the family Bathyergidae. *Proceedings of the Royal Society of London B* 264: 1001–6.

Bennett NC, Faulkes CG, and Jarvis JUM. 1999. Socially induced infertility, incest avoidance and the monopoly of reproduction in cooperatively breeding African mole-rats, family Bathyergidae. *Advances in the Study of Behavior* 28: 75–114.

Bennett NC, Molteno A, and Spinks A. 2000. Pituitary sensitivity to exogenous GnRH in giant Zambian mole-rats, *Cryptomys mechowi* (Rodentia: Bathyergidae): support for the "socially induced infertility continuum." *Journal of Zoology* 252: 447–52.

Berenbaum S. 1999. Effects of early androgens on sex-typed activities and interests in adolescents with congenital adrenal hyperplasia. *Hormones and Behavior* 35: 102–10.

Berenbaum S and Hines M. 1992. Early androgens are related to childhood sex-typed toy preferences. *Psychological Science* 3: 203–6.

Berga S. 1995. Stress and amenorrhea. *Endocrinologist* 5: 416–21.

Bernard D, Eens M, and Ball G. 1996. Age and behavior related variation in the volumes of song control nuclei in male European starlings. *Journal of Neurobiology* 30: 329–39.

Bernstein IS, Gordon TP, and Rose RM. 1983. The interaction of hormones, behavior, and social context in nonhuman primates. In: *Hormones and Aggressive Behavior.* Svare B (ed.). Plenum, pp, 535–61.

Bertram B. 1970. The vocal behavior of the Indian Hill mynah. *Animal Behaviour Monographs* 3: 81–192.

———. 1975. Social factors influencing reproduction in wild lions. *Journal of Zoology* 177: 463–82.

———. 1976. Kin selection in lions and in evolution. In: *Growing Points in Ethology.* Bateson P and Hinde R (eds.). Cambridge University Press, pp. 281–301.

Betzig L. 1989. Rethinking human ethology: a response to some recent critiques. *Ethology and Sociobiology* 10: 315–24.

Beyer C, Eusterschulte B, Pilgrim C, et al. 1992. Sex steroids do not alter sex differences in tyrosine hydroxylase activity of dopaminergic neurons in vitro. *Cell Tissue Research* 270: 547–52.

Biegler R, McGregor A, Krebs J, et al. 2001. A larger hippocampus is associated with longer-lasting spatial memory. *Proceedings of the National Academy of Sciences, USA* 98: 6941–44.

Bingman V, Ioale P, Casini G, et al. 1989. Behavioral and anatomical studies of the avian hippocampus. In: *The Hippocampus, New Vistas*. Chan-Palay V and Koehler C (eds.). Liss, pp. 379–94.

Bird R. 1999. Cooperation and conflict: the behavioral ecology of the sexual division of labor. *Evolutionary Anthropology* 98: 65–75.

Bird R, Smith E, and Bird D. 2001. The hunting handicap: costly signalling in human foraging strategies. *Behavioral Ecology and Sociobiology* 50: 9–19.

Bjorksten T, Fowler K, and Pomiankowski A. 2000. What does sexual trait FA tell us about stress? *Trends in Ecology and Evolution* 15: 163–66.

Blanchard D, Spencer R, Weiss S, et al. 1995. Visible burrow system as a model of chronic social stress: behavioral and neuroendocrine correlates. *Psychoneuroendocrinology* 20: 117–34.

Blinkhorn S. 2001. Yes, but what is it for? *Nature* 412: 771.

Blumenschine RJ and Cavallo JA. 1992. Scavenging and human evolution. *Scientific American* October: 90–96.

Bohner J. 1990. Early acquisition of song in the zebra finch, *Taeniopygia guttata*. *Animal Behaviour* 39: 369–74.

Boles DB. 1980. X-linkage of spatial ability: a critical review. *Child Development* 51: 625–35.

Bolhuis J and Macphail E. 2001. A critique of the neuroecology of learning. *Trends in Cognitive Science* 5: 426–33.

Borg K, Esbenshade K, Johnson B, et al. 1992. Effects of sexual experience, season, and mating stimuli on endocrine concen-

trations in the adult ram. *Hormones and Behavior* 26: 87–109.

Borgerhoff-Mulder M. 1991. Human behavioral ecology: studies in foraging and reproduction. In: *Behavioral Ecology: An Evolutionary Approach*. Krebs J and Davies N (eds.). Blackwell.

Borgia G. 1979. Sexual selection and the evolution of mating systems. In: *Sexual Selection and Reproduction Competition in Insects*. Blum M and Blum N (eds.). Academic Press, pp. 19–80.

Bortolotti GR, Negro JJ, Tella JL, et al. 1996. Sexual dichromatism in birds independent of diet, parasites and androgens. *Proceedings of the Royal Society of London, B* 263: 1171–76.

Bottjer S and Sengelaub D. 1989. Cell death during development of a forebrain nucleus involved with vocal learning in zebra finches. *Journal of Neurobiology* 20: 609–18.

Bottjer S, Glaessner S, and Arnold A. 1985. Ontogeny of brain nuclei controlling song learning and behavior in zebra finches. *Journal of Neuroscience* 5: 1556–62.

Bottjer SW, Miesner EA, and Arnold AP. 1986. Changes in neuronal number, density and size account for increases in volume of song-control nuclei during song development in zebra finches. *Neuroscience Letters* 67: 263–68.

Bouloux PM, Munroe P, Kirk J, et al. 1992. Sex and smell—an enigma resolved. *Journal of Endocrinology* 133: 323–26.

Bouton N, Witte J, van Alphen A, et al. 1999. Local adaptations in populations of rock-dwelling haplochromines (Pisces: Cichlidae) from southern Lake Victoria. *Proceedings of the Royal Society of London, B* 266: 355–60.

Bradford J. 1988. Treatment of sexual offenders with cyproterone acetate. In: *Handbook of Sexology: The Pharmacology and Endocrinology of Sexual Function*. Sitse J (ed.). Elsevier. 6: 526.

Brainard M and Doupe A. 2000. Interruption of a basal ganglia-forebrain circuit prevents plasticity of learned vocalizations. *Nature* 404: 762–66.

Brandon P, Newton B, and Hammond O. 1987. Children's mathematics achievement in Hawaii: sex differences favoring girls. *American Educational Research Journal* 24: 437–61.

Brann D and Mahesh V. 1991. Role of corticosteroids in female reproduction. *Federation of American Societies for Experimental Biology* 5: 2691–98.

Brantley R and Bass A. 1994. Alternative male spawning tactics and acoustic signalling in the plainfin midshipman fish, *Porichthys notatus*. *Ethology* 96: 213–32.

Breedlove SM. 1992. Sexual dimorphism in the vertebrate nervous system. *Journal of Neuroscience* 12: 4133–42.

———. 1994. Sexual differentiation of the human nervous system. *Annual Reviews in Psychology* 45: 389–418.

———. 1997. Sex on the brain. *Nature* 389: 801.

Breedlove S and Arnold A. 1981. Sexually dimorphic motor nucleus in the rat lumbar spinal cord: response to adult hormone manipulation, absence in androgen-insensitive rats. *Brain Research* 225: 297–305.

Brenowitz E and Arnold A. 1986. Interspecific comparisons of the size of neural song control regions and song complexity in duetting birds: evolutionary implications. *Journal of Neuroscience* 6: 2875–79.

Brenowitz EA and Kroodsma DE. 1996. The neuroethology of birdsong. In: *The Ecology and Evolution of Acoustic Communication in Birds*. Kroodsma D and Miller E (eds.). Cornell University Press, pp. 285–304.

Brenowitz E, Arnold A, and Levin R. 1985. Neural correlates of female song in tropical duetting birds. *Brain Research* 343: 104–12.

Brenowitz EA, Lent K, and Kroodsma DE. 1995. Brain space for learned song in birds develops independently of song learning. *Journal of Neuroscience* 15: 6281–86.

Brenowitz EA, Margoliash D, and Nordeen KW. 1997. An introduction to birdsong and the avian song system. *Journal of Neurobiology* 33: 495–500.

Brenowitz EA, Nalis B, Wingfield JC, et al. 1991. Seasonal changes in avian song nuclei without seasonal changes in song repertoire. *Journal of Neuroscience* 11: 1367–74.

Brockmann H. 2001. The evolution of alternative strategies and tactics. *Advances in the Study of Behavior* 30: 1–51.

Bronson M. 1998. The implications for academic attainment of perceived gender-appropriateness upon spatial task performance. *British Journal of Educational Psychology* 68: 203–15.

Brooke J. 1991. *Science and Religion*. Cambridge University Press.

Brooks D and McLennan D. 1991. *Phylogeny, Ecology and Behavior: A Research Program in Comparative Biology*. University of Chicago Press.

Brown J. 1997. A theory of mate choice based on heterozygosity. *Behavioral Ecology* 8: 60–65.

Brown W. 2001. Natural selection of mammalian brain components. *Trends in Ecology and Evolution* 16: 471–73.

Bruner J. 1990. *Acts of Meaning*. Harvard University Press.

Bshary R and Grutter A. 2002. Asymmetric cheating opportunities and partner control in a cleaner fish mutualism. *Animal Behaviour* 63: 547–55.

Bshary R and Schaffer D. 2002. Choosy reef fish select cleaner fish that provide high-quality service. *Animal Behaviour* 63: 557–64.

Bshary R and Wurth M. 2001. Cleaner fish *Labroides dimidiatus* manipulate client reef fish by providing tactile stimulation. *Proceedings of the Royal Society of London, B* 1475: 1495–1501.

Buchanan K, Catchpole C, Lewis J, et al. 1999. Song as an indicator of parasitism in the sedge warbler. *Animal Behaviour* 57: 307–14.

Buchanan K, Evans M, Goldsmith A, et al. 2001. Testosterone influences basal metabolic rate in male house sparrows: a new cost of dominance signalling. *Proceedings of the Royal Society of London, B* 268: 1337–44.

Bull JJ. 1980. Sex determination in reptiles. *Quarterly Review of Biology* 55: 3–21.

———. 1983. *Evolution of Sex-Determining Mechanisms*. Benjamin Cummings.

Bull JJ, Vogt RC, and McCoy CJ. 1982. Sex determining temperatures in emydid turtles: a geographic comparison. *Evolution* 36: 326–32.

Buller D. 1999. DeFreuding evolutionary

psychology: adaptation and human motivation. In: *Where Biology Meets Psychology*. Hardcastle V (ed.). MIT Press, pp. 99–114.

Bullough WS. 1947. Hermaphroditism in the lower vertebrates. *Nature* 160: 9–11.

Burda H. 1995. Individual recognition and incest avoidance in eusocial common mole-rats rather than reproductive suppression by parents. *Experientia* 51: 411–13.

Burns RK. 1950. Sex transformation in the opossum: some new results and a retrospect. *Archives of Anatomy and Microscopic Morphology Experimentation* 39: 467–83.

——. 1961. Role of hormones in the differentiation of sex. In: *Sex and Internal Secretions*. Young W (ed.). Williams and Wilkins, pp. 76–160.

Burr D. 1999. Vision: modular analysis—or not? *Current Biology* 9: R90–R92.

Buss D. 1994. *The Evolution of Desire: Strategies of Human Mating*. Basic Books.

——. 1995. Psychological sex differences: origins through sexual selection. *American Psychologist* 50: 164–68.

——. 1999. *Evolutionary Psychology: The New Science of the Mind*. Allyn and Bacon.

Buss L. 1987. *The Evolution of Individuality*. Princeton University Press.

Butler A and Hodos W. 1996. *Comparative Vertebrate Neuroanatomy*. Wiley.

Butlin R. 2002. The costs and benefits of sex: new insights from old asexual lineages. *Nature Reviews Genetics* 3: 311–17.

Byers BE and Kroodsma DE. 1992. Development of two song categories by chestnut-sided warblers. *Animal Behaviour* 44: 799–810.

Byne W, Bleier R, and Houston L. 1988. Variations in human corpus callosum do not predict gender: a study using magnetic resonance imaging. *Behavioral Neuroscience* 102: 222–27.

Byne W, Tobet S, Mattiace L, et al. 2001. The interstitial nuclei of the human anterior hypothalamus: an investigation of variation with sex, sexual orientation, and HIV status. *Hormones and Behavior* 40: 86–92.

Canady RA, Kroodsma DE, and Nottebohm F. 1984. Population differences in complexity of a learned skill are correlated with the brain space involved. *Proceedings of the National Academy of Sciences, USA* 81: 6232–34.

Cant M. 1998. A model for the evolution of reproductive skew without reproductive suppression. *Animal Behaviour* 55: 163–69.

——. 2000. Social control of reproduction in banded mongooses. *Animal Behaviour* 59: 147–58.

Caplan P and Caplan J. 1997. Do sex-related cognitive differences exist, and why do people seek them out? In: *Gender Differences in Human Cognition*. Caplan P, Crawford M, Hyde J, and Richardson J (eds.). Oxford University Press. pp. 52–80.

Caplan PJ, MacPherson GM, and Tobin P. 1985. Do sex-related differences in spatial abilities exist? *American Psychologist* 40: 786–99.

Cardwell J and Liley N. 1987. Hormonal correlates of sex change and colour phase change in the stoplight parrotfish (*Sparisoma viride*). *Proceedings of the Third International Symposium on the Reproductive Physiology of Fishes*. Idler D, Crim L, and Walsh J (eds.). Memorial University.

——. 1991a. Androgen control of social status in males of a wild population of stoplight parrotfish, *Sparisoma viride* (*Scaridae*). *Hormones and Behavior* 25: 1–18.

——. 1991b. Hormonal control of sex and color change in the stoplight parrotfish, *Sparisoma viride*. *General and Comparative Endocrinology* 81: 7–20.

Caro TM and Bateson P. 1986. Organization and ontogeny of alternative tactics. *Animal Behaviour* 34: 1483–99.

Casey M and Brabeck M. 1989. Exceptions to the male advantage on spatial tasks: family handedness and college major as factors identifying women who excel. *Neuropsychologia* 27: 689–96.

——. 1990. Women who excel on a spatial task: proposed genetic and environmental factors. *Brain and Cognition* 12: 73–84.

Casey M, Colon D, and Goris Y. 1992a. Family handedness as predictor of mental rotation ability among minority girls in a math-science training program. *Brain and Cognition* 18: 88–96.

Casey M, Pezaris E, and Nuttall R. 1992b.

Spatial ability as a sex predictor of math achievement: the importance of sex and handedness patterns. *Neuropsychologia* 30: 35–45.

Casserly P. 1980. Factors affecting female participation in advanced placement programs in mathematics, chemistry, and physics. In: *Women and the Mathematical Mystique*. Fox L, Brody L, and Tobin D (eds.). Johns Hopkins University Press, pp. 138–63.

Casto J, Nolan Jr. V, and Ketterson E. 2001. Steroid hormones and immune function: experimental studies in wild and captive dark-eyed juncos (*Junco hyemalis*). *American Naturalist* 157: 408–20.

Catchpole C. 1980. Sexual selection and the evolution of complex songs among warblers of the genus *Acrocephalus*. *Behaviour* 74: 149–66.

———. 1986. Song repertoires and reproductive success in the great reed warbler *Acrocephalus arundinaceus*. *Behavioral Ecology and Sociobiology* 19: 439–45.

———. 1987. Bird song, sexual selection and female choice. *Trends in Ecology and Evolution* 2: 94–97.

———. 1996. Song and female choice: good genes and big brains? *Trends in Ecology and Evolution* 11: 358–60.

———. 2000. Sexual selection and the evolution of song and brain structure in *Acrocephalus* warblers. *Advances in the Study of Behavior* 29: 45–97.

Catchpole C, Dittami J, and Leisler B. 1984. Differential responses to male song repertoires in female songbirds implanted with oestradiol. *Nature* 312: 563–64.

Cavigelli S. 1999. Behavioral patterns associated with faecal cortisol levels in free-ranging female ring-tailed lemurs, *Lemur catta*. *Animal Behaviour* 57: 935–44.

Chagnon N and Irons W (eds.). 1979. *Evolutionary Biology and Human Social Behavior*. Duxbury Press.

Chambers K and Phoenix C. 1981. Diurnal patterns of testosterone, dihydrotestosterone, estradiol, and cortisol in serum of rhesus males: relationship to sexual behavior in aging males. *Hormones and Behavior* 15: 416–26.

Chan STH, Wai-sum O, and Hui WB. 1975. The gonadal and adenohypophysial functions of natural sex reversal. In: *Intersexuality in the Animal Kingdom*. Reinboth R (ed.). Springer-Verlag, pp. 201–21.

Chan STH, Ng TB, O WS, et al. 1977. Hormones and natural sex succession in *Monopterus*. In: *Biological Research in Southeast Asia*. Lofts B and Chan STH (eds.). University of Hong Kong. pp. 101–20.

Charnier M. 1966. Action de la température sur la sex-ratio chez l'embryon d'*Agama agama* (Agamidae, Lacertilien). *Society for Biology Ouest Africa* 160: 620–22.

Charnov EL. 1982. *The Theory of Sex Allocation*. Princeton University Press.

Chase ID. 1974. Models of hierarchy formation in animal societies. *Behavioral Sciences* 19: 374–82.

———. 1982. Dynamics of hierarchy formation: the sequential development of dominance relationships. *Behaviour* 80: 218–39.

———. 1986. Explanations of hierarchy structure. *Animal Behaviour* 34: 1265–67.

Chase I, Bartolomeo C, and Dugatkin L. 1994. Aggressive interactions and inter-contest interval: how long do winners keep winning? *Animal Behaviour* 48: 393–400.

Chase I, Tovey C, Spangler-Martin D, et al. 2002. Individual differences versus social dynamics in the formation of animal dominance hierarchies. *Proceedings of the National Academy of Sciences* 99: 5744–49.

Cheverud J, Dow M, and Leutenegger W. 1985. The quantitative assessment of phylogenetic constraints in comparative analyses: sexual dimorphism in body weight among primates. *Evolution* 39: 1335–41.

Chisholm A. 1932. Vocal mimicry among Australian birds. *Ibis* 13: 605–25.

Chomsky N. 1959. Review of "Verbal Behavior." *Language* 35: 26–58.

———. 1965. *Aspects of a Theory of Syntax.* MIT Press.

———. 1968. *Language and Mind*. Harcourt Brace & World.

———. 1978. *Rules and Representations*. Columbia University Press.

———. 1980. Rules and representations. *Behavioral and Brain Sciences* 3: 1–61.

——. 1988. *Language and the Problems of Knowledge.* MIT Press.

Christian J. 1964. Actions of ACTH in intact and corticoid-maintained adrenalectomized female mice with emphasis on the reproductive tract. *Endocrinology* 75: 653–69.

Chun MM and Phelps EA. 1999. Memory deficits for implicit contextual information in amnesic subjects with hippocampal damage. *Nature Neuroscience* 2: 844–47.

Clark D, Mitra P, and Wang W-Y. 2001. Scalable architecture in mammalian brains. *Nature* 411: 189–93.

Clark M and Galef B. 1988. Effects of uterine position on sexual development in female Mongolian gerbils. *Physiology and Behavior* 42: 15–18.

Clark M, Spencer B, and Galef B. 1986. Reproductive life history correlates of early and late sexual maturation in female Mongolian gerbils (*Meriones unguiculatus*). *Animal Behaviour* 34: 551–58.

——. 1995. Prenatal influences on reproductive life history strategies. *Trends in Ecology and Evolution* 10: 151–53.

Clark MM, Karpiuk P, and Galef BG. 1993. Hormonally mediated inheritance of acquired characteristics in Mongolian gerbils. *Nature* 364: 712.

Clark M, Bennett G, and Galef J. 2001. Socially induced infertility: familial effects on reproductive development of female Mongolian gerbils. *Animal Behaviour* 62: 897–903.

Clarke FM and Faulkes CG. 1997. Dominance and queen succession in captive colonies of the eusocial naked mole-rat, *Heterocephalus glaber. Proceedings of the Royal Society of London B* 264: 993–1000.

——. 1998. Hormonal and behavioural correlates of male dominance and reproductive status in captive colonies of the naked mole-rat, *Heterocephalus glaber. Proceedings of the Royal Society of London, B* 265: 1391–99.

——. 2001. Intracolony aggression in the eusocial naked mole-rat, *Heterocephalus glaber. Animal Behaviour* 61: 311–24.

Clarke F, Miethe G, and Bennett N. 2001. Reproductive suppression in female Damaraland mole-rats *Cryptomys damarensis*: dominant control or self-restraint? *Proceedings of the Royal Society of London, B* 268: 899–909.

Clarke M. 1974. *Paley: Evidences for the Man.* University of Toronto Press.

Clarke S, Kraftsik R, vanderLoos H, et al. 1989. Forms and measures of adult and developing human corpus callosum: is there sexual dimorphism? *Journal of Comparative Neurology* 280: 213–80.

Clayton N. 1994. The role of age and experience in the behavioural development of food-storing and retrieval in marsh tits, *Parus palustris. Animal Behaviour* 47: 1435–44.

——. 1995a. Development of memory and the hippocampus: comparison of food-storing and nonstoring birds on a one-trial associative memory task. *Journal of Neuroscience* 15: 2796–807.

——. 1995b. The neuroethological development of food-storing memory: a case of use it, or lose it. *Behavioural Brain Research* 70: 95–102.

——. 1995c. Memory and the hippocampus in food-storing birds: a comparative approach. *Journal of Neuropharmacology* 37: 441–52.

——. 1996. Development of food-storing and the hippocampus in juvenile marsh tits (*Parus palustris*). *Behavioral Brain Research* 74: 153–59.

——. 1998. Memory and the hippocampus in food-storing birds. In: *Animal Cognition in Nature.* Balda R, Pepperberg I, and Kamil A (eds.). Academic Press, pp. 99–118.

Clayton NS and Dickinson A. 1999. Motivational control of caching behaviour in the scrub jay, *Aphelocoma coerulescens. Animal Behaviour* 57: 435–44.

Clayton NS and Krebs JR. 1994. Hippocampal growth and attrition in birds affected by experience. *Proceedings of the National Academy of Sciences, USA* 91: 7410–14.

——. 1995. Memory in food-storing birds: from behaviour to brain. *Current Opinion in Neurobiology* 5: 149–54.

Clayton N, Griffiths D, Emergy N, et al. 2001. Elements of episodic-like memory in animals. *Proceedings of the Royal Society of London, B* 356: 1483–91.

Clayton N, Reboreda J and Kacelnik A.

1997. Seasonal changes of hippocampal volume in parasitic cowbirds. *Behavioral Processes* 41: 237–43.

Clayton N and Soha J. 1999. Memory in avian food caching and song learning: a general mechanism or different processes? *Advances in the Study of Behavior* 28: 115–73.

Clemens HP and Inslee T. 1968. The production of unisexual broods by *Tilapia mossambica* sex-reversed with methyl testosterone. *Transactions of the American Fisheries Society* 97: 18–21.

Clutton-Brock T. 1998. Reproductive skew, concessions and limited control. *Trends in Ecology and Evolution* 13: 288–91.

Clutton-Brock T, Harvey P, and Rudder B. 1977. Sexual dimorphism, socioeconomic sex ratio and body weight in primates. *Nature* 269: 797–800.

Clutton-Brock T, Brotherton P, Russell A, et al. 2001. Cooperation, control, and concession in meerkat groups. *Science* 291: 478–81.

Cohen N and Eichenbaum H. 1991. The theory that wouldn't die: a critical look at the spatial mapping theory of hippocampal function. *Hippocampus* 1: 265–68.

Cole CJ. 1975. Evolution of parthenogenetic species of reptiles. *Intersexuality in the Animal Kingdom*. Reinboth R (ed.). Springer-Verlag.

Cole KS. 1989. Patterns of gonad structure in hermaphroditic gobies (Pisces: Gobiidae). *Environmental Biology of Fishes* 28: 125–42.

Collaer ML and Hines M. 1995. Human behavioral sex differences: a role for gonadal hormones during early development? *Psychological Bulletin* 118: 55–107.

Collias NE. 1943. Statistical analysis of factors which make for success in initial encounters between hens. *American Naturalist* 77: 519–38.

Commins D and Yahr P. 1984. Lesions of the sexually dimorphic area disrupt mating and marking in male gerbils. *Brain Research Bulletin* 13: 185.

Conte F, Grumbach M, Ito Y, et al. 1994. A syndrome of female pseudohermaphroditism, hypergonadotropic hypogonadism, and multicystic ovaries associated with missense mutations in the gene encoding aromatase (P450 arom). *Journal of Clinical Endocrinology and Metabolism* 78: 1287–92.

Cosmides L. 1989. The logic of social exchange: has natural selection shaped how humans reason? *Cognition* 31: 187–276.

Cosmides L and Tooby J. 1987. From evolution to behavior: evolutionary psychology as the missing link. In: *The Latest on the Best*. Dupre J (ed.). MIT Press.

——. 1989. Evolutionary psychology and the generation of culture, part 2. *Ethology and Sociobiology* 10: 51–97.

——. 1992. Cognitive adaptations for social exchange. In: *The Adapted Mind: Evolutionary Psychology and the Generation of Culture*. Barkow J, Cosmides L, and Tooby J (eds.). Oxford University Press, pp. 163–228.

Coyne JA and Sohn JJ. 1978. Interspecific brood care in fishes: Reciprocal altruism or mistaken identity? *American Naturalist* 112: 447–50.

Crawford C. 1993. The future of sociobiology: counting babies or studying proximate mechanisms. *Trends in Ecology and Evolution* 8: 183–86.

Creel S and Creel N. 1991. Energetics, reproductive suppression and obligate communal breeding in carnivores. *Behavioral Ecology and Sociobiology* 28: 263–70.

Creel S, Creel N, Wildt DE, et al. 1992. Behavioural and endocrine mechanisms of reproductive suppression in Serengeti dwarf mongooses. *Animal Behaviour* 43: 231–45.

Creel S, Creel N, Monfort S. 1996. Social stress and dominance. *Nature* 379: 212.

——. 1997. Radiocollaring and stress hormones in African wild dogs. *Conservation Biology* 11: 544–48.

Crews D. 1975. Psychobiology of reptilian reproduction. *Science* 189: 1059–65.

——. 1983. Control of male sexual behavior in the Canadian red-sided garter snake. In: *Hormones and Behavior in Higher Vertebrates*. Balthazart J, Prove E, and Gilles R (eds.). Plenum, pp. 398–408.

——. 1984. Gamete production, sex hormone secretion, and mating behavior uncoupled. *Hormones and Behavior* 18: 22–28.

——. 1985. Effects of early hormonal treatment on courtship behavior and sexual activity in the red-sided garter snake, *Thamnophis sirtalis parietalies*. *Physiology and Behavior* 35: 569–75.

——. 1987. Diversity and evolution of behavioral controlling mechanisms. In: *Psychobiology of Reproductive Behavior: An Evolutionary Perspective*. Crews D (ed.). Prentice Hall, pp. 88–119.

——. 1993. The organizational concept and vertebrates without sex chromosomes. *Brain, Behavior and Evolution* 42: 202–14.

Crews D and Gartska W. 1982. The ecological physiology of garter snakes. *Scientific American* 247: 158–68.

Crews D and Sakata J. 2000. Evolution of brain mechanisms controlling sexual behavior. In: *Sexual Differentiation of the Brain*. Matsumoto A (ed.). CRC Press, pp. 113–30.

Crews D, Bergeron JM, Bull JJ, et al. 1994. Temperature-dependent sex determination in reptiles: proximate mechanisms, ultimate outcomes, and practical applications. *Developmental Genetic* 15: 297–312.

Crews D, Grassman M, Lindzey J. 1986. Behavioral facilitation of reproduction in sexual and parthenogenetic whiptail (*Cnemidophorus*) lizards. *Proceedings of the National Academy of Sciences USA* 83: 9547–50.

Crews D and Silver R. 1985. Reproductive physiology and behavioral interactions in nonmammalian vertebrates. In: *Handbook of Behavioral Neurobiology*. Adler N, Pfaff DW, Goy RW (eds.). Plenum, pp. 101–82.

Crews D, Wade J, and Wilczynski W. 1990. Sexually dimorphic areas in the brain of whiptail lizards. *Brain, Behaviour and Evolution* 36: 262–70.

Crucian G and Berenbaum S. 1998. Sex differences in right hemisphere tasks. *Brain and Cognition* 36: 377–89.

Daly M and Wilson M. 1988. *Homocide*. Aldine de Gruyter.

——. 1999. Human evolutionary psychology and animal behaviour. *Animal Behaviour* 57: 509–19.

Danielou AT (ed.). 1994. *The Complete Kama Sutra*. Park Street Press.

Dart R. 1953. The predatory transition from ape to man. *International Anthropological and Linguistic Review* 1: 301–13.

Darwin C. 1958. *The Autobiography of Charles Darwin*. Barlow N (ed.). Collins.

Davies P. 1999. The conflict of evolutionary psychology. In: *Where Biology Meets Psychology*. Hardcastle V (ed.). MIT Press, pp. 67–81.

Davies PS, Fetzer JH, and Foster TR. 1995. Logical reasoning and domain specificity: a critique of the social exchange theory of reasoning. *Biology and Philosophy* 10: 1–37.

Davis MR and Fernald RD. 1990. Social control of neuronal soma size. *Journal of Neurobiology* 21: 1180–88.

Davis P and Takashima. 1980. Sex differentiation in common carp, *Cyprinus carpio*. *Journal of Tokyo University Fisheries* 66: 191–99.

Dawkins R. 1986. *The Blind Watchmaker*. Longmans.

Dawkins R and Krebs J. 1978. Animal signals: information or manipulation. In: *Behavioural Ecology: An Evolutionary Approach*. Krebs J and Davies N (eds.). Blackwell, pp. 282–309.

de Courten-Myers G. 1999. The human cerebral cortex: gender differences in structure and function. *Journal of Neuropathology and Experimental Neurology* 58: 217–26.

DeJonge F, Louwerse A, Ooms M, et al. 1989. Lesions of the SDN-POA inhibit sexual behavior of male Wistar rats. *Brain Research Bulletin* 23: 483–92.

de Lacoste M, Holloway R, and Woodward D. 1986. Sex differences in the fetal human corpus callosum. *Human Neurobiology* 5: 1–5.

de Lacoste-Utamsing C and Holloway R. 1982. Sexual dimorphism in the human corpus callosum. *Science* 216: 1431.

Delgado RJ and Van Schaik C. 2000. The behavioral ecology and conservation of the orangutan (*Pongo pygmaeus*): a tale of two islands. *Evolutionary Anthropology* 9: 201–18.

Dellovade T, Schwanzel-Fukuda M, Gordan J, et al. 1998. Aspects of GnRH neurobiology conserved across vertebrate forms. *General and Comparative Endocrinology* 112: 276–82.

Dennett D. 1978. *Brainstorms*. MIT Press.
———. 1995. *Darwin's Dangerous Idea*. Simon & Schuster.

Densmore L, Moritz C, Wright JW, and Brown WM. 1989. Mitochondrial DNA analysis and the origin and relative age of parthenogenetic lizards (genus *Cnemidophorus*). IV. Nine *sexlineatus* group unisexuals. *Evolutions* 43: 969–83.

Deutsch C, Crocker D, Costa D, et al. 1994. Sex- and age-related variation in reproductive effort of northern elephant seals. In: *Elephant Seals: Population Ecology, Behavior, and Physiology*. LeBoeuf B and Laws R (eds.). University of California Press, pp. 169–209.

DeVoogd TJ. 1991. Endocrine modulation of the development and adult function of the avian song system. *Psychoneuroendocrinology* 16: 41–66.

DeVoogd T and Nottebohm F. 1981. Gonadal hormones induce dendritic growth in the adult avian brain. *Science* 214: 202–4.

DeVoogd TJ, Krebs JR, Healy SD, et al. 1993. Relations between song repertoire size and the volume of brain nuclei related to song: comparative evolutionary analyses amongst oscine birds. *Proceedings of the Royal Society of London, B* 254: 75–82.

Dewey J. 1910. *The Influence of Darwin on Philosophy, and Other Essays in Contemporary Thought*. Holt, Rhinehart and Winston.

de Winter W and Oxnard C. 2001. Evolutionary radiations and convergences in the structural organization of mammalian brains. *Nature* 409: 710–14.

DiPietro J. 1981. Rough and tumble play: a function of gender. *Developmental Psychology* 17: 50–58.

Dixson A. 1980. Androgens and aggressive behavior in primates: a review. *Aggressive Behavior* 6: 37–67.

Dobkin D. 1979. Functional and evolutionary relationships of vocal copying phenomena in birds. *Zeitschrift für Tierpsychologie* 50: 348–63.

Dodd J. 1960. Genetic and environmental aspects of sex determination in cold-blooded vertebrates. *Memoirs of the Society for Endocrinology* 7: 17–44.

Dominey W. 1980. Female mimicry in male bluegill sunfish—a genetic polymorphism? *Nature* 284: 546–48.
———. 1984. Effects of sexual selection and life history on speciation: species flocks in African cichlids and Hawaiian *Drosophila*. In: *Evolution of Species Flocks*. Echelle A and Kornfield I (eds.). University of Maine, pp. 231–49.

Donaldson EM and Hunter GA. 1982. Sex control in fish with particular reference to salmonids. *Canadian Journal of Fisheries and Aquatic Science* 39: 99–110.

Dorit R. 1990. The correlates of high diversity in Lake Victoria haplochromine cichlids: a neontological perspective. In: *Causes of Evolution: A Paleontological Perspective*. Ross R and Allmon W (eds.). University of Chicago Press, pp. 322–53.

Doupe AJ. 1993. A neural circuit specialized for vocal learning. *Current Opinion in Neurobiology* 3: 104–11.

Downs S. 2001. Some recent developments in evolutionary approaches to the study of cognition and behavior. *Biology and Philosophy* 16: 575–95.

Dowsett-Lemaire F. 1979. The imitative range of the song of the marsh warbler *Acrocephalus palustris*, with special reference to imitations of African birds. *Ibis* 135: 181–89.

Drea CM, Weldele ML, Forger NG, et al. 1998. Androgens and masculinization of genitalia in the spotted hyaena (*Crocuta crocuta*). 2. Effects of prenatal anti-androgens. *Journal of Reproduction and Fertility* 113: 117–27.

Drummond H and Canales C. 1998. Dominance between booby nestlings involves winner and loser effects. *Animal Behaviour* 55: 1669–76.

Duchaine B, Cosmides L, and Tooby J. 2001. Evolutionary psychology and the brain. *Current Opinion in Neurobiology* 11: 225–30.

Duff S and Hampson E. 2001. A sex difference on a novel spatial working memory task in humans. *Brain and Cognition* 47: 470–93.

Duffy D and Ball G. 2002. Song predicts immunocompetence in male European starlings (*Sturnus vulgaris*). *Proceedings of the Royal Society of London, B* 269: 847–52.

Dunbar R. 1983. Life history tactics and alternative strategies of reproduction. In: *Mate Choice*. Bateson P (ed.). Cambridge University Press, pp. 423–33.

Duncan J and Owen A. 2000. Common regions of the human frontal lobe recruited by diverse cognitive demands. *Trends in Neuroscience* 23: 475–83.

Durant W and Durant A. 1965. *The Story of Civilization*. Simon & Schuster.

Eagly A. 1995. The science and politics of comparing women and men. *American Psychologist* 50: 145–58.

Eales L. 1985. Song learning in zebra finches: some effects of song model availability on what is learnt and when. *Animal Behaviour* 33: 1293–300.

———. 1987. Do zebra finch males that have been raised by another species still tend to select a conspecific song tutor? *Animal Behaviour* 35: 1347–55.

Eals M and Silverman I. 1994. The hunter-gatherer theory of spatial sex differences: proximate factors mediating the female advantage in recall of object arrays. *Ethology and Sociobiology* 15: 95–105.

East ML and Hofer H. 1997. The peniform clitoris of female spotted hyenas. *Trends in Ecology and Evolution* 12: 401–2.

———. 2002. Conflict and cooperation in a female-dominated society: a reassessment of the "hyperaggressive" image of spotted hyenas. *Advances in the Study of Behavior* 31: 1–30.

East ML, Hofer H, and Wickler W. 1993. The erect "penis" is a flag of submission in a female-dominated society: greetings in Serengeti spotted hyenas. *Behavioral and Evolutionary Sociobiology* 33: 355–70.

Eaton W and Enns L. 1986. Sex differences in human motor activity level. *Psychological Bulletin* 100: 19–28.

Ebeling AW and Chen TR. 1970. Heterogamety in teleosten fishes. *Transactions of the American Fisheries Society* 1: 131–38.

Ebrhardt C and Baptista L. 1977. Intraspecific and interspecific song mimesis in California song sparrows. *Bird Banding* 48: 193–205.

Ehrhardt A, Epstein R, and Money J. 1968a. Fetal androgens and female gender identity in the early-treated adrenogenital syn-

drome. *Johns Hopkins Medical Journal* 122: 160–67.

Ehrhardt A, Evers K, and Money J. 1968b. Influence of androgens and some aspects of sexually dimorphic behavior in women with the late-treated adrenogenital syndrome. *John Hopkins Medical Journal* 123: 115–22.

Ehrhardt AA, Meyer-Bahlburg HFL, Rosen LR, et al. 1985. Sexual orientation after prenatal exposure to exogenous estrogen. *Archives of Sexual Behavior* 14: 57–75.

———. 1989. The development of gender-related behavior in females following prenatal exposure to diethylstilbesterol (DES). *Hormones and Behavior* 23: 526–41.

Eibl-Eibesfeldt I. 1971. *Love and Hate: The Natural History of Behavior*. Holt, Rinehart and Winston.

———. 1989. *Human Ethology*. Aldine de Gruyter.

Eichenbaum H. 1996. Is the rodent hippocampus just for "place"? *Current Opinion in Neurobiology* 6: 187–95.

———. 2000. Hippocampus: mapping or memory? *Current Biology* 10: R785–87.

Eichenbaum H, Dudchenko P, Wood E, et al. 1999. The hippocampus, memory, and place cells: is it spatial memory or a memory space? *Neuron* 23: 209–26.

Eichenbaum H, Otto T, and Cohen NJ. 1994. Two component functions of the hippocampal memory system. *Behavioral and Brain Sciences* 17: 449–517.

Eldridge L, Knowlton B, Furmanski C, et al. 2000. Remembering episodes: a selective role for the hippocampus during retrieval. *Nature Neuroscience* 3: 1149–52.

Ellis H. 1894. *Man and Woman: A Study of Human Secondary and Tertiary Sexual Characters*. Scribner's.

Elofsson U, Mayer I, Damsgard B, et al. 2000. Intermale competition in sexually mature Arctic charr: effects on brain monoamines, endocrine stress responses, sex hormone levels, and behavior. *General and Comparative Endocrinology* 118: 450–60.

Emerson S and Hess D. 2001. Glucocorticoids, androgens, testis mass, and the energetics of vocalization in breeding male frogs. *Hormones and Behavior* 39: 59–69.

Emlen S. 1995. Predicting family dynamics in social vertebrates. *Proceedings of the National Academy of Sciences* 92: 8092–99.

Endler J. 1980. Natural selection on color patterns in *Poecilia reticulata*. *Evolution* 34: 76–91.

———. 1992. Some general comments on the evolution and design of animal communication systems. *Philosophical Transactions of the Royal Society B* 340: 215–25.

Engh A, Funk S, Van Horn R, et al. 2002. Reproductive skew among males in a female-dominated mammalian society. *Behavioral Ecology* 13: 193–200.

Enquist M and Arak A. 1993. Selection of exaggerated male traits by female aesthetic senses. *Nature* 372: 169–72.

———. 1994. Symmetry, beauty and evolution. *Nature* 372: 169–72.

Erichsen JT, Bingman VP, and Krebs JR. 1991. The distribution of neuropeptides in the dorsomedial telencephalon of the pigeon (*Columba livia*): a basis for regional subdivisions. *Journal of Comparative Neurology* 314: 478–92.

Ethington C. 1990. Gender differences in mathematics: an international perspective. *Journal of Research in Mathematics Education* 21: 74–80.

Evans RM. 1988. The steroid and thyroid hormone receptor superfamily. *Science* 240: 889–95.

Evans-Pritchard E. 1937. *Witchcraft, Oracles and Magic among the Azande*. Clarendon.

Eviatar Z, Hellige J, and Zaidel E. 1997. Individual differences in lateralization: effects of gender and handedness. *Neuropsychology* 11: 562–76.

Falk D. 2001. The evolution of sex differences in primate brains. In: *Evolutionary Anatomy of the Primate Cerebral Cortex*. Falk D and Gibson K (eds.). Cambridge University Press, pp. 98–112.

Farabaugh S. 1982. Ecological and social significance of duetting. In: *Acoustic Communication in Birds*. Kroodsma D and Miller E (eds.). Academic Press, pp. 85–124.

Farah MJ. 1994. Neuropsychological inference with an interactive brain: a critique of the locality assumption. *Behavioral and Brain Sciences* 17: 43–104.

Faulks CG and Abbott D. 1993. Social control of reproduction in breeding and non-breeding naked mole rats (*Heterocephalus glaber*). *Journal of Reproduction and Fertility* 93: 427–35.

———. 1997. The physiology of a reproductive dictatorship: regulation of male and female reproduction by a single breeding female in colonies of naked mole-rats. In: *Cooperative Breeding in Mammals*. Solomon N and French J (eds.). Cambridge University Press, pp. 302–34.

Faulkes CG, Abbott D, and Jarvis J. 1990a. Social suppression of ovarian cyclicity in captive and wild colonies of female naked mole rats (*Heterocephalus glaber*). *Journal of Reproduction and Fertility* 88: 559–68.

Faulkes CG, Abbott D, Jarvis J, et al. 1990b. LH responses of female naked mole rats, *Heterocephalus glaber*, to single and multiple doses of exogenous GnRH. *Journal of Reproduction and Fertility* 89: 317–23.

Faulkes CG, Abbott DH, and Jarvis JUM. 1991a. Social suppression of reproduction in male naked mole-rats, *Heterocephalus glaber*. *Journal of Reproduction and Fertility* 91: 593–604.

Faulkes CG, Abbott D, Liddell C, et al. 1991b. Hormonal and behavioral aspects of reproductive suppression in female naked mole rats. In: *The Biology of the Naked Mole Rats*. Sherman P, Jarvis J, and Alexander R (eds.). Princeton University Press, pp. 426–45.

Faulkes CG, Trowell SN, Jarvis JUM, et al. 1994. Investigation of numbers and motility of spermatozoa in reproductively active and socially suppressed males of two eusocial African mole-rats, the naked mole-rat (*Heterocephalus glaber*) and the Damaraland mole-rat (*Cryptomys damarensis*). *Journal of Reproduction and Fertility* 100: 411–16.

Faulkes CG, Bennett NC, Bruford M, et al. 1997. Ecological constraints drive social evolution in the African mole rats. *Proceedings of the Royal Society of London, B* 264: 1619–27.

Fausto-Sterling A. 1985. *Myths of Gender: Biological Theories about Men and Women*. HarperCollins.

Feingold A. 1988. Cognitive gender differences are disappearing. *American Psychologist* 43: 95–103.

Ferguson MWJ and Joanen T. 1982. Temperature of egg incubation determines sex in *Alligator mississippiensis*. *Nature* 296: 850–53.

Fernandez X, Meunier-Salaun M, and Momede P. 1994. Agonistic behavior, plasma stress hormones and metabolites in response to dyadic encounters in domestic pigs: interrelationships and effects of dominance status. *Physiology and Behavior* 56: 841–47.

Fernandez-Guasti A, Kruijver F, Fodor M, et al. 2000. Sex differences in the distribution of androgen receptors in the human hypothalamus. *Journal of Comparative Neurology* 425: 422–25.

Fischer E. 1984. Egg trading in the chalk bass, *Serranus tortugarum*, a simultaneous hermaphroditic reef fish. *Zeitschrift für Tierpsychologie* 66: 124–34.

Fishman M. 2001. Zebrafish—the canonical vertebrate. *Science* 294: 1290–91.

Fleagle J, Kay R, and Simons E. 1980. Sexual dimorphism in early anthropoids. *Nature* 287: 328–30.

Fleming IA and Gross MR. 1994. Breeding competition in a Pacific salmon (coho: *Oncorhynchus kisutch*): measures of natural and sexual selection. *Evolution* 48: 637–57.

Fodor J. 1968. *Psychological Explanation*. Random House.

——. 1974. Special sciences. *Synthese* 28: 77–115.

——. 1981. *Representations*. MIT Press.

——. 1983. *The Modularity of Mind*. MIT Press.

——. 1985. Precis of the modularity of mind. *Behavioral and Brain Sciences* 8: 1–42.

——. 2000. *The Mind Doesn't Work That Way: The Scope and Limits of Computational Psychology*. MIT Press.

Foley R. 1996. The adaptive legacy of human evolution: a search for the environment of evolutionary adaptedness. *Evolutionary Anthropology* 5: 194–203.

Folstad I and Karter A. 1992. Parasites, bright males, and the immunocompetence handicap. *American Naturalist* 139: 603–22.

Forester D and Lykens D. 1986. Significance of satellite males in a population of spring peepers (*Hyla crucifer*). *Copeia* 1986: 719–24.

Forger N and Breedlove N. 1986. Sexual dimorphism in human and canine spinal cord: role of early androgens. *Proceedings of the National Academy of Sciences, USA* 83: 7527–31.

——. 1987. Motoneuronal death during human fetal development. *Journal of Comparative Neurology* 264: 118–22.

Forger N, Hodges L, Roberts S, et al. 1992. Regulation of motoneuron death in the spinal nucleus of the bulbocavernosus. *Journal of Neurobiology* 23: 1192–203.

Fortin N, Agster K, and Eichenbaum H. 2002. Critical role of the hippocampus in memory for sequences of events. *Nature Neuroscience* 5: 458–62.

Fortman M, Dellovade T, and Rissman EF. 1992. Adrenal contribution to the induction of sexual behavior in the female musk shrew. *Hormones and Behavior* 26: 76–86.

Fox CA, Wolff HS, and Baker JA. 1970. Measurement of intra-vaginal and intra-uterine pressures during human coitus by radio-telemetry. *Journal of Reproduction and Fertility* 22: 243–351.

Fox L, Benbow C, and Perkins S. 1983. An accelerated mathematics program for girls. In: *Academic Precocity: Aspects of Its Development*. Benbow C and Stanley J (eds.). Johns Hopkins University Press, pp. 113–39.

Fox H, White S, Kao M, et al. 1997. Stress and dominance in a social fish. *Journal of Neuroscience* 17: 6463–69.

Fraley NB and Fernald RD. 1982. Social control of developmental rate in the African cichlid, *Haplochromis burtoni*. *Zeitschrift für Tierpsychologie* 60: 66–82.

Frame L, Malcolm J, Frame G, et al. 1979. Social organization of African wild dogs (*Lycaon pictus*) on the Serengeti plains, Tanzania (1967–1978). *Zeitschrift für Tierpsychologie* 50: 225–49.

Francis RC. 1983. Experiential effects on agonistic behavior in the paradise fish, *Macropodus opercularis*. *Behaviour* 85: 292–313.

——. 1984. The effects of bidirectional selection for social dominance on agonistic behavior and sex ratios in the paradise fish (*Macropodus opercularis*). *Behaviour* 90: 25–45.

——. 1987. The interaction of genotype and experience in the dominance success of paradise fish (*Macropodus opercularis*). *Biology and Behavior* 12: 1–11.

——. 1988a. On the relationship between aggression and social dominance. *Ethology* 78: 223–37.

——. 1988b. Socially mediated variation in growth rate of the Midas cichlid: the primacy of early size differences. *Animal Behaviour* 36: 1844–45.

——. 1992. Sexual lability in teleosts: developmental factors. *Quarterly Review of Biology* 67: 1–18.

——. 1995. Evolutionary neurobiology. *Trends in Ecology and Evolution* 10: 276–81.

Francis RC and Barlow GW. 1993. Social control of primary sex differentiation in the Midas cichlid. *Proceedings of the National Academy of Sciences, USA* 90: 10673–75.

Francis RC, Jacobson B, Wingfield JC, et al. 1992a. Castration lowers aggression but not social dominance in male *Haplochromis burtoni* (Cichlidae). *Ethology* 90: 247–55.

——. 1992b. Hypertrophy of gonadotropin releasing hormone-containing neurons after castration in the teleost, *Haplochromis burtoni. Journal of Neurobiology* 23: 1084–93.

Francis RC, Soma K, and Fernald RD. 1993. Social regulation of the brain-pituitary-gonadal axis. *Proceedings of the National Academy of Sciences, USA* 90: 7794–98.

Frank LG. 1986. Social organization of the spotted hyaena *Crocuta crocuta*. II. Dominance and reproduction. *Animal Behaviour* 34: 1510–27.

——. 1997. Evolution of genital masculinization: why do female hyenas have such a large "penis"? *Trends in Ecology and Evolution* 12: 58–62.

Frank LG and Glickman SE. 1994. Giving birth through a penile clitoris: parturition and dystocia in the spotted hyaena. *Journal of Zoology London* 234: 659–90.

Frank L, Glickman S, and Licht P. 1991. Fatal sibling aggression, precocial development and androgens in the neonatal spotted hyaenas. *Science* 252: 702–4.

Frank LG, Weldele ML, and Glickman SE. 1995. Masculinization costs in hyaenas. *Nature* 377: 584–85.

Frank L, Brown E, and Wilson M. 2000. Trajectory encoding in the hippocampus and entorhinal cortex. *Neuron* 27: 169–78.

Fredrickson B, Roberts T, Noll S, et al. 1998. That swimsuit becomes you: sex differences in self-objectification, restrained eating, and math performance. *Journal of Personal and Social Psychology* 75: 269–84.

French J, Inglett B, and Dethlefs T. 1989. The reproductive status of nonbreeding group members in golden lion tamarin social groups. *American Journal of Primatology* 18: 73–86.

Fricke H and Fricke S. 1977. Monogamy and sex change by aggressive dominance in coral reef fish. *Nature* 266: 830–32.

Friedl T and Klump G. 2002. The vocal behaviour of male European treefrogs (*Hyla arborea*): implications for inter- and intrasexual selection. *Behaviour* 139: 113–36.

Froehler M and Duffy C. 2002. Cortical neurons encoding path and place: where you go is where you are. *Science* 295: 2462–64.

Frost J, Binder J, Springer J, et al. 1999. Language processing is strongly left lateralized in both sexes. *Brain* 122: 199–208.

Frost P. 1998. Sex differences may indeed exist for 3-D navigational abilities: but was sexual selection responsible? *Behavioral and Brain Sciences* 21: 443–48.

Fryer G and Iles T. 1972. *The Cichlid Fishes of the Great Lakes of Africa: Their Biology and Evolution*. Oliver Boyd.

Furnham A, Reeves E, and Budhani S. 2002. Parents think their sons are brighter than their daughters: sex differences in parental self-estimations and estimations of their children's multiple intelligences. *Journal of Genetic Psychology* 163: 24–39.

Gahr M and Metzdorf R. 1999. The sexually dimorphic expression of androgen receptors in the song nucleus hyperstriatalis ventrale pars caudale of the zebra finch develops independently of gonadal steroids. *Journal of Neuroscience* 19: 2628–36.

Gahr M, Sonnenschein E, and Wickler W. 1998. Sex difference in the size of the neural song control regions in a dueting songbird with similar song repertoire size of males and females. *Journal of Neuroscience* 18: 1124–31.

Galdikas B. 1985. Adult male sociality and reproductive tactics among orangutans at Tanjung Puting. *Folia Primatologica* 45: 9–24.

Galea LAM and Kimura D. 1993. Sex differences in route-learning. *Personality, Individual Differences* 14: 53–65.

Galea L and McEwen B. 1999. Sex and seasonal differences in the rate of cell proliferation in the dentate gyrus of adult wild meadow voles. *Neuroscience* 89: 955–64.

Galis F and Drucker E. 1996. Pharyngeal biting mechanisms in centrarchids and cichlids: insights into a key evolutionary innovation. *Journal of Evolutionary Biology* 9: 641–70.

Galis F and Metz JAJ. 1998. Why are there so many cichlid species? *Trends in Ecology and Evolution* 13: 1–2.

Gallagher A, De Lisi R, Holst P, et al. 2000. Gender differences in advanced mathematical problem solving. *Journal of Experimental Child Psychology* 75: 165–90.

Gallup GG. 1995. Have attitudes toward homosexuals been shaped by natural selection? *Ethology and Sociobiology* 16: 53–70.

Garland Jr. T and Adolph S. 1994. Why not do two-species comparative studies: limitations on inferring adaptation. *Physiological Zoology* 67: 797–828.

Garland Jr. T and Carter P. 1994. Evolutionary physiology. *Annual Review of Physiology* 56: 579–621.

Gartska W. 1981. Female sex pheromone in the skin and circulation of a garter snake. *Science* 214: 681–83.

Gaulin SJC. 1992. Evolution of sex differences in spatial ability. *Yearbook of Physical Anthropology* 35: 125–51.

———. 1995. Does evolutionary theory predict sex differences in the brain? In: *The Cognitive Neurosciences*. Gazzaniga M (ed.). MIT Press, pp. 1211–25.

Gaulin SJC and Fitzgerald RW. 1986. Sex differences in spatial ability: an evolutionary hypothesis and test. *American Naturalist* 127: 74–88.

———. 1988. Home-range size as a predictor of mating systems in *Microtus*. *Journal of Mammalogy* 69: 311–19.

———. 1989. Sexual selection for spatial-learning ability. *Animal Behaviour* 37: 322–31.

Gaunt A, Kikida R, Jehl J, et al. 1990. Rapid atrophy and hypertrophy of an avian flight muscle. *Auk* 107: 649–59.

Geary DC. 1996. Sexual selection and sex differences in mathematical abilities. *Behavior and Brain Sciences* 19: 229–84.

———. 1998. *Male, Female: The Psychology of Human Sex Differences*. American Psychological Association.

———. 1999a. Evolution and developmental sex differences. *Current Directions in Psychological Science* 8: 115–20.

———. 1999b. Sex differences in mathematical abilities: commentary on the math-fact retrieval hypothesis. *Contemporary Educational Psychology* 24: 267–74.

Geary D, Saults S, Liu F, et al. 2000. Sex differences in spatial cognition, computational fluency, and arithmetic reasoning. *Journal of Experimental Child Psychology* 77: 337–53.

Genner MJ, Turner GF, and Hawkins SJ. (1999). Foraging of rocky habitat cichlid fishes in Lake Malawi: coexistence through niche partitioning? *Oecologia* 121: 283–92.

Georgopoulos A, Whang K, Georgopoulos M, et al. 2001. Functional magnetic resonance imaging of visual object construction and shape discrimination: relations among task, hemispheric lateralization, and gender. *Journal of Cognitive Neuroscience* 13: 72–89.

Gerhart J and Kirschner M. 1997. *Cells, Embryos and Evolution*. Blackwell.

Geschwind N and Galaburda A. 1987. *Cerebral Lateralization: Biological Mechanisms, Associations and Pathology*. Bradford Books, MIT Press.

Ghiselin M. 1969. The evolution of hermaphroditism among animals. *Quarterly Review of Biology* 44: 189–208.

——. 1974. *The Economy of Nature and the Evolution of Sex.* University of California Press.

——. 1988. The evolution of sex: A history of competing points of view. In *The Evolution of Sex.* Michod RE and Levin BR (eds.). Sinauer, pp. 7–23.

——. 1996. Differences in male and female cognitive abilities: Sexual selection or division of labor? *Behavioral and Brain Sciences* 19: 254–55.

Gibson B. 2001. Cognitive maps not used by humans (*Homo sapiens*) during a dynamic navigational task. *Journal of Comparative Psychology* 115: 397–402.

Gibson B and Kamil A. 2001. Tests for cognitive mapping in Clark's nutcrackers (*Nucifraga columbiana*). *Journal of Comparative Psychology* 115: 403–17.

Gibson K. 2001. Introduction to Part I. In: *Evolutionary Anatomy of the Primate Cerebral Cortex.* Falk D and Gibson K (eds.). Cambridge University Press, pp. 3–13.

Gibson K, Rumbaugh D, and Beran M. 2001. Bigger is better: primate brain size in relationship to cognition. In: *Evolutionary Anatomy of the Primate Cerebral Cortex.* Falk D and Gibson K (eds.). Cambridge University Press, pp. 79–96.

Gibson M, Kreiger D, Charlton H, et al. 1984. Mating and pregnancy can occur in genetically hypogonadal mice with preoptic area brain grafts. *Science* 225: 949–51.

Gilbert S and Bolker J. 2001. Homologies of process and modular elements of embryonic construction. *Journal of Experimental Zoology* 291: 1–12.

Gilbert SF, Opitz JM, and Raff RA. 1996. Resynthesizing evolutionary and developmental biology. *Developmental Biology* 173: 357–72.

Gilliard E. 1958. *Living Birds of the World.* Doubleday.

Gillner S and Mallot H. 1998. Navigation and acquisition of spatial knowledge in a virtual maze. *Journal of Cognitive Neuroscience* 10: 445–63.

Ginsburg B and Allee W. 1942. Some effects of conditioning on social dominance and subordination in inbred strains of mice.

Journal of Physiological Zoology 15: 485–506.

Gladue BA and Bailey JM. 1995. Spatial ability, handedness, and human sexual orientation. *Psychoneuroendocrinology* 20: 487–97.

Glantz K and Pearce J (eds.). 1989. *Exiles from Eden: Psychotherapy from an Evolutionary Perspective.* Norton.

Glickman SE, Frank LG, Davidson JM, et al. 1987. Androstenedione may organize or activate sex-reversed traits in female spotted hyenas. *Proceedings of the National Academy of Sciences, USA* 84: 3444–47.

Glickman SE, Frank LG, Holekamp KE, et al. 1993. Costs and benefits of "androgenization" in the female spotted hyena: the natural selection of physiological mechanisms. In: *Perspectives in Ethology: Behavior and Evolution.* Bateson PPG (ed.). Plenum, pp. 87–117.

Glickman SE, Coscia EM, Frank LG, et al. 1998. Androgens and masculinization of genitalia in the spotted hyaena (*Crocuta crocuta*). 3. Effects of juvenile gonadectomy. *Journal of Reproduction and Fertility* 113: 129–35.

Godwin J. 1994. Behavioural aspects of protandrous sex change in the anemonefish, *Amphiprion melanopus*, and endocrine correlates. *Animal Behaviour* 48: 551–67.

Godwin J and Crews D. 1995. Sex differences in estrogen and progesterone receptor messenger ribonucleic acid regulation in the brain of little striped whiptail lizards. *Neuroendocrinology* 62: 293–300.

Godwin J, Crews D, and Warner RR. 1996. Behavioural sex change in the absence of gonads in a coral reef fish. *Proceedings of the Royal Society of London, B* 263: 1683–88.

Gold J. 1979. Cytogenetics. In: *Fish Physiology.* Hoar W, Randall DJ, Donaldson EM (eds.). Academic Press, pp. 353–405.

Goldman SA and Nottebohm F. 1983. Neuronal production, migration, and differentiation in a vocal control nucleus of the adult female canary brain. *Proceedings of the National Academy of Sciences, USA* 80: 2390–94.

Goldschmidt T and de Visser J. 1990. On the possible role of egg mimics in speciation. *Acta Biotheretica* 38: 125–34.

Goldstein D, Haldane D, and Mitchell C. 1990. Sex differences in visual-spatial ability: the role of performance factors. *Memory and Cognition* 18: 546–50.

Gonzales C, Coe C, and Levine S. 1982. Cortisol responses under different housing conditions in female squirrel monkeys. *Psychoneuroendocrinology* 7: 209–16.

Goodfellow P and Lovell-Badge R. 1993. SRY and sex determination in mammals. *Annual Review of Genetics* 27: 71–92.

Gorbman A. 1990. Sex differentiation in the hagfish, *Eptatretus stouti*. *General and Comparative Endocrinology* 77: 309–23.

Gorlick D, Atkins PD and Losey GS. 1978. Cleaning stations as water holes, garbage dumps, and sites for the evolution of reciprocal altruism? *American Naturalist* 112: 341–53.

Gorski R, Gordon J, Shryne J, et al. 1978. Evidence for a morphological sex difference within the medial preoptic area of the rat brain. *Brain Research* 148: 333.

Gothard K, Skaggs W, Moore K, et al. 1996. Binding of hippocampal CA1 neural activity to multiple reference frames of a landmark-based navigation task. *Journal of Neuroscience* 16: 823–35.

Gould J. 1982. *Ethology*. Norton.

Gould SJ. 1977. *Ontogeny and phylogeny*. Harvard University Press.

———. 1981a. Hyena myths and realities. *Natural History* 90: 16–24.

———. 1981b. *The Mismeasure of Man*. Norton.

———. 1987. Freudian slip. *Natural History* 96: 14–21.

Gould SJ and Lewontin R. 1979. The spandrels of San Marco and the Panglossian paradigm: a critique of the adaptationist programme. *Proceedings of the Royal Society of London, B* 205: 581–98.

Gould SJ and Vrba E. 1982. Exaptation—a missing term in the science of form. *Paleobiology* 8: 4–15.

Gould-Beierle K. 2000. A comparison of four corvid species in a working and reference memory task using a radial maze. *Journal of Comparative Psychology* 114: 347–56.

Goymann W, East M, and Hofer H. 2001. Androgens and the role of female "hyperaggressiveness" in spotted hyenas (*Crocuta

crocuta*). *Hormones and Behavior* 39: 83–92.

Grafen A. 1990. Biological signals as handicaps. *Journal of Theoretical Biology* 144: 517–46.

Graham A and Begbie J. 2000. Neurogenic placodes: a common front. *Trends in Neurosciences* 23: 313–16.

Grantham T and Nichols S. 1999. Evolutionary psychology: ultimate explanations and Panglossian predictions. In: *Where Biology Meets Psychology*. Hardcastle V (ed.). MIT Press, pp. 47–66.

Grassman M, Crews D. 1986. Hormonal mediation of male- and female-like behaviors in an all-female lizard species. *Hormones and Behavior* 20: 327–35.

———. 1987. Dominance and reproduction in a parthenogenetic lizard. *Behavioral Ecology and Sociobiology* 21: 141–47.

Gray DA and Hagelin JC. 1996. Song repertoires and sensory exploitation: reconsidering the case of the common grackle. *Animal Behaviour* 52: 795–800.

Gray R. 1987. Faith and foraging: A critique of the "paradigm argument from design." In *Foraging Ecology*. Kamil A and Krebs J (eds.). Plenum, pp. 69–142.

Greenberg N and Wingfield JC. 1987. Stress and reproduction: reciprocal relationships. In: *Reproductive Endocrinology of Fish, Amphibians and Reptiles*. Norris D and Jones R (eds.). Plenum, pp. 461–503.

Greenough W, Black J, and Wallace C. 1987. Experience and brain development. *Child Development* 58: 539–59.

Greenwood P. 1991. Speciation. In: *Cichlid Fishes: Behaviour, Ecology and Evolution*. Keenleyside M (ed.). Chapman & Hall, pp. 86–102.

Gresik EW. 1973. Fine structural evidence for the presence of nerve terminals in the testis of the teleost, *Oryzias latipes*. *General and Comparative Endocrinology* 21: 210–13.

Grether G. 2000. Carotenoid limitation and mate preference evolution: a test of the indicator hypothesis in guppies (*Poecilia reticulata*). *Evolution* 54: 1712–24.

Grether G, Hudon J, and Endler J. 2001. Carotenoid scarcity, synthetic pteridine pigments and the evolution of sexual coloration in guppies (*Poecilia reticulata*). *Proceedings of the Royal Society of London, B 268: 1245–53.

Griffith SC, Owens IPF, and Burke T. 1999. Environmental determination of a sexually selected trait. *Nature* 400: 358–59.

Griffo W and Lee CT. 1973. Progesterone antagonism of androgen-dependent marking in gerbils. *Hormones and Behavior* 4: 351–58.

Grimshaw G, Sitarenios G, and Finegan J. 1995. Mental rotation at 7 years: relations with prenatal testosterone levels and spatial play experiences. *Brain and Cognition* 29: 85–100.

Grober M, Jackson I, and Bass A. 1991. Gonadal steroids affect LHRH preoptic cell number in a sex-role changing fish. *Journal of Neurobiology* 22: 734–41.

Grober MS, Fox SH, Laughlin C, et al. 1994. GnRH cell size and number in a teleost fish with two male reproductive morphs: sexual maturation, final sexual status and body size allometry. *Brain, Behavior and Evolution* 43: 61–78.

Gron G, Wunderlich A, Spitzer M, et al. 2000. Brain activation during human navigation: gender-different neural networks as substrate of performance. *Nature Neuroscience* 3: 404–8.

Gross MR. 1982. Sneakers, satellites, and parentals: polymorphic mating strategies in North American sunfishes. *Zeitschrift für Tierpsychologie* 60: 1–26.

———. 1984. Sunfish, salmon, and the evolution of alternative reproductive strategies and tactics in fishes. In: *Fish Reproduction: Strategies and Tactics*. Potts G and Wooton R (eds.). Academic Press, pp. 55–75.

———. 1985. Disruptive selection for alternative life histories in salmon. *Nature* 313: 47–48.

———. 1991. Evolution of alternative reproductive strategies: frequency-dependent sexual selection in male bluegill sunfish. *Philosophical Transactions of the Royal Society B* 332: 59–66.

———. 1996. Alternative reproductive strategies and tactics: diversity within sexes. *Trends in Ecology and Evolution* 11: 92–97.

Gross M and Charnov EL. 1980. Alternative male life histories in bluegill sunfish. *Pro-*

ceedings of the National Academy of Sciences, USA* 77: 6937–40.

Gross M and Repka J. 1998. Game theory and inheritance in the conditional strategy. In: *Game Theory and Animal Behavior*. Dugatkin L and Reeve H (eds.). Oxford University Press, pp. 168–97.

Grossman C. 1984. Regulation of the immune system by sex steroids. *Endocrine Reviews* 5: 435–55.

———. 1985. Interactions between gonadal steroids and the immune system. *Science* 227: 257–61.

Grutter A. 1995. Relationship between cleaning rates and ectoparasite loads in coral reef fishes. *Marine Ecology Progress Series* 118: 51–58.

———. 2001. Parasite infection rather than tactile stimulation is the proximate cause of cleaning behavior in reef fish. *Proceedings of the Royal Society of London, B* 1475: 1362–65.

Guerrero RD. 1975. Use of androgens for the production of all-male *Tilapia aurea* (Steindachner). *Transactions of the American Fisheries Society* 102: 342–48.

Gur R, Turetsky B, Matsui M, et al. 1999. Sex differences in brain gray and white matter in healthy young adults. *Journal of Neuroscience* 19: 4065–72.

Gur R, Alsop D, Glahn D, et al. 2000. An fMRI study of sex differences in regional activation to a verbal and a spatial task. *Brain and Language* 74: 157–70.

Gurney ME and Konishi M. 1980. Hormone-induced sexual differentiation of brain and behavior in zebra finches. *Science* 208: 1380–82.

Haley M, Deutsch C, and Le Boeuf B. 1994. Size dominance and copulatory success in male northern elephant seals, *Mirounga angustirostris*. *Animal Behaviour* 48: 1249–60.

Hall C, Davis N, Bolen L, et al. 1999. Gender and racial differences in mathematical performance. *Journal of Social Psychology* 139: 677–89.

Hall J and Kimura D. 1995. Sexual orientation and performance on sexually dimorphic motor tasks. *Archives of Sexual Behavior* 24: 395–407.

Halpern D. 1986. A different answer to the

question, "do sex related differences in spatial abilities exist?" *American Psychologist* 41: 1014–15.

——. 1992. *Sex Differences in Cognitive Ability*. Laurence Erlbaum.

——. 1997. Sex differences in intelligence: implications for education. *American Psychologist* 52: 1091–102.

Halpern-Sebold LR and Schreibman MP. 1983. Ontogeny of centers containing luteinizing hormone-releasing hormone in the brain of platyfish (*Xiphophorus maculatus*) as determined by immunocytochemistry. *Cell and Tissue Research* 229: 75–84.

Halpern-Sebold LR, Schreibman MP, and Margolis-Nunno H. 1986. Differences between early- and late-maturing genotypes of the platyfish (*Xiphophorus maculatus*) in the morphometry of their immunoreactive luteinizing hormone releasing hormone-containing cells: a developmental study. *Journal of Experimental Zoology* 240: 245–57.

Haltenorth T and Diller H. 1988. *The Collins Field Guide to the Mammals of Africa (including Madagascar)*. Stephen Greene Press.

Hamberg K. 2000. Gender in the brain: a critical scrutiny of the biological gender differences. *Lakartidningen* 8: 5130–35.

Hamilton E. 1940. *Mythology*. Little, Brown.

Hamilton W. 1980. Sex versus non-sex versus parasites. *Oikos* 35: 282–90.

Hamilton W and Brown S. 2001. Autumn tree colors as a handicap signal. *Proceedings of the Royal Society of London, B* 269: 1490–94.

Hamilton W and Poulin R. 1997. The Hamilton and Zuk hypothesis revisited: a meta-analytical approach. *Behaviour* 134: 299–320.

Hamilton W, Tilson R, and Frank L. 1986. Sexual monomorphism in spotted hyenas, *Crocuta crocuta*. *Ethology* 71: 63–73.

Hamilton WD and Zuk M. 1982. Heritable true fitness and bright birds: a role for parasites? *Science* 218: 384–87.

Hammer C and Dusek R. 1996. Brain differences, anthropological stories, and educational implications. *Behavioral and Brain Sciences* 19: 257.

Hampson E. 1990. Estrogen-related varia-

tions in human spatial and articulatory-motor skills. *Psychoneuroendocrinology* 15: 97–111.

Hampson E, Rovet J, and Altmann D. 1998. Spatial reasoning in children with congenital adrenal hyperplasia due to 21-hydroxylase deficiency. *Developmental Neuropsychology* 14: 299–320.

Hampson R, Simeral J, and Deadwyler S. 1999. Distribution of spatial and nonspatial information in dorsal hippocampus. *Nature* 402: 610–14.

Hampton R and Shettleworth S. 1996a. Hippocampal lesions impair memory for location but not color in passerine birds. *Behavioral Neuroscience* 110: 831–35.

——. 1996b. Hippocampus and memory in a food-storing and in a nonstoring bird species. *Behavioral Neuroscience* 110: 946–64.

Hampton R, Sherry D, Shettleworth S, et al. 1995. Hippocampal volume and food storing behavior are related in parids. *Brain, Behavior and Evolution* 45: 54–61.

Hanna G. 1989. Mathematics achievement of girls and boys in grade eight: results from twenty countries. *Educational Studies in Mathematics* 20: 225–32.

Hannes R-P and Franck D. 1983. The effect of social isolation on androgen and corticosteroid levels in male swordtails (*Xiphophorus helleri*) and cichlid fish (*Haplochromis burtoni*). *Hormones and Behavior* 17: 292–301.

Hannes R-P, Franck D, and Liemann F. 1984. Effects of rank-order fights on whole-body and blood concentrations of androgens and corticosteroids in the male swordtail (*Xiphophorus helleri*). *Zeitschrift für Tierpsychologie* 65: 53–65.

Hardcastle V (ed). 1999. *Where Biology Meets Psychology: Philosophical Essays*. MIT Press.

Hardisty MW. 1965a. Sex differentiation and gonadogenesis in lampreys. Part I. The ammocoete gonads of the brook lamprey, *Lampetra planeri*. *Journal of Zoology* 146: 305–45.

——. 1965b. Sex differentiation and gonadogenesis in lampreys. Part II. The ammocoete gonads of the landlocked sea lamprey, *Petromyzon marinus*. *Journal of Zoology* 146: 346–87.

Hardy A. 1960. Was man more aquatic in the past? *New Scientist* 7: 642–45.

Harrington RW. 1961. Oviparous hermaphroditic fish with internal self-fertilization. *Science* 134: 1749–50.

——. 1971. How ecological and genetic factors interact to determine when self-fertilizing hermaphrodites of *Rivulus marmoratus* change into functional secondary males, with a reappraisal of the modes of intersexuality among fishes. *Copeia* 1971: 389–432.

——. 1975. Sex determination and differentiation among uniparental homozygotes of the hermaphroditic fish *Rivulus marmoratus* (Cyprinodontidae: Atheriniformes). In: *Intersexuality in the Animal Kingdom.* Reinboth R (ed.). Springer-Verlag, pp. 249–62.

Harrington RW and Rivas LR. 1958. The discovery in Florida of the cyprinodont fish, *Rivulus marmoratus,* with a redescription and ecological notes. *Copeia* 1958: 125–30.

Harris L. 1978. Sex differences in spatial cognition: possible environmental, genetic, and neurological factors. In: *Asymmetrical Functions of the Brain.* Kinsbourne M (ed.). Cambridge University Press.

Hartman V and Crews D. 1996. Sociosexual stimuli affect ER- and PR-mRNA abundance in the hypothalamus of all-female whiptail lizards. *Brain Research* 741: 344–47.

Harvey P and Pagel M. 1991. *The Comparative Method in Evolutionary Biology.* Oxford University Press.

Hasselquist D, Bensch S, and Schantz T. 1996. Correlation between male song repertoire, extra-pair paternity and offspring survival in the great reed warbler. *Nature* 381: 229–32.

Hasselquist D, Marsh J, Sherman P, et al. 1999. Is avian humoral immunocompetence suppressed by testosterone? *Behavioral Ecology and Sociobiology* 45: 167–75.

Hastings PA and Petersen CW. 1986. A novel sexual pattern in serranid fishes: simultaneous hermaphrodites and secondary males in *Serranus fasciatus. Environmental Biology of Fishes* 15: 59–68.

Hausfater G. 1975. Dominance and reproduction in baboons (*Papio cyanoephalus*). *Contributions to Primatology* 7: 1–150.

Hausmann M, Slabbekoorn D, van Goozen S, et al. 2000. Sex hormones affect spatial ability during the menstrual cycle. *Behavioral Neuroscience* 114: 1245–50.

Hawkes K, O'Connell J, and Rogers L. 1997. The behavioral ecology of modern hunter-gatherers, and human evolution. *Trends in Ecology and Evolution* 12: 29–32.

Healy S. 1995. Memory for objects and positions—delayed non-matching to sample in storing and nonstoring tits. *Quarterly Journal of Experimental Psychology* 48: 179–91.

Healy S and Braithwaite V. 2000. Cognitive ecology: a field of substance? *Trends in Ecology and Evolution* 15: 22–25.

Healy S and Krebs J. 1992a. Comparing spatial memory in two species of tit: recalling a single positive location. *Animal Learning and Behavior* 20: 121–26.

——. 1992b. Delayed-matching-to-sample by marsh tits and great tits. *Quarterly Journal of Experimental Psychology* 45B: 33–47.

——. 1992c. Food storing and the hippocampus in corvids: amount and volume are correlated. *Proceedings of the Royal Society of London,* B 248: 241–45.

——. 1993. Development of hippocampal specialisation in a food-storing bird. *Behavioral Brain Research* 53: 127–31.

——. 1996. Food storing and the hippocampus in Paridae. *Brain, Behavior and Evolution* 47: 195–99.

Healy S and Suhonen J. 1996. Memory for locations of stored food in willow tits and marsh tits. *Behaviour* 133: 71–80.

Healy S, Braham S, and Braithwaite V. 1999. Spatial working memory in rats: no differences between the sexes. *Proceedings of the Royal Society of London,* B 266: 2303–8.

Healy S, Clayton NS, and Krebs JR. 1994. Development of hippocampal specialisation in two species of tit (*Parus* sp.). *Behavioral Brain Research* 61: 23–28.

Helleday J, Bartfai A, Ritzen EM, et al. 1994. General intelligence and cognitive profile in women with congenital adrenal hyperplasia (CAH). *Psychoneuroendocrinology* 19: 343–56.

Heller M, Calcaterra J, Green S, et al. 1999.

Perception of the horizontal and vertical in tangible displays: minimal gender differences. *Perception* 28: 387–94.

Hendriks-Jansen H. 1996. *Catching Ourselves in the Act*. MIT Press.

Henshaw J. 1971. Antlers—the unbrittle bones of contention. *Nature* 231: 469–71.

Hews D and Moore M. 1997. Hormones and sex-specific traits: critical questions. In: *Parasites and Pathogens: Effects on Host Hormones and Behavior*. Beckage N (ed.). Chapman & Hall, pp. 143–55.

Hews DK, Knapp R, and Moore MC. 1994. Early exposure to androgens affects adult expression of alternative male types in tree lizards. *Hormones and Behavior* 28: 96–115.

Hill K, Hurtado H and Hurtado A. 1996. *Ache Life History: The Ecology and Demography of a Foraging People*. Aldine de Gruyter.

Hillgarth N and Wingfield JC. 1997a. Parasite-mediated sexual selection: endocrine aspects. In: *Host-Parasite Evolution: General Principles and Avian Models*. Clayton D and Moore J (eds.). Oxford University Press, pp. 78–104.

——. 1997b. Testosterone and immunosuppression in vertebrates: implications for parasite mediated sexual selection. In: *Parasites and Pathogens: Effects on Host Hormones and Behavior*. Beckage N (ed.). Chapman & Hall, pp. 143–55.

Hillgarth N, Ramenofsky M, and Wingfield JC. 1997. Testosterone and sexual selection. *Behavioral Ecology* 8: 108–9.

Hilton S and Krebs J. 1990. Spatial memory of four species of *Parus*: performance in an open-field analogue of a radial maze. *Quarterly Journal of Experimental Psychology* 45B: 33–47.

Hindmarsh A. 1984. Vocal mimicry in starlings. *Behaviour* 90: 302–24.

——. 1986. Functional significance of vocal mimicry in song. *Behaviour* 92: 87–100.

Hines M. 2000. Gonadal hormones and sexual differentiation of human behavior: effects on psychosexual and cognitive development. In: *Sexual Differentiation of the Brain*. Matsumoto A (ed.). CRC Press, pp. 257–78.

Hines M and Sandberg EC. 1996. Sexual differentiation of cognitive abilities in women exposed to diethylstilbestrol (DES) prenatally. *Hormones and Behavior* 30: 354–63.

Hines M and Shipley C. 1984. Prenatal exposure to diethylstilbestrol (DES) and the development of sexually dimorphic cognitive abilities and cerebral lateralization. *Developmental Psychology* 20: 81–94.

Hines M, Davis F, Coquelin A, et al. 1985. Sexually dimorphic regions in the medial preoptic area and the bed nucleus of the stria terminalis of the guinea pig brain: a description and an investigation of their relationship to gonadal steroids in adulthood. *Journal of Neuroscience* 5: 40.

Hoelzel AR, Halley J, O'Brien SJ, et al. 1993. Elephant seal genetic variation and the use of simulation models to investigate historical population bottlenecks. *Journal of Heredity* 84: 443–49.

Hoelzel AR, Le Boeuf BJ, Reiter J, et al. 1999. Alpha-male paternity in elephant seals. *Behavioral Ecology and Sociobiology* 46: 298–306.

Hofer H and East M. 1997. Skewed offspring sex ratios and sex composition of twin litters in Serengeti spotted hyenas (*Crocuta crocuta*) are a consequence of siblicide. *Applied Animal Behaviour Science* 51: 307–16.

Hoffman L and Polich J. 1999. P300, handedness, and corpus callosal size: gender, modality and task. *International Journal of Psychophysiology* 31: 163–74.

Hofmann HA, Benson ME, and Fernald RD. 1999. Social status regulates growth rate: consequences for life-history strategies. *Proceedings of the National Academy of Sciences, USA* 96: 14171–76.

Holekamp KE and Smale L. 1993. Ontogeny of dominance in free-living hyaenas: juvenile rank relations with other immature individuals. *Animal Behaviour* 46: 451–66.

——. 1998. Dispersal status influences hormones and behavior in the male spotted hyena. *Hormones and Behavior* 33: 205–16.

Holekamp KE, Smale L, and Szykman M. 1996. Rank and reproduction in the female spotted hyena. *Journal of Reproduction and Fertility* 108: 229–37.

Honeycutt R, Allard M, Edwards S, et al.

1991. Systematics and evolution of the family Bathyergidae. In: *The Biology of the Naked Mole Rat*. Sherman P, Jarvis J, and Alexander R (eds.). Princeton University Press, pp. 45–65.

Hori M and Watanabe K. 2000. Aggressive mimicry in the intra-populational color variation of the Tanganyikan scale-eater *Perissodus microlepis*. *Environmental Biology of Fishes* 59: 111–15.

Houde AE. 1987. Mate choice based upon naturally occurring color-pattern variation in a guppy population. *Evolution* 41: 1–10.

———. 1997. *Sex, Color and Mate Choice in Guppies*. Princeton University Press.

Houde A and Torio A. 1992. Effects of parasitic infection on male color pattern and female choice in guppies. *Behavioral Ecology* 3: 346–51.

Houle D and Kondrashov A. 2001. Coevolution of costly mate choice and condition-dependent display of good genes. *Proceedings of the Royal Society of London, B* 269: 97–104.

Houtman AM. 1992. Female zebra finches choose extra-pair copulations with genetically attractive males. *Proceedings of the Royal Society of London, B* 249: 3–6.

Hover E. 1985. Differences in aggressive behavior between throat color morphs in a lizard, *Urosaurus ornatus*. *Copeia* 1985: 933–40.

Howard R. 1974. The influence of sexual selection and interspecific competition on mockingbird song (*Mimus polyglottos*). *Evolution* 28: 428–38.

———. 1978. The evolution of mating strategies in bullfrogs, *Rana catesbeiana*. *Evolution* 32: 850–71.

Hrdy S. 1974. Male-male competition and infanticide among the langurs (*Presbytis entellus*) of Abu. *Folia Primatologica* 22: 19–58.

———. 1977. *The Langurs of Abu*. Harvard University Press.

———. 1981. *The Woman That Never Evolved*. Harvard University Press.

———. 1988. Empathy, polyandry and the myth of the coy female. In: *Feminist Approaches to Science*. Bleier R (ed.). Pergamon Press.

Hrdy S and Hausfater G. 1984. Comparative and evolutionary perspectives on infanticide: Introduction and overview. In: *Infanticide: Comparative and Evolutionary Perspectives*. Hausfater G and Hrdy S (eds.). Aldine, pp. xiii–xxxv.

Hsu Y and Wolf LL. 1999. The winner and loser effect: integrating multiple experiences. *Animal Behaviour* 57: 903–10.

———. 2001. The winner and loser effect: what fighting behaviors are influenced? *Animal Behaviour* 61: 777–86.

Hubbs C. 1932. Apparent parthenogenesis in nature in a form of fish of hybrid origin. *Science* 76: 628–30.

Hugie D and Lank D. 1997. The resident's dilemma: a female choice mode for the evolution of alternative mating strategies in lekking male ruffs (*Philomachus pugnax*). *Behavioral Ecology* 8: 218–225.

Hume D. 1990. *Dialogues concerning Natural Religion*. Bell M (ed.). Penguin.

Hunter GA and Donaldson EM. 1983. Hormonal sex control and its application to fish culture. In: *Fish Physiology*. Hoar W, Randall D, and Donaldson E (eds.). Academic Press, pp. 223–303.

Hunter GA, Donaldson EM, Goetz FW, et al. 1982. Production of all female and sterile groups of coho salmon (*Oncorhyncus kisutch*) and experimental evidence for male heterogamety. *Transactions of the American Fisheries Society* 111: 367–72.

Hutchings J and Myers R. 1988. Mating success of alternative maturation phenotypes in male Atlantic salmon, *Salmo salar*. *Oecologia* 75: 169–74.

Hutchison RE, Opromolla G, and Hutchison JB. 1996. Environmental stimuli influence oestrogen-dependent courtship transitions and brain aromatase activity in male ring doves. *Behaviour* 133: 199–219.

Hyde J. 1990. Meta-analysis and the psychology of gender differences. *Signs* 16: 55–73.

Hyde J and McKinley N. 1997. Gender differences in cognition: results from meta-analyses. In: *Gender Differences in Human Cognition*. Caplan P, Crawford M, Hyde J, and Richardson J (eds.). Oxford University Press, pp. 30–51.

Hyde J and Plant E. 1995. Magnitude of psychological gender differences: another side of the story. *American Journal of Psychology* 50: 159–61.

Hyde JS, Fennema E, and Lamon SJ. 1990. Gender differences in mathematics performance: a meta-analysis. *Psychological Bulletin* 107: 139–55.

Immelmann K. 1969. Song development in the zebra finch and other estrildid finches. In: *Bird Vocalizations*. Hinde R (ed.). Cambridge University Press, pp. 61–74.

Imperato-McGinley J, Pichardo M, Gautier T, et al. (1991). Cognitive abilities in androgen-insensitive subjects: comparison with control males and females from the same kindred. *Clinical Endocrinology* 34: 341–47.

Irwin RE. 1988. The evolutionary importance of behavioural development: the ontogeny and phylogeny of bird song. *Animal Behaviour* 36: 814–24.

Ito Y, Fisher C, Conte F, et al. 1993. Molecular basis of aromatase deficiency in an adult female with sexual infantilism and polycystic ovaries. *Proceedings of the National Academy of Sciences, USA* 90: 673–77.

Iwaniuk AN, Pellis SM, and Whishaw IQ. 1999. Brain size is not correlated with forelimb dexterity in fissiped carnivores (Carnivora): a comparative test of the principle of proper mass. *Brain, Behavior and Evolution* 54: 167–70.

Iwasa Y and Pomiankowski A. 1994. The evolution of mate preferences for multiple sexual ornaments. *Evolution* 48: 853–67.

———. 1999. Good parent and good genes models of handicap evolution. *Journal of Theoretical Biology* 200: 97–109.

Jacob F. 1977. Evolution and tinkering. *Science* 196: 1161–66.

Jacobs J and Eccles J. 1985. Science and the media: Benbow and Stanley revisited. *Educational Researcher* 14: 20–25.

Jacobs LF. 1996. Sexual selection and the brain. *Trends in Ecology and Evolution* 11: 82–86.

Jacobs LF and Spencer WD. 1994. Natural space-use patterns and hippocampal size in kangaroo rats. *Brain, Behavior and Evolution* 44: 125–32.

Jacobs LF, Gaulin SJC, Sherry DF, et al. 1990. Evolution of spatial cognition: sex-specific patterns of spatial behavior predict hippocampal size. *Proceedings of the National Academy of Sciences, USA* 87: 6349–52.

Janowsky J, Chavex B, Zamboni B, et al. 1998. The cognitive neuropsychology of sex hormones in men and women. *Developmental Neuropsychology* 14: 421–40.

Jansson L, Forkman B, and Enquist M. 2002. Experimental evidence of receiver bias of symmetry. *Animal Behaviour* 63: 617–21.

Jardine R and Martin NG. 1983. Spatial ability and throwing accuracy. *Behavior Genetics* 13: 331–40.

Jarvis J. 1981. Eusociality in a mammal: cooperative breeding in naked mole rat colonies. *Science* 212: 571–73.

———. 1991. Reproduction of naked mole rats. In: *The Biology of the Naked Mole Rat*. Sherman P, Jarvis J, and Alexander R (eds.). Princeton University Press, pp. 384–426.

Jarvis J and Bennett NC. 1993. Eusociality has evolved independently in two genera of bathyergid mole rats—but occurs in no other subterranean mammal. *Behavioral Ecology and Sociobiology* 33: 353–60.

Jenks SM, Weldele ML, Frank LG, et al. 1995. Acquisition of matrilineal rank in captive spotted hyenas: emergence of a natural social system in peer-reared animals and their offspring. *Animal Behaviour* 50: 893–904.

Jerison H. 1973. *Evolution of the Brain and Intelligence*. Academic Press.

Johnson M. 2001. Functional brain development in humans. *Nature Reviews Neuroscience* 2: 475–83.

Johnston TD. 1987. The persistence of dichotomies in the study of behavioral development. *Developmental Review* 7: 149–82.

———. 1988. Developmental explanation and the ontogeny of birdsong: nature/nurture redux. *Behavioral and Brain Sciences* 11: 617–63.

Johnstone R and Dugatkin L. 2000. Coalition formation in animals and the nature of winner and loser effects. *Proceedings of the Royal Society of London, B* 267: 17–21.

Johnstone R, Simpson TH, and Youngson AF. 1978. Sex reversal in salmonid culture. *Aquaculture* 13: 115–34.

Jones A, ten Cate C, and Bijleveld C. 2001. The interobserver reliability of scoring sonograms by eye: a study on methods, illustrated on zebra finch songs. *Animal Behaviour* 62: 791–901.

Joseph J and Willingham D. 2000. Effect of sex and joystick experience on pursuit tracking in adults. *Journal of Motor Behavior* 32: 45–56.

Joseph R. 2000. The evolution of sex differences in language, sexuality, and visual-spatial skills. *Archives of Sexual Behavior* 29: 35–66.

Juraska J and Kopcik J. 1988. Sex and environmental influences on the size and ultrastructure of the rat corpus callosum. *Brain Research* 450: 1–8.

Kaas J and Collins C. 2001. Evolving ideas of brain evolution. *Nature* 411: 141–42.

Kahn M. 1951. The effects of severe defeat at various age levels on the aggressive behavior of mice. *Journal of Genetic Psychology* 79: 117–31.

Kallman K. 1970. Sex determination and the restriction of pigment pattern to the X and Y chromosomes in populations of the poeciliid fish, *Xiphophorus maculatus*, from Belize and Sibun rivers of British Honduras. *Zoologica* 51: 107–35.

——. 1973. The sex-determining mechanism of the platyfish, *Xiphophorus maculatus*. In: *Genetics and Mutagenesis of Fish*. Schroder J (ed.). Springer-Verlag, pp. 19–28.

——. 1975. The platyfish *Xiphophorus maculatus*. In: *Handbook of Genetics*. Kin R (ed.). Plenum, pp. 81–132.

——. 1989. Genetic control of size at maturity in *Xiphophorus*. In: *Ecology and Evolution of Livebearing Fishes (Poeciliidae)*. Meffe G and Snelson Jr F. (eds.). Prentice Hall, pp. 163–84.

Kallman K and Schreibman M. 1973. A sex-linked gene controlling gonadotrope differentiation and its significance in determining the age of sexual maturation and size of platyfish. *General and Comparative Endocrinology* 21: 287–304.

Kallmann F, Schoenfeld W, and Barrera S. 1944. The genetic aspects of primary eunuchoidism. *American Journal of Mental Deficiencies* 48: 203–36.

Kalra S, Allen L, Sahu A, et al. 1988. Go-nadal steroids and neuropeptide Y-opioid-LHRH axis: interactions and diversities. *Journal of Steroid Biochemistry* 30: 185–93.

Kamil AC. 1988. A synthetic approach to the study of animal intelligence. In: *Nebraska Symposium on Motivation 1987: Comparative Perspectives in Modern Psychology*. Leger DW (ed.). University of Nebraska Press, pp. 257–308.

Kaplan B and Weisberg F. 1987. Sex differences and practice effects on two visual-spatial tasks. *Perceptual and Motor Skills* 64: 139–42.

Kaplan J, Adams M, Kortinik D, et al. 1986. Adrenal responsiveness and social status in intact and ovariectomized *Macaca fascicularis*. *American Journal of Primatology* 11: 181–93.

Karadi K, Szabo I, Szepesi T, et al. 1999. Sex differences on the hand mental rotation task for 9-yr.-old children and adults. *Perceptual and Motor Skills* 89: 969–72.

Karmiloff-Smith A. 1992. *Beyond Modularity: A Developmental Perspective on Cognitive Science*. MIT Press.

Kaskan P and Finlay B. 2001. Encephalization and its developmental structure: how many ways can the brain get big? In: *Evolutionary Anatomy of the Primate Cerebral Cortex*. Falk D and Gibson K (eds.). Cambridge University Press, pp. 14–28.

Keegan-Rogers V and Schultz RJ. 1988. Sexual selection among clones of unisexual fish (*poeciliopsis: poeciliidae*): Genetic factors and rare-female advantage. *American Naturalist* 132: 846–68.

Keightley P and Eyre-Walker, A. 2000. Deleterious mutations and the evolution of sex. *Science* 290: 331–33.

Keller L and Reeve H. 1994. Partitioning of reproduction in animal societies. *Trends in Ecology and Evolution* 9: 98–102.

Kelley D. 1986. The genesis of male and female brains. *Trends in Neuroscience* 10: 499–502.

Kendrick AM, Rand MS, and Crews D. 1995. Electrolytic lesions to the ventromedial hypothalamus abolish receptivity in female whiptail lizards, *Cnemidophorus uniparens*. *Brain Research* 680: 226–28.

Kenyon R. 1972. Polygyny among superb lyrebirds. *Emu* 72: 70–76.

Kerns KA and Berenbaum SA. 1991. Sex differences in spatial ability in children. *Behavior Genetics* 21: 383–96.

Kester P, Green R, Finch S, et al. 1980. Prenatal "female hormone" administration and psychosexual development in human males. *Psychoneuroendocrinology* 5: 269–85.

Keverne E, Fran L, and Nevison C. 1996. Primate brain evolution: generic and functional considerations. *Proceedings of the Royal Society of London, B* 262: 689–96.

Kim J and Baxter M. 2001. Multiple brain-memory systems: the whole does not equal the sum of its parts. *Trends in Neuroscience* 24: 324–30.

Kimball M. 1989. A new perspective on women's math achievement. *Psychological Bulletin* 105: 198–214.

Kimball RT and Ligon JD. 1999. Evolution of avian plumage dichromatism from a proximate perspective. *American Naturalist* 154: 182–93.

Kime N, Rand A, Kapfer M, et al. 1998. Consistency of female choice in the Tungara frog: a permissive preference for complex characters. *Animal Behaviour* 55: 641–49.

Kimura D. 1992. Sex differences in the brain. *Scientific American* September: 119–25.

——. 1996. Sex, sexual orientation and sex hormones influence human cognitive function. *Current Opinion in Neurobiology* 6: 259–63.

Kimura D and Hampson E. 1994. Cognitive pattern in men and women is influenced by fluctuations in sex hormones. *Current Directions in Psychological Science* 3: 57–61.

Kimura K, Shimizu K, Hayaski M, et al. 2000. Pituitary-adrenocortical responses to the first dyadic encounters in male rhesus monkeys: effect of dominance relationship. *American Journal of Primatology* 50: 247–56.

King A and West M. 1983. Epigenesis of cowbird song—a joint endeavor of males and females. *Nature* 305: 704–6.

Kinsey AC, Pomeroy R, and Martin C. 1953. *Sexual Behavior in the Human Female.* W. B. Saunders.

Kirn JR, Clower RP, Kroodsma DE, et al. 1989. Song-related brain regions in the red-winged blackbird are affected by sex and season but not repertoire size. *Journal of Neurobiology* 20: 139–63.

Kirn JR, Alvarez-Buylla A, and Nottebohm F. 1991. Production and survival of projection neurons in a forebrain vocal center of adult male canaries. *Journal of Neuroscience* 11: 1756–62.

Kirpatrick M. 1982. Sexual selection and the evolution of female choice. *Evolution* 36: 1–12.

——. 1986. The handicap mechanism of sexual selection does not work. *American Naturalist* 127: 222–40.

Klein C. 1989. About girls and science. *Science and Children* 27: 28–31.

Knapp R and Moore MC. 1997. Male morphs in tree lizards have different testosterone responses to elevated levels of corticosterone. *General and Comparative Endocrinology* 107: 273–79.

Knight M and Turner G. 1999. Reproductive isolation among closely related Lake Malawi cichlids: can males recognize females by visual cues? *Animal Behaviour* 58: 761–68.

Kodric-Brown A. 1989. Dietary carotenoids and male mating success in the guppy: an environmental component to female choice. *American Naturalist* 124: 309–23.

Kodric-Brown A and Brown J. 1984. Truth in advertising: the kinds of traits favored by sexual selection. *American Naturalist* 124: 309–23.

Koenig A and Borries C. 2001. Socioecology of Hanuman langurs: the story of their success. *Evolutionary Anthropology* 10: 122–37.

Kolakowski D and Malina R. 1974. Spatial ability, throwing accuracy and man's hunting heritage. *Nature* 251: 410–12.

Kolluru GR and Reznick DN. 1996. Genetic and social control of male maturation in *Phallichthys quadripunctatus* (Pisces: Poiciliidae). *Journal of Evolutionary Biology* 9: 695–715.

Kondrashov A. 2001. Sex and U. *Trends in Genetics* 17: 75–78.

Kondrashov F and Kondrashov A. 2001. Multidimensional epistasis and the disadvantage of sex. *Proceedings of the National Academy of Sciences, USA* 98: 12089–92.

Konishi M. 1964. Effects of deafening on song development in two species of juncos. *Condor* 66: 85–102.

———. 1965. The role of auditory feedback in the control of vocalization in the white-crowned sparrow. *Zeitschrift für Tierpsychologie* 22: 770–83.

———. 1989. Birdsong for neurobiologists. *Neuron* 3: 541–49.

———. 1994. An outline of recent advances in birdsong neurobiology. *Brain, Behavior and Evolution* 44: 279–85.

Konishi M and Akutagawa E. 1985. Neuronal growth, atrophy and death in a sexually dimorphic song nucleus in the zebra finch brain. *Nature* 315: 145–47.

———. 1990. Growth and atrophy of neurons labeled at their birth in a song nucleus of the zebra finch. *Proceedings of the National Academy of Sciences, USA* 87: 3538–41.

Kopcik J, Seymoure P, Schneider S, et al. 1992. Do callosol projection neurons reflect sex differences in axon number? *Brain Research Bulletin* 29: 493–97.

Kornfield I and Smith P. 2000. African cichlid fishes: model systems for evolutionary biology. *Annual Review of Ecology and Systematics* 31: 163–96.

Kotiaho J. 2001. Costs of sexual traits: a mismatch between theoretical considerations and empirical evidence. *Biological Review* 76: 365–76.

Koulish S and Kramer CR. 1989. Human chorionic gonadotropin (hCG) induces gonad reversal in a protogynous fish, the bluehead wrasse, *Thalassoma bifasciatum* (Teleostei, Labridae). *Journal of Experimental Zoology* 252: 156–68.

Kourada S. 1980. Social behavior of the pygmy chimpanzee. *Primates* 21: 181–97.

Kramer CR, Caddell MT, and Bubenheimer-Livolsi L. 1993. sGnRH-A [(D-Arg6,Pro9, NEt-)LHRH] in combination with domperidone induces gonad reversal in a protogynous fish, the bluehead wrasse, *Thalassoma bifasciatum*. *Journal of Fish Biology* 42: 185–95.

Kramer G and Smith R. 2001. An investigation of gender differences in the components influencing the difficulty of spatial ability items. *Journal of Applied Measurement* 2: 65–77.

Krebs JR. 1976. Habituation and song repertoires in the great tit. *Behavioral Ecology and Sociobiology* 1: 215–27.

———. 1977. The significance of song repertoires: the beau geste hypothesis. *Animal Behaviour* 25: 475–78.

———. 1990. Food storing birds: adaptive specialization in brain and behavior? *Philosophical Transactions of the Royal Society of London B* 329: 55–62.

Krebs JR and Davies N. 1997. The evolution of behavioral ecology. In: *Behavioural Ecology*. Krebs J, and Davies N (eds.). Blackwell Scientific, pp. 3–18.

Krebs JR, Erichsen JT, Webber MI, et al. 1977. Optimal prey selection in the great tit (*Parus major*). *Animal Behaviour* 25: 30–38.

Krebs JR, Kacelnik A, and Taylor P. 1978. Test of optimal sampling by foraging great tits. *Nature* 275: 27–31.

Krebs JR, Sherry DF, Healy SD, et al. 1989. Hippocampal specialization of food-storing birds. *Proceedings of the National Academy of Sciences, USA* 86: 1388–92.

Krebs JR, Clayton N, Hampton R, et al. 1995. Effects of photoperiod on food-storing and the hippocampus in birds. *Neuroreport* 6: 1701–04.

Krebs JR, Clayton N, Healy S, et al. 1996. The ecology of the avian brain: food storing memory and the hippocampus. *Ibis* 138: 34–46.

Kreiger D, Perlow M, Gibson M, et al. 1982. Brain grafts reverse hypogonadism of gonadotropin releasing hormone deficiency. *Nature* 298: 468–71.

Kroodsma D. 1982. Song repertoires: problems in their definition and use. In: *Acoustic Communication in Birds*. Kroodsma D and Miller E (eds.). Academic Press, pp. 125–46.

Kroodsma D and Byers B. 1998. Songbird song repertoires: an ethological approach to studying cognition. In: *Animal Cognition in Nature*. Balda R, Pepperberg I, and Kamil A (eds.). Academic Press, pp. 305–36.

Kroodsma D and Canady R. 1985. Differences in repertoire size, singing behavior, and associated neuroanatomy among marsh wren populations have a genetic basis. *Auk* 102: 439–46.

Kroodsma DE and Konishi M. 1991. A suboscine bird (eastern phoebe, *Sayornis phoebe*) develops normal song without auditory feedback. *Animal Behaviour* 42: 477–87.

Kruuk H. 1972. *The Spotted Hyena*. University of Chicago Press.

Kuhn T. 1962. *The Structure of Scientific Revolutions*. University of Chicago Press.

Kulynych J, Vladar K, Jones D, et al. 1994. Gender differences in the normal lateralization of the supratemporal cortex: MRI surface-rendering mophometry of Heschl's gyrus and the planum temporale. *Cerebral Cortex* 4: 107–18.

Kurtz J and Sauer K. 1999. The immunocompetence handicap hypothesis: testing the genetic predictions. *Proceedings of the Royal Society of London, B* 266: 2515–22.

Lacey E and Sherman P. 1991. Social organization of the naked mole-rat: evidence for divisions of labour. In: *The Biology of the Naked Mole Rat*. Sherman P, Jarvis J, and Alexander R (eds.). Princeton University Press, pp. 274–37.

Lacey E, Alexander R, Braude S, et al. 1991. An ethogram for the naked mole rat: nonvocal behaviors. In: *The Biology of the Naked Mole Rat*. Sherman P, Jarvis J, and Alexander R (eds.). Princeton University Press, pp. 209–42.

Lamb D, Weigel N, and Marcelli M. 2001. Androgen receptors and their biology. *Vitamins and Hormones* 62: 199–230.

Lampe H and Saetre F. 1995. Female pied flycatchers prefer males with larger song repertoires. *Proceedings of the Royal Society of London, B* 262: 163–67.

Landau H. 1951. On dominance relations and the structure of animal societies: I. Effect of inherent characteristics. *Bulletin of Mathematical Biophysics* 13: 1–19.

Lande R. 1989. Sexual dimorphism, sexual selection, and adaptation in polygenic characters. *Evolution* 34: 292–305.

———. 1981. Models of speciation by sexual selection on polygenic traits. *Proceedings of the National Academy of Sciences, USA* 78: 3721–25.

Landry C, Garant P, Duchesne P, et al. 2001. "Good genes as heterozygosity": the major histocompatibility complex and mate choice in Atlantic salmon (*Salmo salar*). *Proceedings of the Royal Society of London, B* 268: 1279–85.

Lank DB, Smith CM, Hanotte O, et al. 1995. Genetic polymorphism for alternative mating behaviour in lekking male ruff *Philomachus pugnax*. *Nature* 378: 59–62.

———. 2002. High frequency of polyandry in a lek mating system. *Behavioral Ecology* 13: 209–15.

Laskey A. 1944. A mockingbird acquires his song repertory. *Auk* 61: 211–19.

Lavenex P, Steele M, and Jacobs L. 2000. Sex differences, but no seasonal variations in the hippocampus of food-caching squirrels: a stereological study. *Journal of Comparative Neurology* 425: 152–66.

Laws R. 1994. History and present status of southern elephant seal populations. In: *Elephant Seals: Population Ecology, Behavior, and Physiology*. Le Boeuf B and Laws R (eds.). University of California Press, pp. 49–65.

Lea J, Dyson M, Halliday T. 2001. Calling by male midwife toads stimulates females to maintain reproductive condition. *Animal Behaviour* 61: 373–77.

Leamy L. 1997. Is developmental stability heritable? *Journal of Evolutionary Biology* 10: 21–29.

Le Boeuf B. 1974. Male-male competition and reproductive success in elephant seals. *American Zoologist* 14: 163–74.

Le Boeuf B and Reiter J. 1988. Lifetime reproductive success in northern elephant seals. In: *Reproductive Success: Studies of Individual Variation in Contrasting Breeding Systems*. Clutton-Brock T (ed.). University of Chicago Press, pp. 344–62.

Lee K and Devore I (eds.). 1968. *Man the Hunter*. Aldine.

Lehne G. 1988. Treatment of sex offenders with methoxyprogesterone acetate. In: *Handbook of Sexology: The Pharmacology and Endocrinology of Sexual Function*. Sitse J (ed.). Elsevier. 6: 516.

Lehrman D. 1965. Interaction between internal and external environments in the regulation of the reproductive cycle of the ring dove. In: *Sex and Behavior*. Beach F (ed.). Wiley, pp. 355–80.

Leibniz G. 1988. *Theodicy: Essays on the Goodness of God the Freedom of Man and the Origin of Evil*. Huggard ET (ed.). Open Court.

Leitner S, Voigt C, Garcia-Segura L-M, et al. 2001. Seasonal activation and inactivation of song motor memories in wild canaries is not reflected in neuroanatomical changes of forebrain song areas. *Hormones and Behavior* 40: 160–68.

LeMahieu D. 1976. *The Mind of William Paley*. University of Nebraska Press.

Lens L, Van Dongen S, Klark S, et al. 2002. Fluctuating asymmetry as an indicator of fitness: can we bridge the gap between studies? *Biological Review* 77: 27–38.

Lepori NG. 1980. *Sex Differentiation, Hermaphroditism and Intersexuality in Vertebrates including Man*. Piccin Medical Books, pp. 297–333.

Lepri J and Vandenbergh J. 1986. Puberty in pine voles, *Microtus pinetorum*, and the influence of chemosignals on female reproduction. *Biology of Reproduction* 34: 370–77.

Levin D. 1975. Pest pressure and recombination in plants. *American Naturalist* 109: 437–51.

Levin R. 1996. Song behaviour and reproductive strategies in a duetting wren, *Thryothorus nigricapillus*. I. Playback experiments. *Animal Behaviour* 52: 1107–17.

Levine S, Huttenlocher J, Taylor A, et al. 1999. Early sex differences in spatial skills. *Developmental Psychology* 35: 940–49.

Levy J and Heller W. 1992. Gender differences in human neuropsychological function. In: *Handbook of Behavioral Neurobiology: Sexual Differentiation*. Gerall A, Moltz H, and Ward I (eds.). Plenum, pp. 245–73.

Liben L, Susman E, Finkelstein J, et al. 2002. The effects of sex steroids on spatial performance: a review and an experimental clinical investigation. *Developmental Psychology* 38: 236–53.

Licht P, Hayes T, Tsai P, et al. 1998. Androgens and masculinization of genitalia in the spotted hyaena (*Crocuta crocuta*). 1. Urogential morphology and placental an-

drogen production during fetal life. *Journal of Reproduction and Fertility* 113: 105–16.

Liem K. 1973. Evolutionary strategies and morphological innovations: cichlid pharyngeal jaws. *Systematic Zoology* 22: 425–41.

——. 1980. Adaptive significance of intra- and interspecific differences in the feeding repertoires of cichlid fishes. *American Zoologist* 20: 295–314.

Lillie F. 1916. The theory of the freemartin. *Science* 43: 611–13.

Lima NRW, Kobak CJ, and Vrijenhoek RC. 1996. Evolution of sexual mimicry in sperm-dependent all-female forms of *Poeciliopsis* (*Pisces: poeciliidae*). *Journal of Evolutionary Biology* 9: 185–203.

Limbaugh C. 1961. Cleaning symbiosis. *Scientific American* 205: 42–49.

Lincoln G. 1994. Teeth, horns and antlers: the weapons of sex. In: *The Differences between the Sexes*. Short R and Balaban E (eds.). Cambridge University Press, pp. 131–58.

Lindeque M and Skinner JD. 1982. Fetal androgens and sexual mimicry in spotted hyaenas (*Crocuta crocuta*). *Journal of Reproduction and Fertility* 65: 405–10.

Lindeque M, Skinner JD, and Millar RP. 1986. Adrenal and gonadal contribution to circulating androgens in spotted hyaenas (*Crocuta crocuta*) as revealed by LHRH, HCG, and ACTH stimulation. *Journal of Reproduction and Fertility* 78: 211–17.

Lindzey J, and Crews D. 1986. Hormonal control of courtship and copulatory behavior in male *Cnemidophorus inornatus*, a direct sexual ancestor of a unisexual parthenogenetic lizard. *General and Comparative Endocrinology* 64: 411–18.

——. 1988. Psychobiology of sexual behavior in a whiptail lizard, *Cnemidophorus inornatus*. *Hormones and Behavior* 22: 279–93.

——. 1992. Interactions between progesterone and androgens in the stimulation of sex behaviors in male little striped whiptail lizards, *Cnedmidophorus inornatus*. *General and Comparative Endocrinology* 86: 52–58.

——. 1993. Effects of progesterone and dihydrotestosterone on stimulation of androgen-dependent sex behavior, accessory sex structures, and in vitro binding characteristics of cytosolic androgen receptors in male whiptail lizards (*Cnemidophorus inornatus*). *Hormones and Behavior* 27: 269–81.

Linn M and Hyde J. 1989. Gender, mathematics, and science. *Educational Researcher* 18: 22–27.

Linn MC and Petersen AC. 1985. Emergence and characterization of sex differences in spatial ability: a meta-analysis. *Child Development* 56: 1479–98.

Liu D, Diorio J, Tannenbaum B, et al. 1997. Hypothalamic-pituitary-adrenal response to stress. *Science* 277: 1659–62.

Lloyd EA. 1993. Pre-theoretical assumptions in evolutionary explanations of female sexuality. *Philosophical Studies* 69: 139–53.

——. 1999. Evolutionary psychology: the burdens of proof. *Biology and Philosophy* 14: 211–33.

Logan C, Hyatt L, and Gregorcyk L. 1990. Song playback initiates nest building during clutch overlap in mockingbirds, *Mimus polyglottus*. *Animal Behaviour* 39: 943–53.

Lorenz K. 1941. Vergleichende bewegungsstudien an Anatinen. *Journal für Ornithologie* 79 (suppl.): 221–68.

——. 1950. Part and parcel in animal and human societies. In: *Studies in Animal and Human Behavior*. Martin R (ed.). Harvard University Press, pp. 115–95.

——. 1954. Psychology and phylogeny. In: *Studies in Animal and Human Behavior*. Lorenz K (ed.). Harvard University Press, pp. 196–245.

Losey G. 1972. The ecological importance of cleaning symbiosis. *Copeia* 1972: 820–33.

Lovejoy O. 1981. The origin of man. *Science* 211: 341–50.

Lowe TP and Larkin JR. 1975. Sex reversal in *Betta splendens* Regan with emphasis on the problem of sex determination. *Journal of Experimental Zoology* 191: 25–32.

Lowry C, Burke K, Renner K, et al. 2001. Rapid changes in monoamine levels following administration of corticotropin-releasing factor or corticosterone are localized in the dosomedial hypothalamus. *Hormones and Behavior* 39: 195–205.

Lubbock R. 1980. Why are clownfish not stung by sea anemones? *Proceedings of the Royal Society of London, B* 207: 35–61.

Lubinski D, Benbow C, Shea D, et al. 2001. Men and women at promise for scientific excellence: similarity not dissimilarity. *Psychological Science* 12: 309–17.

Lunneborg P. 1982. Sex differences in self-assessed, everyday spatial abilities. *Perceptual and Motor Skills* 55: 200–202.

——. 1984. Sex differences in self-assessed, everyday spatial abilities: differential practice or self-esteem. *Perceptual and Motor Skills* 58: 213–14.

Maccoby E. 1988. Gender as a social category. *Developmental Psychology* 24: 755–765.

Maccoby E and Jacklin C. 1974. *The Psychology of Sex Differences*. Stanford University Press.

MacDonald K. 1991. A perspective on Darwinian psychology: the importance of domain-general mechanisms, plasticity, and individual differences. *Ethology and Sociobiology* 12: 449–80.

MacDougall-Shackleton S, Hulse S, and Ball G. 1998. Neural correlates of singing behavior in male zebra finches (*Taeniopygia guttata*). *Journal of Neurobiology* 36: 421–30.

MacDougall-Shackleton SA, and Ball GF. 1999. Comparative studies of sex differences in the song-control system of songbirds. *Trends in Neurosciences* 22: 432–36.

MacNab B and Eisenberg J. 1989. Brain size and its relation to the rate of metabolism in mammals. *American Naturalist* 133: 157–67.

Macphail E and Bolhuis J. 2001. The evolution of intelligence: adaptive specializations versus general processes. *Biological Review* 76: 341–64.

Madden J. 2000. Sex, bowers and brains. *Proceedings of the Royal Society of London, B* 268: 833–38.

Maggioncalda A and Sapolsky R. 2002. Disturbing behaviors of the orangutan. *Scientific American* 286: 60–65.

Maggioncalda A, Sapolsky R, and Czekala N. 1999. Reproductive hormone profiles in captive male orangutans: implications for understanding developmental arrest.

American Journal of Physical Anthropology 109: 19–32.

Maguire E, Burgess N, and O'Keefe J. 1999. Human spatial navigation: cognitive maps, sexual dimorphism, and neural substrates. *Current Opinion in Neurobiology* 9: 171–77.

Maguire E, Gadian D, Johnsrude I, et al. 2000. Navigation-related structural change in the hippocampi of taxi drivers. *Proceedings of the National Academy of Sciences, USA* 97: 4398–403.

Malinowski J. 2001. Mental rotation and real-world wayfinding. *Perceptual and Motor Skills* 92: 19–30.

Mameli M. 2001. Modules and mindreaders. *Biology and Philosophy* 16: 377–93.

Manger T and Eikkeland O. 1998. The effects of spatial visualization and students' sex on mathematical achievement. *British Journal of Psychology* 89: 17–25.

Mann VA, Sasanuma S, Sakuma N, et al. 1990. Sex differences in cognitive abilities: a cross-cultural perspective. *Neuropsychologia* 28: 1063–77.

Manning J, Trivers R, Singh D, et al. 1999. The mystery of female beauty. *Nature* 399: 214–15.

Margulis L and Sagan D. 1986. *Origins of Sex: Three Billion Years of Recombination.* Yale University Press.

Markow T and Clarke G. 1997. Meta-analysis of the heritability of developmental stability: a giant step backward. *Journal of Evolutionary Biology* 10: 31–37.

Marler P. 1997. Three models of song learning: evidence from behavior. *Journal of Neurobiology* 33: 501–16.

Marler P and Nelson DA. 1993. Action-based learning: a new form of developmental plasticity in bird song. *Netherlands Journal of Zoology* 43: 91–103.

Marler P and Peters S. 1982. Developmental overproduction and selective attrition: new processes in the epigenesis of birdsong. *Developmental Psychobiology* 15: 369–78.

Marler P and Peters S. 1988. Sensitive periods for song acquisition from tape recordings and live tutors in the swamp sparrow, *Melospiza georgiana*. *Ethology* 77: 76–84.

Marler P and Tamura M. 1962. Song "dialects" in three populations of white-crowned sparrows. *Condor* 64: 368–77.

Martin J. 1989. *Neuroanatomy: Text and Atlas.* Appleton and Lange.

Martin R. 1996. Scaling of the mammalian brain: the maternal energy hypothesis. *News in Physiological Sciences* 11: 149–56.

Martins E. 2000. Adaptation and the comparative method. *Trends in Ecology and Evolution* 15: 296–99.

Masica D, Money J, and Ehrhardt A. 1971. Fetal feminization and female gender identity in the testicular feminizing syndrome of androgen insensitivity. *Archives of Sexual Behavior* 1: 131–42.

Mason AJ, Hayflick JS, Zoeller RT, et al. 1986a. A deletion truncating the gonadotropin-releasing hormone gene is responsible for hypogonadism in the HPG mouse. *Science* 234: 1366–71.

Mason A, Pitts S, Nikolics K, et al. 1986b. The hypogonadal mouse: reproduction functions restored by gene therapy. *Science* 234: 1372–78.

Mason A, Pitts S, Nikolics K, et al. 1987. Gonadal development and gametogenesis in the hypogonadal mouse are restored by gene transfer. *Annals of the New York Academy of Sciences* 513: 16–26.

Mason R and Crews D. 1985. Female mimicry in garter snakes. *Nature* 316: 59–60.

Mason R and Crews D. 1986. Pheromonal mimicry in garter snakes. In *Chemical Signals in Vertebrates.* Duval D and Muller-Schwartz D (eds.). Plenum.

Masters M. 1998. The gender difference on the mental rotations test is not due to performance factors. *Memory and Cognititon* 26: 444–48.

Masters MS and Sanders B. 1993. Is the gender difference in mental rotation disappearing? *Behavior Genetics* 23: 337–41.

Matthews L. 1939. Reproduction of the spotted hyaena (*Crocuta crocuta* Erxleben). *Philosophical Transactions* 230: 1–78.

Maynard-Smith J. 1969. The status of neo-Darwinism. In: *Towards a Theoretical Biology.* Waddington C (ed.). Edinburgh University Press.

———. 1974. The theory of games and the evolution of animal conflicts. *Journal of Theoretical Biology* 47: 209–21.

——. 1976. Sexual selection and the hand-icap principle. *Journal of Theoretical Biology* 57: 239–42.

——. 1978. *The Evolution of Sex*. Cambridge Univesity Press.

——. 1982. *Evolution and the Theory of Games*. Cambridge University Press.

——. 1986. The evolution of recombination. In: *The Evolution of Sex*. Michod RE and Levin BR (eds.). Sinauer, pp. 106–25.

Maynard Smith J and Parker G. 1976. The logic of asymmetric contests. *Animal Behaviour* 24: 159–75.

Maynard-Smith J and Price G. 1973. The logic of animal conflict. *Nature* 246: 15–18.

Mayr E. 1961. Cause and effect in biology. *Science* 134: 1501–6.

——. 1974. Teleological and teleonomic: a new analysis. *Boston Studies in the Philosophy of Science* 14: 91–117.

——. 1982. *The Growth of Biological Thought*. Harvard University Press.

McBurney D, Gaulin S, Devineni T, et al. 1997. Superior spatial memory of women: stronger evidence for the gathering hypothesis. *Ethology and Sociobiology* 17: 165–73.

McCarrey JR and Abbott UK. 1979. Mechanisms of genetic sex determination, gonadal sex differentiation and germ-cell development in mammals. *Advances in Genetics* 20: 217–90.

McClelland J and Rumelhart O. 1986. *Parallel Distributed Processing*. Vol. 2: *Psychological and Biological Foundations*. MIT Press.

McCormick C and Teillon S. 2001. Menstrual cycle variation in spatial ability: relation to salivary cortisol levels. *Hormones and Behavior* 39: 29–38.

McCormick CM and Witelson SF. 1991. A cognitive profile of homosexual men compared to heterosexual men and women. *Psychoneuroendocrinology* 16: 459–73.

McEwen B, Lieberburg I, Chaptal C, et al. 1977. Aromatization: important for sexual differentiation of the neonatal rat brain. *Hormones and Behavior* 9: 249–63.

McGlone J. 1980. Sex differences in human brain asymmetry: a critical survey. *Behavioral Brain Science* 3: 215–63.

McGuire L, Ryan K, and Omenn G. 1975.

Congenital adrenal hyperplasia II: cognitive and behavioral studies. *Behavior Genetics* 5: 175–88.

McGuire M, Brammer G, and Raleigh MJ. 1986. Restiong cortisol levels and emergence of dominant status among male vervet monkeys. *Hormones and Behavior* 20: 106–17.

McHenry H. 1992. Body size and proportions in early hominids. *American Journal of Physical Anthropology* 87: 407–31.

McKay F. 1971. Behavioral aspects of population dynamics in unisexual-bisexual *Poeciliopsis* (*Pisces: Poeciliidae*). *Ecology* 52: 778–90.

McKaye KR. 1977. Defense of a predator's young by a herbivorous fish: An unusual strategy. *American Naturalist* 111: 301–15.

——. 1979. Defense of a predator's young revisited. *American Naturalist* 114: 595–601.

McKenzie WJ, Crews D, Kallman K, et al. 1983. Age, weight and the genetics of sexual maturation in the platyfish, *Xiphophorus maculatus*. *Copeia* 1983: 770–74.

McLennan D and Brooks D. 1991. Parasites and sexual selection: A macroevolutionary perspective. *Quarterly Review of Biology* 66: 255–86.

McLeod P, Moser WH, Ryan J. et al. 1996. The relation between urinary cortisol levels and social behaviour in captive timber wolves. *Journal of Zoology* 74: 209–16.

McManus IC and Bryden MP. 1991. Geschwind's theory of cerebral lateralization: developing a formal, causal model. *Psychological Bulletin* 110: 237–53.

Mendl M, Adroaldo J, and Broom B. 1992. Physiological and reproductive correlates of behavioural strategies in female domestic pigs. *Animal Behaviour* 44: 1107–21.

Meurling A, Tonning-Olsson I, and Levander S. 2000. Sex differences in strategy and performance on computerized neuropsychological tests as related to gender identity and age at puberty. *Scandanavian Jouranl of Psychology* 41: 81–90.

Meyer A. 1993. Phylogenetic relationships and evolutionary processes in East African cichlid fishes. *Trends in Ecology and Evolution* 8: 279–84.

Meyer A, Kocher T, Basasibwaki P, et al. 1990. Monophyletic origin of Lake Victoria cichlid fishes suggested by mitochondrial DNA sequences. *Nature* 347: 550–53.

Meyer-Bahlburg H, Gruen RS, New MI, et al. 1995. Prenatal estrogens and the development of homosexual orientation. *Developmental Psychology* 31: 12–21.

——. 1996. Gender change from female to male in classical congenital adrenal hyperplasia. *Hormones and Behavior* 30: 319–32.

Michener C. 1969. Comparative social behavior of bees. *Annual Review of Entomology* 14: 277–42.

Michod R and Levin B. 1988. *The Evolution of Sex: An Examination of Current Ideas.* Sinauer.

Mizcek K, Thompson M, and Tornatzky W. 1991. Subordinate animals: behavioral and physiological adaptations and opioid tolerance. In: *Stress: Neurobiology and Neuroendocrinology.* Brown M, Koob G, and Rivier C (eds.). Dekker, pp. 323–57.

Miller G. 2001. *The Mating Mind: How Sexual Choice Shaped the Evolution of Human Nature.* Anchor Books.

Miller R and Schultz R. 1959. All-female strains of the teleost fishes of the genus *Poeciliopsis. Science* 130: 1656–57.

Milner B. 1966. Amnesia following operation on the temporal lobes. In: *Amnesia.* Whitty C and Zangwill O (eds.). Butterworths.

Milton K. 1999. A hypothesis to explain the role of meat-eating in human evolution. *Evolutionary Anthropology* 8: 11–21.

Mitani J. 1985. Mating behaviour of male orangutans in the Kutai Game Reserve, Indonesia. *Animal Behaviour* 33: 392–402.

Mitton J. 1993. Enzyme heterozygosity, metabolism, and developmental stability. *Genetica* 89: 47–65.

Mitton JB and Grant MC. 1984. Associations among protein heterozygosity, growth rate, and developmental homeostasis. *Annual Review of Ecology and Systematics* 15: 479–99.

Mock D. 1984. Siblicidal aggression and resource monopolization in birds. *Science* 225: 731–33.

Moehlman P and Hofer H. 1997. Cooperative breeding, reproductive suppression, and body mass in canids. In: *Cooperative Breeding in Mammals.* Solomon N and French J (eds.). Cambridge University Press, pp. 76–128.

Moffat SD and Hampson E. 1996. A curvilinear relationship between testosterone and spatial cognition in humans: possible influence of hand preference. *Psychoneuroendocrinology* 21: 323–37.

Moffat SD, Hampson E, and Hatzipantelis M. 1998. Navigation in a "virtual" maze: sex differences and correlation with psychometric measures of spatial ability in humans. *Evolution and Human Behavior* 19: 73–87.

Moguilewsky M and Philibert D. 1985. Biochemical profile of RU 486. In: *The Antiprogesterone Steriod RU 486 and Human Fertility Control.* Baulier E (ed.). Plenum, pp. 87–97.

Møller A. 1991. Parasite loads reduce song output in a passerine bird. *Animal Behaviour* 41: 723–30.

——. 1992. Parasites differentially increase fluctuating asymmetry in secondary sexual characteristics. *Journal of Evolutionary Biology* 5: 691–99.

——. 1993. Patterns of fluctuating asymmetry in sexual ornaments predict female choice. *Journal of Evolutionary Biology* 6: 481–91.

——. 1995. Hormones, handicaps and bright birds. *Trends in Ecology and Evolution* 10: 121.

——. 1997. Immune defence, extra-pair paternity, and sexual selection in birds. *Proceedings of the Royal Society of London, B* 264: 561–66.

Møller A and Petrie M. 2001. Condition dependence, multiple sexual signals, and immunocompetence in peacocks. *Behavioral Ecology* 13: 248–53.

Møller A and Thornhill R. 1997. A metaanalysis of the heritability of developmental stability. *Journal of Evolutionary Biology* 10: 1–16.

——. 1998. Bilateral symmetry and sexual selection: a meta-analysis. *American Naturalist* 151: 174–92.

Møller A, Henry P-Y and Erritzoe J. 2000. The evolution of song repertoires and immune defence in birds. *Proceedings of the Royal Society of London, B* 267: 165–69.

Møller A, Kimball R, and Erritzoe J. 1996. Sexual ornamentation, condition, and immune defence in the house sparrow, *Passer domesticus*. *Behavioral Ecology and Sociobiology* 39: 317–22.

Molteno A and Bennett N. 2002. Rainfall, dispersal and reproductive inhibition in eusocial Damaraland mole-rats (*Cryptomys damarensis*). *Journal of Zoology* 256: 445–48.

Money J and Dalery J. 1977. Hyperadrenocortical 46, XX hermaphroditism with penile urethra: psychological studies in seven cases, three reared as boys, four as girls. In: *Congenital Adrenal Hyperplasia*. Lee P, Plotnick L, Kowarski A, and Migeon C (eds.). Univesity Park Press.

Money J and Ogunro C. 1974. Behavioral sexology: ten cases of genetic male intersexuality with impaired prenatal and pubertal androgenization. *Archives of Sexual Behavior* 3: 181–205.

Montagnese CM, Krebs JR, Szekely AD, et al. 1993. A subpopulation of large calbindin-like immunopositive neurones is present in the hippocampal formation in food-storing but not in non-storing species of bird. *Brain Research* 614: 291–300.

Moore CL, Dou H, and Juraska JM. 1992. Maternal stimulation affects the number of motor neurons in a sexually dimorphic nucleus of the lumbar spinal cord. *Brain Research* 72: 52–56.

Moore, MC. 1991. Application of organization-activation theory to alternative male reproductive strategies: a review. *Hormones and Behavior* 25: 154–79.

Moore M and Crews D. 1986. Sex steroid hormones in natural populations of a sexual whiptail lizard, *Cnemidophorus inornatus*, a direct sexual ancestor of a unisexual parthenogenic lizard. *General and Comparative Endocrinology* 63: 424–30.

Moore M, Whittier JM, Billy AJ, Crews D. 1985. Male-like behavior in an all-female lizard: relationship to ovarian cycle. *Animal Behaviour* 33: 284–89.

Morgan E. 1982. *The Aquatic Ape*. Souvenir.

Mori A. 1984. An ethological study of pygmy chimpanzees in Wambe Zaire: A comparison with chimpanzees. *Primates* 25: 255–78.

Moore W. 1975. Stability of unisexual-bisexual populations of *Poeciliopsis* (*Pisces: Poeciliidae*). *Ecology* 56: 791–808.

——. 1976. Components of fitness in the unisexual fish *Poeciliopsis monacha-occidentalis*. *Evolution* 30: 564–78.

Moore W and McKay F. 1971. Coexistence in unisexual-bisexual species complexes of *Poeciliopsis* (*Pisces: Poeciliidae*). *Ecology* 52: 791–99.

Morton E. 1982. Grading discreteness, redundancy, and motivation-structural rules. In: *Acoustic Communication in Birds*. Kroodsma D and Miller E (eds.). Academic Press, pp. 183–212.

Mosconi G, Carnevali O, Franzoni M, et al. 2002. Environmental estrogens and reproductive biology in amphibians. *General and Comparative Endocrinology* 126: 125–29.

Moser E and Paulsen O. 2001. New excitement in cognitive space: between place cells and spatial memory. *Current Opinion in Neurobiology* 11: 745–51.

Mrowka W. 1987. Oral fertilization in a mouth-brooding cichlid fish. *Ethology* 74: 293–96.

Mueller R-A. 1996. Innateness, autonomy, universality? Neurobiological approaches language. *Behavioral Brain Sciences* 19: 611–75.

Muller H. 1932. Some genetic aspects of sex. *American Naturalist* 66: 118–38.

Muller M and Wrangham R. 2002. Sexual mimicry in hyenas. *Quarterly Review of Biology* 77: 3–16.

Muller R, Bostock E, Taube J, et al. 1994. On the directional firing properties of hippocampal place cells. *Journal of Neuroscience* 14: 7235–51.

Mulligan J. 1966. Singing behavior and its development in the song sparrow, *Melospiza melodia*. *University of California Publications in Zoology* 81: 1–76.

Muma K and Weatherhead P. 1989. Male traits expressed in females: direct or indirect sexual selection? *Behavioral Ecology and Sociobiology* 25: 23–31.

Munz H, Stumpf WE, and Jennes L. 1981. LHRH systems in the brain of platyfish. *Brain Research* 221: 1–13.

Muske LE. 1993. Evolution of gonadotropin-releasing hormone (GnRH) neuronal systems. *Brain, Behavior and Evolution* 42: 215–30.

Nachman G and Heller NE. 1999. Fluctuating asymmetry as an index of fitness: causality or statistical artifact? *Oikos* 86: 357–65.

Nadel L. 1991. The hippocampus and space revisited. *Hippocampus* 1: 221–29.

Nagl S, Tichy H, Mayer W, et al. 2000. The origin and age of haplochromine fishes in Lake Victoria, East Africa. *Proceedings of the Royal Society of London, B* 267: 1049–61.

Naish K and Ribbink A. 1990. A preliminary investigation of sex change in *Pseudotropheus lombardoi* (Pisces: Cichlidae). *Evironmental Biology of Fishes* 28: 285–94.

Nakamura M and Iwahashi M. 1982. Studies on the practical masculinization in *Tilapia nilotica* by the oral administration of androgen. *Bulletin of the Japanese Society for Scientific Fisheries* 486: 763–69.

Nash S. 1975. The relationship among sex-role stereotyping, sex-role preference, and the sex difference in spatial visualization. *Sex Roles* 1: 15–32.

Nass R and Baker S. 1991. Androgen effects on cognition: congenital adrenal hyperplasia. *Psychoneuroendocrinology* 16: 189–201.

Nealen P and Perkel D. 2000. Sexual dimorphism in the song system of the Carolina wren *Thryothorus ludovicianus*. *Journal of Comparative Neurology* 418: 346–60.

Neave N, Menaged M, and Weightman D. 1999. Sex differences in cognition: the role of testosterone and sexual orientation. *Brain and Cognition* 41: 245–62.

Neaves WB, Griffin JE, and Wilson JD. 1980. Sexual dimorphism of the phallus in spotted hyaena (*Crocuta crocuta*). *Journal of Reproduction and Fertility* 59: 509–13.

Noor M. 2000. On the evolution of female mating preferences as pleiotropic by-products of adaptive evolution. *Adaptive Behavior* 8: 3–12.

Nordeen E and Nordeen K. 1988. Sex and regional differences in the incorporation of neurons born during song learning in zebra finches. *Journal of Neuroscience* 8: 2869–74.

Nordeen KW and Nordeen EJ. 1988. Projection neurons within a vocal motor pathway are born during song learning in zebra finches. *Nature* 334: 149–51.

Nordeen K, Marler P, and Nordeen E. 1989. Addition of song-related neurons in swamp sparrows coincides with memorization, not production, of learned songs. *Journal of Neurobiology* 20: 651–61.

Northcutt RG and Muske LE. 1994. Multiple embryonic origins of gonadotropin-releasing hormone (GnRH) immunoreactive neurons. *Developmental Brain Research* 78: 279–90.

Nottebohm F. 1981. A brain for all seasons: cyclical anatomical changes in song control nuclei of the canary brain. *Science* 214: 1368–70.

———. 1989. From bird song to neurogenesis. *Scientific American* 260: 74–79.

Nottebohm F and Arnold A. 1976. Sexual dimorphism in vocal control areas of the songbird brain. *Science* 194: 211–13.

Nottebohm F, Kasparian S, and Pandazis C. 1981. Brain space for a learned task. *Brain Research* 213: 99–109.

Nottebohm F, Nottebohm M, and Crane L. 1986. Developmental and seasonal changes in canary song and their relation to changes in the anatomy of song-control nuclei. *Behavioral Neural Biology* 46: 445–71.

Nowicki S, Hasselquist D, Bensch S, et al. 2000. Nestling growth and song repertoire size in great reed warblers: evidence for song learning as an indicator mechanism in mate choice. *Proceedings of the Royal Society of London, B* 267: 2419–24.

Nunez JL and Juraska JM. 1998. The size of the splenium of the rat corpus callosum: influence of hormones, sex ratio, and neonatal cryoanesthesia. *Developmental Psychobiology* 33: 295–303.

O'Connor D, Archer J, Hair W, et al. 2001. Activational effects of testosterone on cognitive function in men. *Neuropsychologia* 39: 1385–94.

Ohlsson T, Smith H, Råberg L, et al. 2001. Pheasant sexual ornaments reflect nutritional conditions during early growth. *Proceedings of the Royal Society of London, B* 269: 21–27.

Oka S, Miyamoto O, Jamjua N, et al. 1999.

Re-evaluation of sexual dimorphism in human corpus callosum. *Neuroreport* 10: 937–40.

O'Keefe J and Burgess N. 1996. Geometric determinants of the place fields of hippocampal neurons. *Nature* 381: 425–28.

O'Keefe J and Dostrovsky J. 1971. The hippocampus as a spatial map: preliminary evidence from unit activity in the freely-moving rat. *Brain Research* 34: 171–75.

O'Keefe and Nadel L. 1978. *The Hippocampus As a Cognitive Map*. Clarendon.

O'Keefe J, Burgess N, Donnett JG, et al. 1998. Place cells, navigational accuracy, and the human hippocampus. *Philosophical Transactions of the Royal Society of London*, B 353: 1333–40.

Olson D and Eliot J. 1986. Relationships between experiences, processing style, and sex-related differences in performance on spatial tests. *Perceptual and Motor Skills* 62: 447–60.

Olton D, Becker J, and Handelmann G. 1979. Hippocampus, space, and memory. *Behavioral Brain Science* 2: 313–65.

O'Riain MJ, Jarvis JUM, and Faulkes CG. 1996. A dispersive morph in the naked mole-rat. *Nature* 380: 619–21.

Orchinik M, Licht P, and Crews D. 1988. Plasma steroid concentrations change in response to sexual behavior in *Bufo marinus*. *Hormones and Behavior* 22: 338–50.

Osborne J. 2001. Testing stereotype threats: does anxiety explain race and sex differences in achievement? *Contemporary Educational Psychology* 26: 291–310.

Otto S and Lenormand T. 2002. Resolving the paradox of sex and recombination. *Nature Reviews Genetics* 3: 252–61.

Owens IPF and Wilson K. 1999. Immunocompetence: a neglected life history trait or conspicuous red herring? *Trends in Ecology and Evolution* 14: 170–72.

Packard J, Mech L, and Seal U. 1983. Social influences on reproduction in wolves. *Canadian Wildlife Service Reports Series* 45: 78–85.

Packer C, Pusey A, and Eberly L. 2001. Egalitariansim in female African lions. *Science* 293: 690–93.

Paley W. 1825. *Natural Theology: Evidence of the Existence and Attributes of the Deity, Collected from the Appearance of Nature*. 12th ed. Paley E (ed.). Classworks.

Palmer A. 1994. Fluctuating asymmetry analyses: a primer. In: *Developmental Instability: Its Origins and Evolutionary Implications*. Markow T (ed.). Kluwer, pp. 335–64.

——. 1996. Waltzing with asymmetry. *BioScience* 46: 518–32.

——. 1999. Detecting publication bias in meta-analyses: a case study of fluctuating asymmetry and sexual selection. *American Naturalist* 154: 220–33.

——. 2000. Quasireplication and the contract of error: lesson from sex ratios, heritabilities and fluctuating asymmetry. *Annual Review of Ecology and Systematics* 31: 441–80.

Palmer A and Strobeck C. 1986. Fluctuating asymmetry: measurement, analysis, patterns. *Annual Review of Ecology and Systematics* 17: 391–421.

——. 1997. Fluctuating asymmetry and developmental stability: heritability of observable variation vs heritability of inferred cause. *Journal of Evolutionary Biology* 10: 39–49.

Pankhurst NW, vanderKraak G, and Peter RE. 1995. Evidence that the inhibitory effects of stress on reproduction in teleost fish are not mediated by the action of cortisol on ovarian steroidogenesis. *General and Comparative Endorcrinology* 99: 249–57.

Parker G. 1974. Assessment strategy and the evolution of animal conflicts. *Journal of Theoretical Biology* 47: 223–43.

Paulescu E, McCrory E, Fazio F, et al. 2000. A cultural effect on brain function. *Nature Neuroscience* 3: 92–95.

Payne R. 1982. Ecological consequences of song matching: breeding success and intraspecific song mimicry in indigo buntings. *Ecology* 63: 401–11.

Perlman S. 1973. Cognitive abilities of children with hormone abnormalities: screening by psychoeducational tests. *Journal of Learning Disabilities* 6: 21–29.

Perrill S and Magier M. 1988. Male mating behavior in *Acris crepitans*. *Copeia* 1988: 245–48.

Perrill S, Magier M, and Daniel R. 1982. Mating strategy shifts in male green treefrogs (*Hyla cinerea*): an experimental study. *Animal Behaviour* 30: 43–48.

Peters A. 2000. Testosterone treatment is immunosuppressive in superb fairy-wrens, yet free-living males with high testosterone are more immunocompetent. *Proceedings of the Royal Society of London*, B 267: 883–89.

Peters H. 1975. Hermaphroditism in cichlid fishes. In: *Intersexuality in the Animal Kingdom*. Reinboth R (ed.). Springer-Verlag, pp. 228–35.

Petersen C. 1983. Reproductive strategies in a simultaneously hermaphroditic reef fish, *Serranus fasciatus*. *American Zoologist* 23: 880.

———. 1987. Reproductive behavior and gender allocation in a simultaneously hermaphroditic reef fish, *Serranus fasciatus*. *Animal Behaviour* 35: 1601–14.

Petrie M. 1992. Peacocks with low mating success are more likley to suffer predation. *Animal Behaviour* 44: 585–86.

———. 1994. Improved growth and survival of offspring of peacocks with more elaborate trains. *Nature* 371: 598–99.

Petrie M and Kempenaers B. 1998. Extrapair paternity in birds: explaining variation between species and populations. *Trends in Ecology and Evolution* 13: 52–57.

Petrie M, Halliday T, and Sanders C. 1991. Peahens prefer peacocks with elaborate trains. *Animal Behaviour* 41: 323–31.

Petrinovich L. 1984. A two-factor dual-process theory of habituation and sensitization. In: *Habituation, Sensitization, and Behavior*. Peeke HVS and Petrinovich L (eds.). Academic Press, pp. 17–55.

Pezaris E and Casey M. 1991. Girls who use "masculine" problem-solving strategies on a spatial task: proposed genetic and environmental factors. *Brain and Cognition* 17: 1–22.

Pfaff D, McCarthy M, Schwartz-Giblin S, et al. 1994. Female reproductive behavior. In *The Physiology of Reproduction*. Knobil E and Neill J (eds.). Raven. 2: 107–220.

Pfaff D and Schwartz-Giblin S 1988. Cellular mechanisms of female reproductive behaviors. In: *The Physiology of Reproduction*.

Knobil E and Neill J (eds.). Raven Press, pp. 1487–568.

Phelps SM and Ryan MJ. 1998. Neural networks predict response biases of female tungara frogs. *Proceedings of the Royal Society of London*, B 265: 279–85.

Phelps S, Ryan M, and Rand A. 2001. Vestigial preference functions in neural networks and tungara frogs. *Proceedings of the National Academy of Sciences, USA* 98: 13161–66.

Pieau C. 1972. Effets de la température sur le développment des glandes genitales chez les embryons de deux Cheloniens, *Emys orbicularis* L. et *Testudo graeca* L. C. R. *Academy of Science Paris* (D) 274: 719–22.

———. 1975. Temperature and sex differentiation in embryos of two chelonians, *Emys obicularis* L. and *Testudo graeca* L. In: *Intersexuality in the Animal Kingdom*. Reinboth R (ed.). Springer-Verlag, pp. 332–39.

———. 1996. Temperature variation and sex determination in reptiles. *BioEssays* 18: 19–26.

Pieau C, Mignot T-M, Dorizzi M, et al. 1982. Gonadal steroid levels in the turtle *Emys obicularis* L.: a preliminary study in embryos, hatchlings, and young as a function of the incubation temperature of eggs. *General and Comparative Endocrinology* 47: 392–98.

Piersma T and Lindstrom A. 1997. Rapid reversible changes in organ size as a component of adaptive behavior. *Trends in Ecology and Evolution* 12: 134–38.

Pilastro A and Bisazza A. 1999. Insemination efficiency of two alternative male mating tactics in the guppy (*Poecilia reticulata*). *Proceedings of the Royal Society of London*, B 266: 1887–91.

Pilgrim C and Reisert I. 1992. Differences between male and female brains—developmental mechanisms and implications. *Hormones and Metabolic Research* 24: 353–59.

Pinker S. 1994. *The Language Instinct*. Morrow.

Pinker S and Bloom P. 1990. Natural language and natural selection. *Behavioral and Brain Sciences* 13: 707–84.

Plavcan J and Van Schaik C. 1994. Canine

dimorphism. *Evolutionary Anthropology* 4: 208–14.

Poiani A, Goldsmith A, and Evans M. 2000. Ectoparasites of house sparrows (*Passer domesticus*): an experimental test of the immunocompetence handicap hypothesis and a new model. *Behavioral Ecology and Sociobiology* 47: 230–42.

Policansky D. 1982. Sex change in plants and animals. *Annual Review of Ecology and Systematics* 13: 471–95.

Pomiankowski A. 1990. How to find the top male. *Nature* 347: 616–17.

Pomiankowski A and Iwasa Y. 1998. Runaway ornament diversity caused by Fisherian sexual selection. *Proceedings of the National Academy of Sciences* 95: 5106–11.

Pomiankowski A and Møller A. 1995. A resolution of the lek paradox. *Proceedings of the Royal Society of London, B* 260: 21–29.

Pontius A. 1989. Color and spatial error in block design in stone-age Auca Indians: Ecological underuse of occipital-parietal system in men and of frontal lobes in women. *Brain and Cognition* 10: 54–75.

Pope A. 1994. *Essay on Man and Other Poems*. Dover.

Porteus S. 1965. *Porteus Maze Test: Fifty Years of Applications*. Pacific Books.

Postma A, Izendoorn P, and De Haan E. 1998. Sex differences in object location memory. *Brain and Cognititon* 36: 334–45.

Postma A, Winkel J, Tuiten A, et al. 1999. Sex differences and menstrual cycle effects in human spatial memory. *Psychoneuroendocrinology* 24: 175–92.

Potter S and Graves R. 1988. Is interhemispheric transfer related to handedness and gender? *Neuropsychologia* 26: 319–25.

Pottinger T, Campbell P, and Sumpter J. 1991. Stress-induced disruption of the salmonid liver-gonad axis. In: *Reproductive Physiology of Fish Symposium*. Scott A, Sumpter J, Kime D, and Rolfe M (eds.). Sheffield University, pp. 114–16.

Potts G. 1973. The ethology of *Labroides dimidiatus* (Cuv. & Val.) (Labroidae, Pisces) on Aldabra. *Animal Behaviour* 21: 250–91.

Poulin R and Grutter AS. 1996. Cleaning symbioses: proximate and adaptive explanations. *BioScience* 46: 512–17.

Poulin R, Marshall L, and Spencer H. 2000.

Genetic variation and prevalence of blood parasites do not correlate among bird species. *Journal of Zoology* 252: 381–88.

Pravosudov V and Clayton N. 2000. Effects of demanding foraging conditions on cache retrieval accuracy in food-caching mountain chickadees (*Poecile gambeli*). *Proceedings of the Royal Society of London, B* 268: 363–68.

Pressley P. 1981. Pair formation and joint territoriality in a simultaneous hermaphrodite: the coral reef fish, *Serranus tigrinus*. *Zeitschrift für Tierpsychologie* 56: 33–46.

Quinnett P. 1996. *Darwin's Bass: The Evolutionary Psychology of Fishing Man*. Keokee Co. Publishing.

Råberg L, Grahn M, Hasselquist D, et al. 1998. On the adaptive significance of stress-induced immunosuppression. *Proceedings of the Royal Society of London, B* 265: 1637–41.

Råberg L, Vestberg M, Hasselquist D, et al. 2002. Basal metabolic rate and the evolution of the adaptive immune system. *Proceedings of the Royal Society of London, B* 269: 817–21.

Racey P and Skinner J. 1979. Endocrine aspects of sexual mimicry in spotted hyenas *Crocuta crocuta*. *Journal of Zoology* 187: 315–26.

Raff R. 1996. *The Shape of Life*. University of Chicago Press.

Ralls K. 1976. Mammals in which females are larger than males. *Quarterly Review of Biology* 51: 245–76.

Rand MS and Crews D. 1994. The bisexual brain: Sex behavior differences and sex differences in parthenogenetic and sexual lizards. *Brain Research* 663: 163–67.

Rassmussen D, Liu J, Wolf P, et al. 1983. Endogenous opioid regulation of gonadotropin-releasing hormone release from the human fetal hypothalamus in vitro. *Journal of Clinical Endocrinology and Metabolism* 57: 881–85.

Rawlins J. 1985. Associations across time: the hippocampus as a temporary memory store. *Behavioral Brain Science* 8: 479–96.

Read A. 1988. Sexual selection and the role of parasites. *Trends in Ecology and Evolution* 3: 97–101.

Read A and Weary D. 1990. Sexual selection and the evolution of bird song: a test of the Hamilton-Zuk hypothesis. *Behavioral Ecology and Sociobiology* 26: 47–56.

——. 1992. The evolution of bird song: comparative analyses. *Philosophical Transactions of the Royal Society of London, B* 338: 166–86.

Reboreda JC, Clayton NS, and Kacelnik A. 1996. Species and sex differences in hippocampus size in parasitic and non-parasitic cowbirds. *NeuroReport* 7: 505–8.

Rechten C. 1978. Interspecific mimicry in birdsong: does the Beau Geste hypothesis apply? *Animal Behaviour* 26: 304–12.

Redies C and Puelles L. 2001. Modularity in vertebrate brain development and evolution. *BioEssays* 23: 1100–11.

Reeve H. 1998. Game theory, reproductive skew and nepotism. In: *Game Theory and Animal Behavior*. Dugatkin L and Reeve H (eds.). Oxford University Press, pp. 118–45.

Reeve J and Fairbairn D. 1996. Sexual size dimorphism as a correlated response to selection on body size: an empirical test of the quantitative genetic model. *Evolution* 50: 1927–38.

Reeve J and Fairbairn D. 2001. Predicting the evolution of sexual size dimorphism. *Journal of Evolutionary Biology* 14: 244–54.

Regal P. 1998. Violence and sex: a review of demonic males: apes and the origins of human violence. *Quarterly Review of Biology* 73: 473–76.

Reinboth R. 1967. Zum problem der amphisexuallen Fische. *Verhalten Deutsche Zoologie Gesellschaft* 1967: 316–25.

——. 1988. Physiological problems of teleost ambisexuality. *Environmental Biology of Fishes* 22: 249–59.

Resnick S. 1993. Sex differences in mental rotations: an effect of time limits. *Brain and Cognition* 21: 71–79.

Resnick S, Berenbaum S, Gottesman I, et al. 1986. Early hormonal influences on cognitive functioning in congential adrenal hyperplasia. *Developmental Psychology* 22: 191–98.

Rhen T. 2000. Sex-limited mutations and the evolution of sexual dimorphism. *Evolution* 54: 37–43.

Rice WR. 1983. Parent-offspring transmission: A selective agent promoting sexual reproduction. *American Naturalist* 121: 187–203.

——. 2002. Experimental tests of the adaptive significance of sexual recombination. *Nature Reviews Genetics* 3: 241–51.

Richards CM and Nace GW. 1978. Gynogenetic and hormonal sex reversal in tests of the XX-XY hypothesis of sex determination in *Rana pipiens*. *Growth* 42: 319–31.

Richardson J. 1994. Gender differences in mental rotations. *Perceptual and Motor Skills* 78: 435–48.

——. 1997. Introduction to the study of gender differences in cognition. In: *Gender Differences in Human Cognition*. Caplan P, Crawford M, Hyde J, and Richardson J (eds.). Oxford University Press, pp. 3–29.

Rissman EF and Bronson F. 1987. Role of the ovary and adrenal gland in the sexual behavior of the musk shrew (*Suncus murinus*). *Biology of Reproduction* 36: 664–68.

Roberts J and Bell M. 2000a. Sex differences on a computerized mental rotation task disappear with computer familiarization. *Perceptual and Motor Skills* 91: 1027–34.

——. 2000b. Sex differences on a mental rotation task: variations in electroencephalogram hemispheric activation between children and college students. *Developmental Neuropsychology* 17: 199–223.

Robertson D. 1972. Social control of sex reversal in a coral reef fish. *Science* 177: 1007–9.

Rodd F, Hughes K, Grether G, et al. 2002. A possible non-sexual origin of mate preference: are male guppies mimicking fruit? *Proceedings of the Royal Society of London, B* 269: 475–81.

Rohwer S. 1975. The social significance of avian plumage variability. *Evolution* 29: 593–610.

Ronnekleiv O and Resko J. 1990. Ontogeny of gonadotropin-releasing hormone-containing neurons in early fetal development of Rhesus macaques. *Endocrinology* 126: 498.

Rood J. 1980. Mating relationships and breeding suppression in the dwarf mongoose. *Animal Behaviour* 28: 143–50.

Rosenblum P and Peter R. 1989. Evidence for the involvement of endogenous opioids in the regulation of gonadotropin secretion in male goldfish, *Carassius auratus*. *General and Comparative Endocrinology* 73: 21–27.

Roth G. 1987. *Visual Behavior in Salamanders*. Springer.

Roth G, Blanke J, and Wake DB. 1994. Cell size predicts morphological complexity in the brains of frogs and salamanders. *Proceedings of the National Academy of Sciences, USA* 91: 4796–800.

Roth G, Naujoks-Manteuffel C, and Grunwald W. 1990a. Cytoarchitecture of the tectum mesencephali in salamanders: a Golgi and HRP study. *Journal of Comparative Neurology* 291: 27–42.

Roth G, Nishikawa KC, Naujoks-Manteuffel C, et al. 1993. Paedomorphosis and simplification in the nervous system of salamanders. *Brain, Behavior and Evolution* 42: 137–70.

Roth G, Rottluff B, and Grunwald W. 1990b. Miniaturization in plethodontid salamanders (Caudata: Plethodontidae) and its consequences for the brain and visual system. *Biological Journal of the Linnaen Society* 40: 165–90.

Roth G, Rottluff B, and Linke R. 1988. Miniaturization, genome size and the origin of functional constraints in the visual system of salamanders. *Naturwissenschaften* 75: 297–304.

Rothstein S and Fleischer R. 1987. Vocal dialects and their possible relation to honest status signalling in the brown-headed cowbird. *Condor* 889: 1–23.

Rothstein SI and Robinson SK. 1994. Conservation and coevolutionary implications of brood parasitism by cowbirds. *Trends in Evolution and Ecology* 9: 162–64.

Rubin DA. 1985. Effect of pH on sex ratio in cichlids and a poecilliid (Teleostei). *Copeia* 1985: 233–34.

Rumelhart D. 1989. The architecture of mind. In: *Foundations of Cognitive Science*. Posner M (ed.). MIT Press, pp. 133–59.

Rumelhart D and McClelland J. 1986. *Parallel Distributed Processing*. Vol. 1: *Foundations*. MIT Press.

Russell E. 1916. *From and Function: A Contribution to the History of Animal Morphology*. University of Chicago Press.

Russo P, Persegani C, Papeschi L, et al. 1999. Sex differences in hemisphere preference as assessed by a paper-pencil test. *International Jouranl of Neuroscience* 100: 29–37.

Ruvinsky A. 1997. Sex, meiosis and multicellularity. *Acta Biotheoretica* 45:127–41.

Ryan M. 1989. Alternative mating behavior in the swordtails *Xiphophorus nigrensis* and *X. pygmaeus*. *Behavioral Ecology and Sociobiology* 24: 341–48.

———. 1998a. Principle with a handicap. *Quarterly Review of Biology* 73: 477–79.

———. 1998b. Sexual selection, receiver biases, and the evolution of sex differences. *Science* 281: 1999–2002.

Ryan M and Getz W. 2000. Signal decoding and receiver evolution: an analysis using an artificial neural network. *Brain, Behavior and Evolution* 56: 45–62.

Ryan MJ and Keddy-Hector A. 1992. Directional patterns of female mate choice and the role of sensory biases. *American Naturalist* 139: S4–S35.

Ryan M and Rand A. 1993a. Sexual selection and signal evolution: the ghost of biases past. *Philosophical Transactions of the Royal Society of London, B* 340: 187–95.

———. 1993b. Species recognition and sexual selection as a unitary problem in animal communication. *Evolution* 47: 647–57.

———. 1995. Female responses to ancestral advertisement calls in tungara frogs. *Science* 269: 390–92.

———. 1999. Phylogenetic influence on mating call preferences in female tungara frogs, *Physalaemus pustulosus*. *Animal Behaviour* 57: 945–58.

Ryan M, Fox J, Wilczynski W, et al. 1990. Sexual selection for sensory exploitation in the frog *Physalaemus pustulosus*. *Nature* 343: 66–67.

Sachs B and Meisel R. 1988. The physiology of male sexual behaviour. In: *The Physiology of Reproduction*. Knobil E and Neill J (eds.). Raven Press, pp. 1393–485.

Sachser N, Lick C, and Stanzel K. 1994. The environment, hormones, and aggresive behaviour: a 5-year study in guinea pigs. *Psychoneuroendocrinology* 19: 697–707.

Saino N and Møller A. 1994. Secondary sexual characters, parasites and testosterone in the barn swallow, *Hirundo rustica. Animal Behaviour* 48: 1325–33.

Saino N, Møller A, and Bolzern A. 1995. Testosterone effects on the immune system and parasite infestations in the barn swallow (*Hirundo rustica*): an experimental test of the immunocompetence hypothesis. *Behavioral Ecology* 6: 397–404.

Saino N, Galeotti P, Sacchi R, et al. 1997. Song and immunological condition in male barn swallows (*Hirundo rustica*). *Behavioral Ecology* 8: 364–71.

Sakthivel M, Patterson P, and Cruz-Neira C. 1999. Gender differences in navigating virtual worlds. *Biomedical Science Instrumentation* 35: 353–59.

Saltzman W, Schultz-Darken N, Wegner F, et al. 1998. Suppression of cortisol levels in subordinate female marmosets: reproductive and social contributions. *Hormones and Behavior* 33: 58–74.

Sanders B, Chen M, and Soares M. 1986. The sex differences in spatial ability: a rejoinder. *American Psychologist* 41: 1015–16.

Sandstrom N, Kaufman J, and Huettel S. 1998. Males and females use different distal cues in a virtual environmental navigation task. *Cognitive Brain Research* 6: 351–60.

Sapolsky RM. 1982. The endocrine stress-response and social status in the wild baboon. *Hormones and Behavior* 16: 279–92.

———. 1983. Endocrine aspects of social instability in the olive baboon (*Papio anubis*). *American Journal of Primatology* 5: 365–79.

———. 1986. Stress-induced elevation of testosterone concentrations in high-ranking baboons: role of catecholamines. *Endocrinology* 118: 1630–35.

———. 1987. Stress, social status, and reproductive physiology in free-living baboons. In: *Psychobiology of Reproductive Behavior: An Evolutionary Perspective*. Crews D (ed.). Prentice Hall, pp. 291–322.

Sapolsky R, Alberts S, and Altmann J. 1997. Hypercortisolism associated with social subordination or social isolation among wild baboons. *Archives of General Psychiatry* 54: 1137–43.

Saucier D, Green S, Leason J, et al. 2002. Are sex differences in navigation caused by sexually dimorphic strategies or by differences in the ability to use the strategies? *Behavioral Neuroscience* 116: 403–10.

Saunders R and Henderson E. 1965. Precocious sexual development in male postsmolt Atlantic salmon reared in the laboratory. *Journal of the Fisheries Research Board of Canada* 22: 1567–70.

Schaefer P and Thomas J. 1998. Difficulty of a spatial task and sex difference in gains from practice. *Perceptual and Motor Skills* 87: 56–58.

Scharloo W. 1991. Canalization: genetic and developmental aspects. *Annual Review of Ecology and Systematics* 22: 65–93.

Scheib JE, Gangestad SW, and Thornhill R. 1999. Facial attractiveness, symmetry and cues of good genes. *Proceedings of the Royal Society of London, B* 266: 1913–17.

Schiml P and Rissman EF. 1999. Cortisol facilitates induction of sexual behavior in the female musk shrew (*Suncus murinus*). *Behavioral Neuroscience* 113: 166–75.

———. 2000. Effects of gonadotropin-releasing hormone, corticotropin-releasing hormone, and vasopressin on female sexual behavior. *Hormones and Behavior* 37: 212–20.

Schiml-Webb P, Temple J, and Rissman EF. 2001. Glucocorticoids affect gonadotropin-releasing hormone immunoreactivity in musk shrew brain. *General and Comparative Endocrinology* 123: 163–69.

Schjelderup-Ebbe T. 1922. Beiträge zur Sozialpsychologie des Haushuhns. *Zeitschrift für Psychologie* 88: 225–52.

Schlicter D. 1972. Chemische Tarnung. Die stoffliche Grundlage der Anpassung von Anemonefischen und Riff anemonen. *Marine Biology* 12: 137–50.

———. 1976. Macromolecular mimicry: substances released by sea anemones and their role in the protection of anemonefishes. In: *Coelenterate Ecology and Behavior*. Mackie G (ed.). Plenum, pp. 433–41.

Schlinger BA. 1994. Estrogens to song: picograms to sonograms. *Hormones and Behavior* 28: 191–98.

———. 1998. Sexual differentiation of avian brain and behavior: current views on gonadal hormone-dependent and independent mechanisms. *Annual Review of Physiology* 60: 407–29.

Schlinger B and Arnold A. 1991. Brain is the major site of estrogen synthesis in the male zebra finch. *Proceedings of the National Academy of Sciences, USA* 88: 4191–94.

Schlinger BA, Greco C, and Bass AH. 1999. Aromatase activity in the hindbrain vocal control region of a teleost fish: divergence among males with alternative reproduction tactics. *Proceedings of the Royal Society of London, B* 266: 131–36.

Schmidt-Nielsen K. 1964. *Desert Animals: Physiological Problems of Heat and Water.* Clarendon.

Schoenemann P, Budinger T, Sarich V, et al. 2000. Brain size does not predict general cognitive ability within families. *Proceedings of the National Academy of Sciences, USA* 97: 4932–37.

Schratz M. 1978. A developmental investigation of sex differences in spatial (visual-analysis) and mathematical skills in three ethnic groups. *Developmental Psychology* 14: 263–67.

Schuett G. 1997. Body size and agonistic experience affect dominance and mating success in male copperheads. *Animal Behaviour* 54: 213–24.

Schwanzel-Fukuda M and Pfaff DW. 1989. Origin of luteinizing hormone-releasing hormone neurons. *Nature* 338: 161–63.

Schwanzel-Fukuda M, Bick D, and Pfaff D. 1989. Luteinizing hormone releasing hormone (LHRH)-expressing cells do not migrate normally in an inherited hypogonadal (Kallmann) syndrome. *Molecular Brain Research* 6: 311–26.

Schwanzel-Fukuda M, Jorgenson KL, Bergen HT, et al. 1992. Biology of normal luteinizing hormone-releasing hormone neurons during and after their migration from olfactory placode. *Endocrine Reviews* 13: 623–34.

Schwegler H, Crusio WE, Lipp H-P, et al. 1988. Water-maze learning in the mouse correlates with variation in hippocampal morphology. *Behavior Genetics* 18: 153–165.

———. 1991. Early postnatal hyperthyroidism alters hippocampal circuitry and improves radial-maze learning in adult mice. *Journal of Neuroscience* 11: 2102–8.

Schwenk K and Wagner G. 2001. Function and the evolution of phenotypic stability: connecting pattern to process. *American Zoologist* 41: 552–63.

Schwier H. 1939. Geschlectsbestimmung und differnzierung bei *Macropodus opercularis*, concolor, chinensis und deren Artbastarden. *Zeitschrift Induktive Abstammungs- und Verbungslehre* 77: 291–335.

Scott JP and Fredericson E. 1951. The causes of fighting in mice and rats. *Physiological Zoology* 24: 273–308.

Scoville W and Milner B. 1957. Loss of memory after bilateral hippocampal lesions. *Journal of Neurology, Neurosurgery & Psychiatry* 20: 11–21.

Searcy W. 1984. Song repertoire size and female preferences in song sparrows. *Behavioral Ecology and Sociobiology* 14: 281–86.

———. 1992. Song repertoire and mate choice in birds. *American Zoologist* 32: 71–80.

Searcy W and Andersson M. 1986. Sexual selection and the evolution of song. *Annual Review of Ecology and Systematics* 17: 507–33.

Seehausen O, van Alpern JJM, and White F. 1997. Cichlid fish diversity is threatened by eutrophication that curbs sexual selection. *Science* 277: 1808–11.

Setchell J and Dixson A. 2001. Changes in the secondary sexual adornments of male mandrillls (*Mandrillus sphinx*) are associated with gain and loss of alpha status. *Hormones and Behavior* 39: 177–84.

Shakakura Y and Noakes D. 2000. Age, growth and sexual development in the self-fertilizing hermaphroditic fish *Rivulus marmoratus*. *Environmental Biology of Fishes* 59: 309–17.

Shapiro DY. 1987. Differentiation and evolution of sex change in fishes. *BioScience* 37: 490–97.

———. 1988. Behavioral influences on gene structure and other new ideas concerning sex change in fishes. *Environmental Biology of Fishes* 23: 283–97.

———. 1989. Inapplicability of the size-advantage model to coral reef fishes. *Trends in Ecology and Evolution* 4: 272–73.

———. 1992. Plasticity of gonadal development and protandry in fishes. *Journal of Experimental Zoology* 261: 194–203.

Shapiro DY, Marconoto A, and Yoshikawa T. 1994. Sperm economy in a coral reef fish. *Ecology* 75: 1334–44.

Shaywitz B, Shaywitz S, Pugh K, et al. 1995. Sex differences in the functional organization of the brain for language. *Nature* 373: 607–9.

Shepard R and Metzler J. 1971. Mental rotation of three-dimensional objects. *Science* 171: 701–3.

Sherman PW. 1988. The levels of analysis. *Animal Behaviour* 36: 616–19.

Sherman P and Reeve H. 1997. Forward and backward: alternative approaches to studying human and social evolution. In: *Human Nature: A Critical Reader*. Betzig L (ed.). Oxford University Press, pp. 147–58.

Sherman P, Jarvis J, and Alexander R (eds.). 1991. *The Biology of the Naked Mole Rat*. Princeton University Press.

Sherry D. 1984. Food storage by black-capped chickadees—memory for the location and contents of caches. *Animal Behaviour* 32: 451–64.

Sherry DF and Vaccarino AL. 1989. Hippocampus and memory for food caches in black-capped chickadees. *Behavioral Neuroscience* 103: 308–18.

Sherry D, Vaccarino A, Buckenham K, et al. 1989. The hippocampal complex of food-storing birds. *Brain, Behavior and Evolution* 34: 308–17.

Sherry DF, Jacobs LF, and Gaulin SJC. 1992. Spatial memory and adaptive specialization of the hippocampus. *Trends in Neurosciences* 15: 298–303.

Sherry DF, Forbes MRL, Khurgel M, et al. 1993. Females have a larger hippocampus than males in the brood-parasitic brown-headed cowbird. *Proceedings of the National Academy of Sciences, USA* 90: 7839–43.

Shettleworth S. 2001. Animal cognition and animal behavior. *Animal Behaviour* 61: 277–86.

Shettleworth S and Hampton R. 1998. Adaptive specializations of spatial cognition in food-storing birds? Approaches to testing a comparative hypothesis. In: *Animal Cognition in Nature*. Balda R, Pepperberg I, and Kamil A (eds.). Academic Press, pp. 65–98.

Shields S. 1975. Functionalism, Darwinism, and the psychology of women. *American Psychologist* 30: 739–54.

Shine R. 1999. Why is sex determined by nest temperature in many reptiles? *Trends in Ecology and Evolution* 14: 186–89.

Shine R, Harlow P, LeMaster M, et al. 2000a. The transvestite serpent: why do male garter snakes court (some) other males? *Animal Behaviour* 59: 349–59.

Shine R, O'Connor D, and Mason R. 2000b. Female mimicry in garter snakes: behavioural tactics of "she-males" and the males that court them. *Canadian Journal of Zoology* 78: 1391–96.

Shively C, Grant K, Ehrenkaufer R, et al. 1997a. Social stress, depression, and brain dopamine in female cynomolgus monkeys. *Annals of the New York Academy of Sciences* 807: 574–77.

Shively C, Laber-Laird K, and Anton R. 1997b. Behavior and physiology of social stress and depression in female cynomolgus monkeys. *Biological Psychology* 41: 871–82.

Shozu M, Akasofu K, Harada T, et al. 1991. A new cause of female pseudohermaphroditism. *Journal of Clinical Endocrinology and Metabolism* 72: 560–66.

Sibley C and Ahlquist J. 1985. The phylogeny and classification of the passerine birds based on comparisons of the genetic material. In: *Proceedings of the XVIII Congress of International Ornithologists Acta*. Ilyichev V and Gavrilov V (eds.). Nauka, pp. 83–121.

Signorella M and Jamison W. 1986. Masculinity, femininity, androgyny, and cognitive performance: a meta-analysis. *Psychological Bulletin* 100: 207–28.

Signorella M, Jamison W, and Krupa M. 1989. Predicting spatial performance from gender stereotyping in activity preferences and in self-concept. *Developmental Psychology* 25: 89–95.

Siller S. 2001. Sexual selection and the maintenance of sex. *Nature* 411: 689–92.

Silverman I and Eals M. 1992. Sex differences in spatial ability: evolutionary theory and data. In: *The Adapted Mind: Evolutionary Psychology and the Generation of Culture.* Barkow J, Cosmides L, and Tooby J (eds.). Oxford University Press, pp. 487–503.

Silverman I, Kastuk D, Choi J, et al. 1999. Testosterone levels and spatial ability in men. *Psychoneuroendocrinology* 24: 813–22.

Silverman I, Choi J, Mackewn A, et al. 2000. Evolved mechanisms underlying wayfinding, further studies on the hunter-gatherer theory of spatial sex differences. *Evolution and Human Behavior* 21: 201–13.

Silverman R, Gibson M, and Silverman A. 1992. Application of a fluorescent dye to study connectivity between third ventricular preoptic area grafts and host hypothalamus. *Journal of Neuroscience Research* 31: 156–65.

Simmons LW, Tomkins JL, Kotiaho JS, et al. 1999. Fluctuating paradigm. *Proceedings of the Royal Society of London, B* 266: 593–95.

Simon NG and Cologer-Clifford A. 1991. In utero contiguity to males does not influence morphology, behavioral sensitivity to testosterone, or hypothalamic androgen binding in cf-1 female mice. *Hormones and Behavior* 25: 518–30.

Simpson E, Clyne C, Rubin G, et al. 2002. Aromatase—a brief overview. *Annual Review of Physiology* 64: 93–127.

Sinervo B and Lively CM. 1996. The rock-paper-scissors game and the evolution of alternative male strategies. *Nature* 380: 240–43.

Sinervo B, Bleay C, and Adamopoulou C. 2001. Social causes of correlational selection and the resolution of a heritable throat color polymorphism in a lizard. *Evolution* 55: 2040–52.

Sinforiani E, Livieri C, Mauri M, et al. 1994. Cognitive and neuroradiological findings in congenital adrenal hyperplasia. *Psychoneuroendocrinology* 19: 55–64.

Sirinathsinghji D. 1986. Regulation of lordosis behavior in the female rat by corticotropin-releasing factor, B-endorphin/corticotropin and luteinizing hormone-releasing hormone neural systems in the medial preoptic area. *Brain Research* 375: 45–56.

Skinner B. 1957. *Verbal Behavior.* Appleton Century Crofts.

Slabbekoorn D, van Goozen S, Sanders G, et al. 2000. The dermatoglyphic characteristics of transexuals: is there evidence for an organizing effect of sex hormones? *Psychoneuroendocrinology* 25: 365–75.

Slotow R, vanDyk G, Poole J, et al. 2000. Older bull elephants control young males. *Nature* 408: 425–26.

Smale L, Frank LG, and Holekamp KE. 1993. Ontogeny of dominance in free-living spotted hyaenas: juvenile rank relations with adult females and immigrant males. *Animal Behaviour* 46: 467–77.

Smale L, Holekamp KE, and White PA. 1999. Siblicide revisited in the spotted hyaena: does it conform to obligate or facultative models? *Animal Behaviour* 58: 545–51.

Smith CL. 1965. The patterns of sexuality and the classification of serranid fishes. *American Museum Novitates* 2207: 1–20.

Smith E. 1998. Is Tibetan polyandry adaptive? Methodological and metatheoretical analyses. *Human Nature* 9: 225–61.

Smith E and Winterhalder B (eds.). 1992. *Evolutionary Ecology and Human Behavior.* Aldine de Gruyter.

Smith E, Borgerhoff-Mulder M, and Hill K. 2001a. Controversies in the evolutionary social sciences: a guide for the perplexed. *Trends in Ecology and Evolution* 16: 128–35.

———. 2001b. Human rape—adaptive or not? *Trends in Ecology and Evolution* 16: 489.

Smith T and French J. 1997. Social and reproductive conditions modulate urinary cortisol excretion in black tufted-ear marmosets (*Callithrix kuhli*). *American Journal of Primatology* 42: 253–67.

Smith TE, Faulkes CG, and Abbott DH. 1997a. Combined olfactory contact with the parent colony and direct contact with nonbreeding animals does not maintain suppression of ovulation in female naked mole-rats (*Heterocephalus glaber*). *Hormones and Behavior* 31: 277–88.

Smith TE, Schaffner CM, and French JA.

1997b. Social and developmental influences on reproductive function in female Wied's black tufted-ear marmosets (*Callithrix kuhli*). *Hormones and Behavior* 31: 159–68.

Smulders TV and Dhondt A. 1997. How much memory do tits need? *Trends in Ecology and Evolution* 12: 417–18.

Smulders TV, Sasson AD, and DeVoogd TJ. 1995. Seasonal variation in hippocampal volume in a food-storing bird, the black-capped chickadee. *Journal of Neurobiology* 27: 15–25.

Smulders T, Shiflett M, Sperling A, et al. 2000. Seasonal changes in neuron numbers in the hippocampal formation of a food-hoarding bird: the black-capped chickadee. *Journal of Neurobiology* 44: 414–22.

Smuts RW. 1991. The present also explains the past: a response to Tooby and Cosmides. *Ethology and Sociobiology* 12: 77–82.

Sober E. 1984. *The Nature of Selection*. MIT Press.

Sohn JJ and Crews D. 1977. Size-mediated onset of genetically determined maturation in the platyfish, *Xiphophorus maculatus*. *Proceedings of the National Academy of Sciences, USA* 74: 4547–48.

Sohrabji F, Nordeen E, and Nordeen K. 1990. Selective impairment of song learning following lesions of a forebrain nucleus in the juvenile zebra finch. *Behavioral Neural Biology* 53: 51–63.

Sola L, Cataudella S, and Capanna E. 1981. New developments in vertebrate cytotaxonomy III. Karyology of bony fishes: a review. *Genetika* 54: 285–328.

Soma KK, Francis RC, Wingfield JC, et al. 1996. Androgen regulation of hypothalamic neurons containing gonadotropin-releasing hormone in a cichlid fish: integration with social cues. *Hormones and Behavior* 30: 216–26.

Soto C, Leatherland J, and Noakes D. 1992. Gonadal histology in the self-fertilizing hermaphroditic fish *Rivulus marmoratus*. *Canadian Journal of Zoology* 70: 2338–47.

Spence J. 1985. Gender identity and its implications for the concepts of masculinity and femininity. In: *Nebraska Symposium on Motivation*. Sonderegger T (ed.). University of Nebraska Press, pp. 59–95.

Sperry R and Gazzaniga M. 1967. Language following surgical disconnection of the hemispheres. In: *Brain Mechanisms Underlying Speech and Language*. Grune and Stratton.

Spinks A, Bennett N, Faulkes CG, et al. 2000. Circulating LH levels and the response to exogenous GnRH in the common mole-rat: implications for reproductive regulation in this social, seasonal breeding species. *Hormones and Behavior* 37: 221–28.

Squire L. 1992. Memory and the hippocampus: a synthesis from findings with rats, monkeys and humans. *Psychological Review* 99: 195–231.

Squire L and Zola S. 1998. Episodic memory, semantic memory, and amnesia. *Hippocampus* 8: 205–11.

Stafford R. 1961. Sex differences in spatial visualization as evidence of sex-linked inheritance. *Perceptual and Motor Skills* 13: 428–31.

Stanley J and Stumpf H. 1996. Able youths and achievement tests. *Behavioral and Brain Sciences* 19: 263–64.

Stavisky R, Adams M, Watson S, et al. 2001. Dominance, cortisol, and behavior in small groups of female cynomolgus monkeys (*Macaca fascicularis*). *Hormones and Behavior* 39: 232–38.

Stavnezer A, McDowell C, Hyde L, et al. 2000. Spatial ability of XY sex-reversed female mice. *Behavioral Brain Research* 112: 135–43.

Stearns S. 1987. *The Evolution of Sex and Its Consequences*. Birkhauser.

Sterelny K. 1995. The adapted mind. *Biology and Philosophy* 10: 365–80.

Stevens T and Krebs J. 1984. Retrieval of stored seeds by marsh tits (*Parus palustris*) in the field. *Ibis* 128: 513–15.

Stewart BS, Yochem P, Huber H, et al. 1994. History and present status of the northern elephant seal. In: *Elephant Seals: Population Ecology, Behavior, and Physiology*. Le Boeuf BJ and Laws R (eds.). University of California Press, pp. 29–48.

Strasser R, Bingman VP, Ioale P, et al. 1998. The homing pigeon hippocampus and the development of landmark navigation. *Developmental Psychobiology* 33: 305–15.

Streisinger G, Walker C, Dower N, et al. 1981. Production of clones of homozygous diploid zebra fish (*Brachydanio rerio*). *Nature* 291: 293–96.

Stumpf H and Klieme E. 1989. Sex-related differences in spatial ability: more evidence for convergence. *Perceptual and Motor Skills* 69: 915–21.

Sussman RW, Cheverud JM, and Bartlett TQ. 1995. Infant killing as an evolutionary strategy: Reality or myth? *Evolutionary Anthropology* 4: 149–51.

Swaab D and Fliers E. 1985. A sexually dimorphic nucleus in the human brain. *Science* 228: 1112–15.

Swaab D, Fliers E, and Partiman T. 1985. The suprachiasmatic nucleus of the human brain in relation to sex, age and senile dementia. *Brain Research* 342: 37–44.

Swaab D and Hofman M. 1988. Sexual differentiation of the human hypothalamus: ontogeny of the sexually dimorphic nucleus of the preoptic area. *Developmental Brain Research* 44: 314–18.

——. 1995. Sexual differentiation of the human hypothalamus in relation to gender and sexual orientation. *Trends in Neurosciences* 18: 264–70.

Swaab DF, Gooren LJG, and Hofman MA. 1992. The human hypothalamus in relation to gender and sexual orientation. *Progress in Brain Research* 93: 205–19.

Swaab D, Chung W, Kruijvjer F, et al. 2001. Structural and functional sex differences in the human hypothalamus. *Hormones and Behavior* 40: 93–98.

Symons D. 1979. *The Evolution of Human Sexuality*. Oxford University Press.

——. 1989. A critique of Darwinian anthropology. *Ethology and Sociobiology* 10: 131–44.

——. 1990. Adaptiveness and adaptation. *Ethology and Sociobiology* 11: 427–44.

——. 1992. On the use and misuse of Darwinism in the study of human behavior. In: *The Adapted Mind: Evolutionary Psychology and the Generation of Culture*. Barkow J, Cosmides L, and Tooby J (eds.). Oxford University Press.

Szekely T, Catchpole CK, DeVoogd A, et al.

1996. Evolutionary changes in a song control area of the brain (HVC) are associated with evolutionary changes in song repertoire among European warblers (Sylviidae). *Proceedings of the Royal Society of London, B* 263: 607–10.

Taborsky M. 1994. Sneakers, satellites, and helpers: parasitic and cooperative behavior in fish reproduction. *Advances in the Study of Behavior* 23: 1–100.

Takahashi H. 1977. Juvenile hermaphroditism in the zebrafish, *Brachydanio rerio*. *Bulletin of the Faculty of Fisheries Hokkaido Univ.* 28: 57–65.

Takahashi H and Shimizu M. 1983. Juvenile intersexuality in a cyprinid fish, the Sumatra barb, *Barbus tetrazona tetrazona*. *Bulletin of the Faculty of Fisheries Hokkaido Univ.* 34: 69–78.

Tang F, Chan STH, and Lofts B. 1974. Effects of steroid hormones on the process of natural sex reversal in the ricefield eel, *Monopterus albus* (Zuiew). *General and Comparative Endocrinology* 24: 227–41.

Tayamen MM and Shelton WL. 1978. Inducement of sex reversal in *Sarotherodon niloticus* L. *Aquaculture* 14: 349–54.

Temeles E, Pan I, Brennan J, et al. 2000. Evidence for ecological causation of sexual dimorphism in a hummingbird. *Science* 289: 441–43.

ten Cate C. 1989. Behavioral development: toward understanding processes. In: *Perspectives in Ethology*. Bateson P and Klopfer P (eds.). Plenum, pp. 243–69.

Thomaz D, Beall E, and Burke T. 1997. Alternative reproductive tactics in Atlantic salmon: factors affecting mature parr success. *Proceedings of the Royal Society of London, B* 264: 219–26.

Thompson C and Moore M. 1991. Throat color reliably signals status in male tree lizards, *Urosaurus ornatus*. *Animal Behaviour* 42: 745–53.

Thornhill R. 1980. Rape in *Panorpa* scorpionflies and a general rape hypothesis. *Animal Behaviour* 28: 52–59.

Thornhill R and Gangestad S. 1993. Human facial beauty: averageness, symmetry and parasite resistance. *Human Nature* 4: 237–69.

Thornhill R and Palmer C. 2000. *A Natural*

History of Rape: Biological Bases of Sexual Coercion. MIT Press.

Thornhill R and Thornhill NW. 1983. Human rape: an evolutionary analysis. *Ethology and Sociobiology* 4: 137–73.

——. 1992. The evolutionary psychology of men's coercive sexuality. *Behavioral and Brain Sciences* 15: 363–421.

Thornhill R, Gangestad SW, and Comer R. 1995. Human female orgasm and mate fluctuating asymmetry. *Animal Behaviour* 50: 1601–15.

Thorpe J. 1975. Early maturity in male Atlantic salmon. *Scottish Fisheries Bulletin* 42: 15–17.

Thorpe JE and Morgan RIG. 1978. Parental influence on growth rate, smolting rate and survival in hatchery reared juvenile Atlantic salmon, *Salmo salar. Journal of Fish Biology* 13: 549–56.

——. 1980. Growth rate and smolting rate of progeny of male Atlantic salmon parr, *Salmo salar L. Journal of Fish Biology* 17: 451–59.

Thorpe W. 1958. The learning of song-patterns in birds with especial reference to the song of the chaffinch, *Fringilla coelebs. Ibis* 100: 535–70.

Tiger L and Fox R. 1971. *The Imperial Animal.* Holt, Rinehart and Winston.

Tinbergen N. 1963. On aims and methods of ethology. *Zeitschrift für Tierpsychologie* 20: 410–33.

Tobet S, Zahniser D, and Baum M. 1986. Sexual dimorphism in the preoptic/anterior hypothalamic area of ferrets: effects of adult exposure to sex steroids. *Brain Research* 364: 249.

Tokarz R, Crews D, and McEwen BS. 1981. Estrogen-sensitive progestin binding in the brain of the lizard, *Anolis carolinensis. Brain Research* 220: 95–105.

Tolman E. 1948. Cognitive maps in rats and men. *Psychological Review* 55: 189–208.

Tomback D. 1980. How nutcrackers find their stores. *Condor* 82: 10–19.

Tomizuka C and Tobias S. 1982. Mathematical ability: is sex a factor? *Science* 212: 114–15.

Tomkins J and Simmons L. 1999. Heritability of size but not symmetry in a sexually selected trait chosen by female earwigs. *Heredity* 82: 151–57.

Tooby J and Cosmides L. 1989. Evolutionary psychology and the generation of culture, part 1. *Ethology and Sociobiology* 10: 29–49.

——. 1990. The past explains the present: emotional adaptations and the structure of ancestral environments. *Ethology and Sociobiology* 11: 375–424.

——. 1992. The psychological foundations of culture. In: *The Adapted Mind: Evolutionary Psychology and the Generation of Culture.* Barkow J, Cosmides L, and Tooby J (eds.). Oxford University Press, pp. 19–136.

Tooby J and DeVore I. 1987. The reconstruction of hominid behavioral evolution through strategic modeling. In: *The Evolution of Human Behavior: Primate Models.* Kinzey WG (ed.). State University of New York Press, 183–237.

Tramontin AD and Brenowitz EA. 1999. A field study of seasonal neuronal incorporation into the song control system of a songbird that lacks adult song learning. *Journal of Neurobiology* 40: 316–26.

Tramontin A, Perfito N, Wingfield JC, et al. 2001. Seasonal growth of song control nuclei precedes seasonal reproductive development in wild adult song sparrows. *General and Comparative Endocrinology* 122: 1–9.

Trine C, Robinson W, and Robinson S. 1998. Consequences of brown-headed cowbird brood parasitism for host population dynamics. In: *Parasitic Birds and Their Hosts: Studies in Coevolution.* Rothstein S and Robinson S (eds.). Oxford University Press, pp. 273–95.

Trivers R. 1972. Parental investment and sexual selection. In: *Sexual Selection and the Descent of Man.* Campbell B (ed.). Aldine, pp. 136–79.

Turke PW. 1990. Which humans behave adaptively, and why does it matter? *Ethology and Sociobiology* 11: 305–39.

Turner G and Burrows M. 1995. A model of sympatric speciation by sexual selection. *Proceedings of the Royal Society of London, B* 260: 287–92.

Tuzel E, Sevim V, and Erzan A. 2001. Evolutionary route to diploidy and sex. *Proceedings of the National Academy of Sciences, USA* 98: 13774–77.

Uecker A and Obrzut J. 1993. Hemisphere and gender differences in mental rotation. *Brain and Cognition* 22: 4–50.

Vandenberg S and Kuse A. 1978. Mental rotation, a group test of three-dimensional spatial visualization. *Perceptual and Motor Skills* 47: 599–604.

Vandenbergh J and Huggett C. 1994. Mother's prior intrauterine position affects the sex ratio of her offspring in house mice. *Proceedings of the National Academy of Sciences, USA* 91: 11055–59.

VanderWall S. 1982. An experimental analysis of cache recovery in Clark's nutcracker. *Animal Behaviour* 20: 84–94.

———. 1990. *Food Hoarding in Animals*. University of Chicago Press.

VanderWall S and Balda R. 1981. Ecology and evolution of food storage behavior in conifer-seed-caching corvids. *Zeitschrift für Tierpsychologie* 56: 217–42.

van der Westhuizen L, Bennett NC, and Jarvis J. 2002. Behavioral interactions, basal plasma luteinizing hormone concentrations and the differential pituitary responsiveness to exogenous gondotrophin-releasing hormone in entire colonies of the naked mole-rat (*Heterocephalus glaber*). *Journal of Zoology* 256: 25–33.

van Jaarsveld AS and Skinner JD. 1991. Plasma androgens in spotted hyaenas (*Crocuta croctua*): influence of social and reproductive development. *Journal of Reproduction and Fertility* 93: 195–201.

Van Oppen MJH, Turner GF, Rico C, et al. 1998. Unusually fine-scale genetic structuring found in rapidly speciating Malawi cichlid fishes. *Proceedings of the Royal Society of London, B* 264: 1803–12.

Van Rhijn J. 1973. Behavioural dimorphism in male ruffs, *Philomachus pugnax* (L). *Behaviour* 47: 153–229.

———. 1991. *The Ruff*. Poyser.

Van Tienhoven A. 1983. *Reproductive Physiology of Vertebrates*. Cornell University Press.

Van Valen L. 1962. A study of fluctuating asymmetry. *International Journal of Organic Evolution* 16: 125–42.

———. 1973. A new evolutionary law. *Evolutionary Theory* 1: 1–30.

Vehrencamp S. 1983. Optimal degree of skew in cooperative societies. *American Zoologist* 23: 327–35.

Verhaegen M, Puech P-F, and Munro S. 2002. Aquaboreal ancestors? *Trends in Ecology and Evolution* 17: 212–17.

Vermeij G. 1974. Adaptation, versatility, and evolution. *Systematic Zoology* 22: 466–77.

Vernon C. 1973. Vocal imitation by South African birds. *Ostrich* 44: 23–30.

Vines G. 1992. Castrated top-dog fish fail to lose their balls. *New Scientist* 134: 15.

Virgin CJ and Sapolsky R. 1997. Styles of male social behavior and their endocrine correlates among low-ranking baboons. *American Journal of Primatology* 42: 25–39.

Voland E. 1998. Evolutionary ecology of human reproduction. *Annual Review of Anthropology* 27: 347–74.

Vollestad L, Hindar K, and Møller A. 1999. A meta-analysis of fluctuating asymmetry in relation to heterozygosity. *Heredity* 83: 206–18.

Voltaire. 1947. *Candide*. Penguin.

vom Saal F. 1983. The interaction of circulating oestrogens and androgens in regulating mammalian sexual differentiation. In: *Hormones and Behavior in Higher Vertebrates*. Balthazart J, Prove E, and Gilles R (eds.). Plenum, pp. 159–77.

———. 1984. Models of early hormonal effects on intrasex aggression in mice. In: *Hormones and Aggressive Behavior*. Svare B (ed.). Plenum, pp. 197–222.

———. 1989. Sexual differentiation in litter-bearing mammals: influence of sex of adjacent fetuses in utero. *Journal of Animal Science* 67: 1824–40.

Von Neumann J and Morgenstern O. 1953. *Theory of Games and Economic Behavior*. Princeton University Press.

Voorzanger B. 1987. No norms and no nature—the moral relevance of evolutionary biology. *Biology and Philosophy* 2: 253–70.

Voyer D. 1997. Scoring procedure, performance factors, and magnitude of sex differences in spatial performance. *American Journal of Psychology* 110: 259–76.

Voyer D and Bryden MP. 1990. Gender, level of spatial ability, and lateralization of mental rotation. *Brain and Cognition* 13: 18–29.

Voyer D, Voyer S, and Bryden MP. 1995.

Magnitude of sex differences in spatial abilities: a meta-analysis and consideration of critical variables. *Psychological Bulletin* 117: 250–70.

Vrijenhoek R. 1989. Genetic and ecological constraints on the origin and establishment of unisexual vertebrates. In: *Evolution and Ecology of Unisexual Vertebrates.* Dawley R and Bogert J (eds.). New York State Museum, pp. 24–31.

———. 1993. The origin and evolution of clones versus the maintenance of sex in Poeciliopsis. *Journal of Heredity* 84: 388–95.

Waddington C. 1953. Genetic assimilation of acquired characters. *Evolution* 7: 118–26.

Wade J. 1999. Sexual dimorphisms in avian and reptilian courtships: two systems that do not play by mammalian rules. *Brain, Behavior and Evolution* 54: 15–27.

Wade J and Crews D. 1991. The relationship between reproductive state and "sexually" dimorphic brain areas in sexually reproducing and parthenogenetic whiptail lizards. *Journal of Comparative Neurology* 309: 507–14.

Wade J, Huang J-M, and Crews D. 1993. Hormonal control of sex differences in the brain, behavior, and accessory sex structures of whiptail lizards (Cnemidophorus species). *Journal of Neuroendocrinology* 5: 81–93.

Wade J, Springer ML, Wingfield JC, et al. 1996. Neither testicular androgens nor embryonic aromatase activity alters morphology of the neural song system in zebra finches. *Biology of Reproduction* 55: 1126–32.

Wade J, Swender D, and McElhinny T. 1999. Sexual differentiation of the zebra finch song system parallels genetic, not gonadal, sex. *Hormones and Behavior* 36: 141–52.

Wagner G. 2000. What is the promise of developmental evolution? Part I. Why is developmental biology necessary to explain evolutionary innovations? *Journal of Experimental Zoology* 288: 95–98.

———. 2001. What is the promise of developmental evolution? Part II. A causal explanation of evolutionary change may be

impossible. *Journal of Experimental Zoology* 291: 305–9.

Wagner H and Zimmerman B. 1986. Identification and fostering of mathematically gifted students. *Educational Studies in Mathematics* 17: 243–59.

Wainer H and Steinberg L. 1992. Sex differences in performance on the mathematics section of the Scholastic Aptitude Test: a bidirectional validity study. *Harvard Educational Review* 62: 323–36.

Wake D. 1987. Adaptive radiation of salamanders in Middle American cloud forests. *Annals of the Missouri Botanical Garden* 74: 242–46.

Walker D, Knapp D, and Hunt E. 2001. Spatial representations of virtual mazes: the role of visual fidelity and individual differences. *Human Factors* 43: 147–58.

Wallenstein GV, Eichenbaum H, and Hassellmo ME. 1998. The hippocampus as an associator of discontiguous events. *Trends in Neurosciences* 21: 317–23.

Ward B, Nordeen E, and Nordeen K. 1998. Individual variation in neuron number predicts differences in the propensity for avian vocal imitation. *Proceedings of the National Academy of Sciences, USA* 95: 1277–82.

Warner R. 1975. The adaptive significance of sequential hermaphroditism in animals. *American Naturalist* 109: 61–82.

———. 1978. The evolution of hermaphroditism and unisexuality in aquatic and terrestrial vertebrates. In: *Contrasts in Behavior*. Reese E and Lighter FJ (eds.). Wiley, pp. 77–101.

———. 1988. Sex change and the size-advantage model. *Trends in Ecology and Evolution* 3: 133–36.

———. 1989. Reply to Shapiro. *Trends in Ecology and Evolution* 4: 272–73.

Warner R and Robertson D. 1978. Sexual patterns in the labroid fishes of the western Caribbean. I. The wrasses (Labridae). *Smithsonian Contributions in Zoology* 254: 1–27.

Warner R, Robertson DR, and Leigh EG. 1975. Sex change and sexual selection. *Science* 190: 633–38.

Wasser S and Barash D. 1983. Reproductive suppression among female mammals: implications for biomedicine and sexual selection theory. *Quarterly Review of Biology* 58: 513–38.

Watson PJ and Thornhill R. 1994. Fluctuating asymmetry and sexual selection. *Trends in Ecology and Evolution* 9: 21–25.

Weatherhead P, Dufour K, Lougheed S, et al. 1999. A test of the good-genes-as-heterozygosity hypothesis using red-winged blackbirds. *Behavioral Ecology* 10: 619–25.

Weber T and Piersma T. 1996. Basal metabolic rate and the mass of tissues differing in metabolic scope: migration-related covariation between individual knots *Calidris canutus*. *Journal of Avian Biology* 27: 215–24.

Wegesin D. 1998a. Event-related potentials in homosexual and heterosexual men and women: sex-dimorphic patterns in verbal asymmetries and mental rotation. *Brain and Cognition* 36: 73–92.

——. 1998b. A neuropsychologic profile of homosexual and heterosexual men and women. *Archives of Sexual Behavior* 27: 91–108.

Weismann A. 1891. *Essays Upon Heredity And Kindred Biological Problems.* Clarendon Press.

Weiss K. 2002. Biology's theoretical kudzu: the irrepressible illusion of teleology. *Evolutionary Anthropology* 11: 4–8.

Wells KD. 1977. The courtship of frogs. In: *The Reproductive Biology of Amphibians.* Taylor DH and Guttman SI (eds.). Plenum, pp. 233–62.

West M and King A. 1985. Learning by performing: an ecological theme for the study of vocal learning. In: *Issues in the Ecological Study of Learning.* Johnston T and Pietrewicz A (eds.). Erlbaum.

——. 1988. Female visual displays affect the development of male song in the cowbird. *Nature* 334: 244–46.

West P and Packer C. 2002. Sexual selection, temperature, and the lion's mane. *Science* 297: 1339–43.

West-Eberhard M. 1979. Sexual selection, social competition and evolution. *Proceedings of the American Philosophical Society* 123: 222–34.

Westlin L, Bennett NC, and Jarvis J. 1994. Relaxation of reproductive suppression in non-breeding naked mole rats. *Journal of Zoology London* 234: 177–88.

Westneat D and Birkhead T. 1998. Alternative hypotheses linking the immune system and mate choice for good genes. *Proceedings of the Royal Society of London, B* 265: 1065–73.

Wetherington J, Schenck RA, and Vrijenhoek RC. 1989. The origins and ecological success of unisexual *Poeciliopsis*: The frozen niche model. In *The Ecology and Evolution of Poeciliid Fishes.* Meffe G and Snelson, Jr., FF (eds.). Prentice Hall, pp. 259–76.

White M. 1973. *Animal Cytology and Evolution.* Cambridge University Press.

Whitlock M and Fowler K. 1997. The instability of studies of instability. *Journal of Evolutionary Biology* 10: 63–67.

Wickler W. 1964. Vom Grupenlebben einiger Säugetiere Afrikas. *Mitt Max-Planck-Ges* 56: 296–309.

——. 1965. Die aüsseren Genitalien als soziale Signalen bei einegen Primaten. *Naturwissenschaften* 52: 268–70.

——. 1968. *Mimicry in Plants and Animals.* McGraw-Hill.

Widemo F. 1998. Alternative reproductive strategies in the ruff, *Philomarchus pugnax*: a mixed ESS? *Animal Behaviour* 56: 329–36.

Widenthal J. 1965. Structure in the primary song of the mockingbird. *Auk* 82: 161–89.

Wiley R, Steadman L, Chadwick L, et al. 1999. Social inertia in white-throated sparows results from recognition of opponents. *Animal Behaviour* 57: 453–63.

Williams CL, Barnett AM, and Meck WH. 1990. Organizational effects of early gonadal secretions on sexual differentiation in spatial memory. *Behavioral Neuroscience* 104: 84–97.

Williams GC. 1966. *Adaptation and Natural Selection.* Princeton University Press.

——. 1970. The question of adaptive sex ratio in outcrossed vertebrates. *Proceedings of the Royal Society of London, B* 205: 567–80.

——. 1975. *Sex and Evolution.* Princeton University Press.

Williams L and MacRoberts M. 1978. Song variation in dark-eyed juncos in Nova Scotia. *Condor* 80: 237–304.

Williams T, Pepitone M, Christensen S, et al. 2000. Finger-length ratios and sexual orientation. *Nature* 404: 455.

Wilson DS. 1994. Adaptive genetic variation and human evolutionary psychology. *Ethology and Sociobiology* 15: 219–35.

Wilson E. 1971. *The Insect Societies.* Harvard University Press.

———. 1975. *Sociobiology.* Harvard University Press.

———. 1978. *On Human Nature.* Harvard University Press.

Wilson J. 2001. Androgens, androgen receptors, and male gender role behavior. *Hormones and Behavior* 40: 358–66.

Winberg S and Nilsson GE. 1993. Roles of brain monoamine neurotransmitters in agonistic behaviour and stress reactions, with particular reference to fish. *Comparative Biochemistry and Physiology* 106C: 597–614.

Winge O. 1932. The nature of sex chromosomes. *Proceedings of the Sixth International Congress of Genetics* 1: 343–55.

———. 1934. The experimental alteration of sex chromosomes into autosomes and vice versa, as illustrated by *Lebistes. C. R. Carlsberg Series in Physiology* 21: 1–49.

Witelson S. 1985. Sex and the single hemisphere. *Science* 194: 425–27.

Witschi E. 1929. Studies on sex differentiation and sex-determination in amphibians. I. Developmental and sexual differentiation of the gonads of *Rana sylvatica. Journal of Experimental Zoology* 52: 235–66.

Witt DM, Young LJ, and Crews D. 1995. Progesterone modulation of androgen-dependent sexual behavior in male rats. *Physiology and Behavior* 57: 307–13.

Wolf O, Preut R, Hellhammer D, et al. 2000. Testosterone and cognition in elderly men: a single testosterone injection blocks the practice effect in verbal fluency, but has no effect on spatial or verbal memory. *Biological Psychiatry* 47: 650–54.

Woodruff-Pak D, Goldenberg G, Downey-Lamb M, et al. 2000. Cerebellar volume in humans related to magnitude of classical conditioning. *Cognitive Neuroscience and Neurophysiology* 11: 609–15.

Woodson J and Gorski R. 2000. Structural sex differences in the mammalian brain: reconsidering the male/female dichotomy. In: *Sexual Differentiation of the Brain.* Matsumoto A (ed.). CRC Press, pp. 230–55.

Woolley H. 1910. Psychological literature: a review of the recent literature on the psychology of sex. *Psychological Bulletin* 7: 335–42.

Wray S, Grant P, and Gainer H. 1989a. Evidence that cells expressing luteinizing hormone-releasing hormone mRNA in the mouse are derived from progenitor cells in the olfactory placode. *Proceedings of the National Academy of Sciences, USA* 86: 8132–36.

Wray S, Nieburgs A, and Elkabes S. 1989b. Spatiotemporal cell expression of luteinizing hormone-releasing hormone in the prenatal mouse: evidence for an embryonic origin in the olfactory placode. *Developmental Brain Research* 46: 309–18.

Yahr P. 1988. Sexual differentiation of behavior in the context of developmental psychobiology. In: *Handbook of Behavioral Neurobiology.* Blass E (ed.). Plenum, pp. 197–243.

Yalcinkaya TM, Sitteri PK, Vigne J-L, et al. 1993. A mechanism for virilization of female spotted hyenas in utero. *Science* 260: 1929–31.

Yamamoto T. 1953. Artificially induced sex reversal in genotypic males of the medaka (*Oryzias latipes*). *Journal of Experimental Zoology* 123: 571–94.

———. 1958. Artificial induction of functional sex-reversal in genotyic females of the medaka (*Oryzias latipes*). *Journal of Experimental Zoology* 137: 227–62.

———. 1961. Progenies of induced sex-reversal females mated with induced sex-reversal males in the medaka, *Oryzias latipes. Journal of Experimental Zoology* 146: 133–54.

———. 1969. Sex differentiation. In: *Fish Physiology.* Hoar WS and Randall DJ (eds.). Academic Press, pp. 117–75.

Yamazaki F. 1976. Application of hormones in fish culture. *Journal of the Fisheries Research Board of Canada* 33: 948–58.

———. 1983. Sex control and manipulation in fish. *Aquaculture* 33: 329–54.

Yen W. 1975. Sex-linked major-gene influences on selected types of spatial performance. *Behavior Genetics* 5: 281–98.

Yntema C. 1976. Effects of incubation temperatures on sexual differentiation in the turtle, *Chelhydra serpentina*. *Journal of Morphology* 150: 453–62.

———. 1979. Temperature levels and periods of sex determination during incubation of eggs of *Chelhydra serpentina*. *Journal of Morphology* 159: 17–27.

Young LJ, Nag PK, and Crews D. 1995a. Species differences in behavioral and neural sensitivity to estrogen in whiptail lizards: Correlation with hormone receptor messenger ribonucleic acid expression. *Neuroendocrinology* 61: 680–86.

———. 1995b. Species differences in estrogen receptor and progesterone receptor-mRNA expression in the brain of sexual and unisexual whiptail lizards. *Journal of Neuroendocrinology* 7: 567–76.

Young R. 1970. *Mind, Brain and Adaptation in the Nineteenth Century*. Oxford University Press.

Yund P. 2000. How severe is sperm limitation in natural populations of marine free-spawners? *Trends in Ecology and Evolution* 15: 10–13.

Zahavi A. 1975. Mate selection—a selection for a handicap. *Journal of Theoretical Biology* 53: 205–14.

———. 1977. The cost of honesty—further remarks on the handicap principle. *Journal of Theoretical Biology* 67: 603–5.

Zahavi A and Zahavi A. 1999. *The Handicap Principle: A Missing Piece of Darwin's Puzzle*. Oxford University Press.

Zamudio K and Sinervo B. 2000. Polygyny, mate-guarding, and posthumous fertilization as alternative male mating strategies. *Proceedings of the National Academy of Sciences, USA* 97: 14427–32.

Zhao X, Lein E, He A, et al. 2001. Transcriptional profiling reveals strict boundaries between hippocampal subregions. *Journal of Comparative Neurology* 441: 187–96.

Zheng L-M, Pfaff D, and Schwanzel-Fukuda M. 1992. Electron microscopic identification of luteinizing hormone releasing hormone-immunoreactive neurons in the medial olfactory placode and basal forebrain in embryonic mice. *Neuroscience* 46: 407–18.

Ziegler T, Scheffler G, and Snowdon C. 1995. The relationship of cortisol levels to social environment and reproductive functioning in female cotton-top tamarins, *Saguinus oedipus*. *Hormones and Behavior* 29: 407–24.

Zimmerer E and Kallman K. 1989. Genetic basis for alternative reproductive tactics in the pygmy swordtail, *Xiphophorus nigrensis*. *Evolution* 43: 1298–307.

Zola-Morgan S, Squire L, and Rasmus S. 1994. Severity of memory impariment in monkeys as a function of locus and extent of damage within the medial temporal lobe memory system. *Hippocampus* 5: 232–39.

Zucker N and Murray L. 1996. Determinants of dominance in the tree lizard *Urosaurus ornatus*: the relative importance of mass, previous experience and coloration. *Ethology* 102: 812–24.

INDEX

aardwolf (*Proteles crisatus*), 177
AC. *See* anterior commissure
Acrocephalus palustris (marsh warbler), 231n.86
action-based learning, 227n.57
activation effects, 146
adaptation: asexual reproduction in, 23–24; Darwin on, 222n.10; in mammals, 67; sexual reproduction and, 23
adaptationists: on alternative reproductive tactics, 52, 60, 62, 73–74; on bird mimicry, 104–5, 112–13, 123; on cowbird brain structure, 125; evolutionary game theory and, 62–65; on female orgasm, 11–15, 17–18; history of, 4–6, 196; on human evolution, 150–54; on hyena genitalia, 176–82, 188–91; on optimal foraging theory, 126–27; on peacock plumage, 107–8; on sex change in fishes, 38; on sexual *vs.* asexual reproduction, 20–24, 29, 33–35; on songbird versatility, 112–13; on spatial cognition in humans, 152–54, 157–66; teleological thinking by, 2–6; on wet dreams, 15–16. *See also* brain ecology; evolutionary psychology; whybiology
adaptive radiation, 75, 217n.1
Aeschylus, 26
African lion (*Panthera leo*), 177, 181, 182–83
African wild dog (*Lycaon pictus*), 80–81, 83, 177
Agelaius phoeniceus (red-winged blackbird), 117, 120, 237n.62
age-limited song learning, 116–17, 120, 228n.64, 230n.81
AH. *See* anterior hypothalamus
AIS. *See* androgen insensitivity syndrome
Alcock, John, 12, 17–18
Alexander, Richard, 248n.82
algorithms, Darwinian, 169–70

alternative reproductive tactics (ARTs): adaptationists on, 52, 60, 62, 73–74; in bullfrogs, 57–58; conditional, 63–66, 67, 83; in European Ruff, 59–60, 64, 65–66; evolutionary game theory and, 62–63; in fishes, 51–54, 62–67, 71–73; fixed *vs.* plastic, 216n.45; in garter snakes, 58–59; how-biology of, 66–72, 73–74; in humans, 154; identification of, 52–56, 64–66; in mammals, 60–61, 66–67, 71–72; in orangutans, 61, 72; in rodents, 60–61; sex change and, 56–57; in side-blotched lizard, 59; in toadfish, 51–52, 53–54, 62–63, 65–66, 73; in treefrogs, 54–55, 65; in tree lizards, 59; why-biology of, 66, 73–74
Amazon molly (*Poecilia formosa*), 25–27, 206n.33
Amazons (tribe), 26
amphimixis, 203n.3
Amphiprion percula (clown anemonefish), 38; and acclimation to anemones, 37, 208n.3; life cycle of, 38–39; mating in, 42; sex change in, 37–38, 39, 42–43, 47; and symbiosis with anemones, 36–37
amygdala, 132f.
androgen insensitivity syndrome (AIS), 144, 147
androgen receptors (AR), 31, 144–45, 147
androgens: in androgen insensitivity syndrome, 144, 147; in congenital adrenal hyperplasia, 163; in female spotted hyenas, 185–86, 190, 251n.19; in reproductive physiology of whiptail lizards, 30–33, 30f.; in sexual differentiation, 143
anemic materialism, 94
anemonefish: acclimation to anemones, 37, 208n.3; life cycle of, 38–39; mating in, 42; sex change in, 37–38, 39, 42–43, 47; and symbiosis with anemones, 36–37
anemone: stinging element of, 37. *See also* anemonefish
anterior commissure (AC), 242n.25

anterior hypothalamus (AH), 30
antlers, natural selection and, 141–42
ants, social suppression of reproduction in, 81
Aphelocoma coerulescens (scrub jay), 134t., 234n.31
Aphelocoma ultramarina (Mexican jay), 134t., 234n.31
aposomatic coloration, 222n.9
aquaculture, sex manipulation in, 44
AR. *See* androgen receptors
Area X, in birdsong, 115, 119
Aristotle, 13, 15, 184–85
aromatase, 144, 251n.19
ARTs. *See* alternative reproductive tactics
asexual reproduction: adaptationists on, 20–24, 29, 33–35; in Amazon mollys, 25–27, 206n.33; defined, 21; disadvantages of, 22; how-biology of, 24, 29, 33–35; inertia of, 22, 24; and parthenogenesis, *see* parthenogenic reproduction; and sperm-dependency, 26–27, 27f., 205n.30, 206n.33; in whiptail lizards, 20, 27–33, 206n.38, 207n.51, 208n.57
asexual vertebrates: in Amazon molly, 25–27, 206n.33; how-biology of, 33–35; meiosis in, 24; origins of, 33; sexual development in, 24. *See also* whiptail lizard
asking directions, male reluctance about, 1, 153
atomism in biological sciences, 95, 100–101
automimicry, 250n.6
automixis, 203n.3
Azande peoples, 192

baboon: mating practices of, 179; social stress in, 90–91; social suppression of reproduction in, 79–81, 83; social system of, 79–80
Bactrian hump, 53, 54f.
Baker, Robin, 13–15, 18
bannertail kangaroo rat (*Dipodomys spectabilis*), 139, 140t.
Barlow, George, 46
Bass, Andrew, 214 nn. 3–5
bay-winged cowbird (*Molothrus badianus*), 140
bay wren (*Thryothorus nigricapillus*), 118–19
Beau Geste hypothesis, 222n.7
behavior: cause and effect in, 135–37, 229n.75

behavioral ecology: brain ecology and, 129; optimal foraging theory and, 126–27; origins of, 125–26; on spatial memory in birds, 127–29
Behaviorism, 167–68, 174
Bell, Graham, 23, 24, 196, 198
Bellis, Mark, 13–15, 18
Benson, Mark, 77
biological impact of social phenomena: barriers to recognition of, 94–95, 99–101; examples of, 235n.35; gene expression and, 99–100. *See also* social suppression of reproduction
biology. *See* evolutionary biology; how-biology; sexual biology; why-biology
biosociology, 100–101
birds: migration of, 127; spatial memory in, 127–29. *See also* song; songbirds; specific species
blackbird (Icteridae), 112
blackbird, red-winged (*Agelaius phoeniceus*), 117, 120, 237n.62
black-capped chickadee (*Parus atricapillus*), 127–28
bluegill sunfish (*Lepomis macrochirus*), mating practices of, 55–56
bluehead wrasse (*Thalassoma bifasciatum*): mating practices of, 56–57; sex change in, 48, 213n.45
blue tit (*Parus caeruleus*), 128, 134, 134t., 135–36
bonobo (*Pan panmiscus*), sexual behavior in, 11
booby (*Sulidae*), mating rituals of, 28
brain ecology: on brain size *vs.* capacity, 133, 199; on causation in behavior, 135–37, 229n.75; on hippocampal function, 129–32; on hippocampal sex differences, 137–49; origins of, 125, 129, 148; questionable assumptions of, 129–37, 148–49; testosterone in, 146–48
brain structure and chemistry: birdsong learning and, 115–16, 117, 120; birdsong sex differences and, 118–19, 123; cause and effect in, 135–37, 229n.75; estrogen in, 144–45, 147, 163–64; human sexual dimorphism in, 23, 24, 156–57, 241 nn. 22 and 25; lateralization, 23, 24, 156, 241 nn. 22 and 25; mammalian sexual dimorphism in, 155–57; in sex change, 47–48; sex steroid hormones and, 30–33, 30f.,

206n.38 (*see also* sex steroid hormones); size *vs.* capacity, 132–35, 199; spatial memory and, 125, 129; testosterone in, 144–48; and use-it-or-lose-it phenomenon, 142–43. *See also* hippocampus/hippocampal formation; hypothalamus

brown-headed cowbird (*Molothrus ater*): brain structure of, 125; human impact on, 124; nesting practices of, 124–25; sex differences in, 125; song repertoire of, 115

brown hyena (*Hyaena brunnea*), 177

buff-breasted wren (*Thryothorus leucotis*), 118–19

bulbocavernosus, spinal nucleus of (SNB), 156

bullfrog (*Rana catesbiana*), alternative reproductive tactics in, 57–58

Burton, Richard, 10

bush shrike, East African (*Laniarius funebris*), 119, 229n.76

caching of food: motivation for, 236n.48; and spatial memory in birds, 127–29

CAH. *See* congenital adrenal hyperplasia

camouflage hypothesis, 189

canary, 103f., 114f., 117, 228n.64

Candide (Voltaire), 194–95

Canidae, 177

Caprimulgus vociferus (whip-poor-will), 104

Caribbean stoplight parrotfish (*Sparisoma viride*), mating practices of, 56–57

Caribou, 238 nn. 64 and 65

carnivores, social suppression of reproduction in, 80–81, 83

Carolina wren (*Thryothorus ludovicianus*), 118–19

Catholic church, on eroticism, 10, 17

causation in behavior, 135–37, 229n.75

CC. *See* corpus callosum

cerebral cortex, 132f.

Cervidae (deer), natural selection and, 141–42

chickadee, black-capped (*Parus atricapillus*), 127–28

chickens, social hierarchy in, 95–97

children, teleological thinking in, 2–3

Chomsky, Noam, 168, 169, 171

chromosomes. *See* sex chromosomes

Cichlasoma citrinellum (Midas cichlid): care of young, 77; sex determination in, 46, 47

cichlid: adaptive radiation in, 75, 217n.1. *See also* mouth-brooders

cichlid, Midas (*Cichlasoma citrinellum*): and care of young, 77; sex determination in, 46, 47

Cistothoras palustris (marsh wren), 117, 118, 229n.75

Clark's nutcracker (*Nucifraga columbiana*), 128, 234n.31

Clayton, Nicola, 135–36, 235n.36, 236 nn. 42–48, 238n.69

cleaner wrasse (*Labroides dimidiatus*), 39–40; mating practices of, 49; sex change in, 40

clitoris: contraction of during orgasm, 17; as homologous to penis, 17; of spotted hyena, 176 (*see also* spotted hyenas)

cloning, of zebrafish, 44–45, 210n.21

clown anemonefish (*Amphiprion percula*), 38; and acclimation to anemones, 37, 208n.3; life cycle of, 38–39; mating in, 42; sex change in, 37–38, 39, 42–43, 47; and symbiosis with anemones, 36–37

Cnemidophorus (whiptail lizards): asexual reproduction in, 20, 27–33; evolution of asexual reproduction in, 28, 29; male role adoption in, 29, 32; reproductive physiology of, 29–33, 30f., 206n.38, 207n.51, 208n.57; sex act in, 19–20

Cnemidophorus burti (rusty rumped whiptail), 28

Cnemidophorus inornatus (little-striped whiptail), 28; hormone activity in, 31, 32; sex act in, 19–20

Cnemidophorus uniparens (desert grassland whiptail): hormone activity in, 31, 32; origin of, 28; sex act in, 20

cognition: sex steroid hormones and, 163–66, 246n.77; sexual selection in, 160–61; social factors in, 165–66, 173. *See also* spatial cognition

cognitive maps, 150–51, 233 nn. 23 and 24

cognitive revolution, 152, 168, 169

Comer, Randall, 15

comparative method, 177

competition-aggression hypothesis, 250n.10

computationalists, 168–69

conditional alternative reproductive tactics, 63–66, 67, 83

conditional traits, selection for, 147

congenital adrenal hyperplasia (CAH), 163, 185–86, 251 nn. 17 and 18

connectionism, 169
convergences, in comparative method, 177
coral reefs, 36
corpus callosum (CC), 242
corticotropin, 89
corticotropin-releasing hormone (CRH), 89
cortisol, 89, 90–91
Corvidae (crow family), spatial memory in, 127–28, 134
Cosmides, Leda, 169, 170, 171
cowbird, spatial memory in, 140
cowbird, bay-winged (*Molothrus badianus*), 140
cowbird, brown-headed (*Molothrus ater*): brain structure of, 125; human impact on, 124; nesting practices, 124–25; sex differences in, 125; song repertoire of, 115
cowbird, screaming (*Molothrus rufoaxillaris*), 140
cowbird, shiny (*Molothrus bonariensis*), 140
Crew, David, 215 nn. 17–20
CRH. *See* corticotropin-releasing hormone
Crocuta crocuta (spotted hyena): adaptationists on, 176–82, 188–91; androgen levels in, 185–86, 190, 251n.19; birthing process in, 186–89, 188f.; female genitalia in, 176, 253n.33; female genitalia disadvantages for, 186–91, 188f.; female genitalia function in, 178–80, 185–86, 188–91; greeting ritual of, 176, 178–81; how-biology of, 184–86, 190–91; hunting practices of, 180; reputation of, 175–76; social structure in, 99, 175, 181–82, 182–84; social suppression of reproduction in, 80–81, 83
cross-cultural studies, 162–63
crossing over, in meiosis, 21
crow family (Corvidae), spatial memory in, 127–28, 134
Cryptomys, 220 nn. 29–31
Cryptomys damarensis (Damaraland mole rat), 82, 220n.30
Cryptomys darlingi (Mashona mole rat), 220n.31
crystallization process in birdsong learning, 114, 114f., 121
cyclostomes, 212n.42

Damaraland mole rat (*Cryptomys damarensis*), 82, 220n.30
damselfish. *See* anemonefish
Darwin, Charles: influences on, 196; on sexual selection, 105–6, 222n.10; teleological thinking and, 5

Darwin, Erasmus, 17, 202n.24
Darwinian algorithms, 169–70
Darwinian paranoic explanatory style, 3–7, 11–13, 83, 95, 129, 166, 169–70, 174, 178, 197–98
Dawkins, Richard, 5, 14, 16, 196–97, 253 nn. 11 and 13
dawn chorus, 103–4
deer (Cervidae), natural selection and, 141–42
Dennett, Daniel, 168, 197–99, 254n.17, 255 nn. 18–20
DES (diethylstilbestrol), 147
Descartes, René, 166, 174
descent with modification, conflict over, 5
desert grassland whiptail (*Cnemidophorus uniparens*): hormone activity in, 31, 32; origin of, 28; sex act in, 20
design in nature. *See* teleology
developmental constraints, 8–9
Devoogd, Timothy, 118
Dialogues Concerning Natural Religion (Hume), 195–96
diethylstilbestrol (DES), 147
Dipodomys merriami (Merriam's kangaroo rat), 139, 140t.
Dipodomys spectabilis (bannertail kangaroo rat), 139, 140t.
divergences, in comparative method, 177
dog, African wild (*Lycaon pictus*), 80–81, 83, 177
double standard, sexual, 14–15
dromedary hump, 53, 54f.
dry biology, 199
duetting in songbirds, 117, 226n.37, 229n.75

Eals, Marion, 158–59
East African bush shrike (*Laniarius funebris*), 119, 229n.76
eastern phoebe (*Sayornis phoebe*), 230n.81
EEA. *See* environment of evolutionary adaptedness
Einstein, Albert, 133
elephant, social suppression of reproduction in, 78–79
elephant seal (*Mirounga*): mating practices of, 49–50, 55, 199; species of, 213 nn. 48 and 49
Ellis, Havelock, 157
Emberizidae (New World sparrow), 113
embryonic stem cells, 145

environment of evolutionary adaptedness (EEA), 171

EPCs. See extra-pair copulations

ER. See estrogen receptors

eroticism, in Christian vs. Hindu thought, 10

ESS. See evolutionarily stable strategy

estradiol, 144–46, 147, 247n.77

estrogen: in brain chemistry, 144–45, 147, 163–64; estrogen receptor levels and, 207n.49; in reproductive physiology of whiptail lizards, 30–33, 30f.; in sexual differentiation, 143

estrogen receptors (ER), 31, 144–45, 147, 207n.49

ethology, 125–26, 152, 227n.57

Ethology and Sociobiology (journal), 239n.4

eunuchs, and biological systems, 184–85

European ruff (*Philomachus pugnax*), alternative reproductive tactics in, 59–60, 64, 65–66

European warbler (Sylviidae), 228n.69

Evans-Pritchard, 192

evil, problem of, 4–5, 194–96

evolution: human, 150–54, 239n.1; of meiosis, 24; of sexual dimorphism, 159–66; of sexual reproduction, 22–25; of vertebrate eye, 254n.16. See also natural selection; sexual selection

Evolution and Human Behavior (journal), 203n.25

evolutionarily stable strategy (ESS), 62–63

evolutionary biology: agenda of, 7; teleological thinking in, 2, 3

evolutionary causation, 201n.11

evolutionary developmental biology, 8

evolutionary game theory: alternative reproductive tactics in, 62–63; critique of, 63–72; development of, 62; principles of, 62

evolutionary history. See phylogeny

evolutionary inertia: in reproduction, 22, 24, 27; in underlying structure, 146–49, 173

evolutionary neurobiology: critique of brain ecology, 140–49; critique of evolutionary psychology, 172–74; focus of, 8, 129; how-biology and, 129, 173

evolutionary psychology: adaptationism in, 2, 3, 6; agenda of, 8; contrasted with evolutionary neurobiology, 172–74; evolution

claims of, 171–72; founders of, 169; as Freudian phrenology, 169–71; history of as a discipline, 166–69; on human evolution, 152–54, 160–61; on human sex dimorphism, 158–66; teleological thinking in, 172–73

The Evolution of Human Sexuality (Symons), 17

experience-expectance, HVC size and, 236n.45

extra-pair copulations (EPCs): female motivations for, 12–14; male motivations for, 12

eye evolution in vertebrates, 254n.16

Felidae, 177

female: definition of, 154; reproductive biology in, 154–55; in spotted hyena social structure, 182–84. See also sexual dimorphism

female human: gender roles of, 152; mate choice in, 152; SAT scoring and, 161–66

female orgasm: adaptationist view of, 11–15, 17–18; clitoral contraction during, 17; Freud on, 16; how-biology on, 18; Kinsey on, 16–18

feminist functionalists, on female orgasm, 12–13, 17

Fernald, Russell, 77, 218n.6, 221n.49

finch, zebra (*Taeniopygia guttata*), 103f., 114f., 116–17, 118–19

finger length in humans, 159

Fisher, R. A., 107

fishes: alternative reproductive tactics in, 51–54, 62–67, 71–73; gonad structure in, 212n.42; sex change in, 37–38, 39–40, 42–48; sex chromosomes in, 43–44, 46; sex manipulation in, 44, 45–46; sexual development of, 43–44, 67, 69–72, 211n.38. See also particular species

flanges, 61

fluctuating asymmetry, 108, 223n.22

Fodor, Jerry, 168–69, 171

folk psychology, 168

food storing: motivation for, 236n.48; and spatial memory in birds, 127–29

foraging theory, 126–27

forest fires, 192–93

fornix, 132f.

Frank, Lawrence, 186–87

freemartin effect, 210n.20

Freud, Sigmund, 16, 167
Freudian paranoic explanatory style, 166–71, 174

Gahr, Manfred, 119
Gall, Franz Joseph, 170
gametes, production and transmission of, 154–55
game theory. *See* evolutionary game theory
Gangestad, Steven, 15
garter snake (*Thamnophis sirtalis*), alternative reproductive tactics in, 58–59
Geary, David, 153, 160–66, 244n.48, 245n.57
genealogy, influence on phylogeny, 8–9
gene combination in sexual reproduction, 21
gene expression: in sexual differentiation, 145–46; social influences on, 99–100, 199
Ghiselin, Michael, 23, 41
Glickman, Stephen, 186–87
glucocorticoids, 89, 220n.46
GnRH. *See* gonadotropin-releasing hormone
God, in natural theology, 4
gonadotropin, 82
gonadotropin-releasing hormone (GnRH): estrogen levels and, 207n.49; in mammalian reproduction, 155; in sexual maturation, 67–72, 72f.; in social suppression of reproduction, 84–85, 85f., 99–100; stress and, 91–93
gonadotropin-releasing hormone (GnRH) neurons: migration of, 70–72; in sexual maturation, 48; in social suppression of reproduction, 84–88, 85f.
good genes, and sexual selection, 107–10, 123, 223n.12
Goodwin, Brian, 34
Gould, James, 110
Gould, Stephen Jay, 17, 24, 185–86, 191, 194
grackle (*Quiscalus quiscula*), 112, 120
Gray, Russell, 126
great tit (*Parus major*), 128, 134, 134t.
Gross, Mart, 64
guppy (*Poecilia reticulata*), 110
Gymnorhinus cyanocephalus (pinyon jay), 134t., 234n.31
gynogenesis, 26–27, 27f, 205n.30, 206n33

H. M., 130–31
hagfish (cyclostomes), 212n.42

Hamilton-Zuk hypothesis, 224 nn. 24, 25, and 26, 226n. 41
handicap principle, 108–9, 223n.23
Haplochromis burtoni (mouth-brooder): mating practices of, 75–76, 76f., 110, 179; sexual competition in, 76–77; social rank in, 84–88; social stress in, 91–93; social suppression of reproduction in, 77–78, 83, 100
hermaphroditism, 211n.38
heterochrony, 212n.39, 230n.80
heteromorphism of sex chromosomes, 209n.12
High Vocal Center. *See* HVC
hippocampus/hippocampal formation, 132f; cost of increased size in, 238n.68; function of, 130–32; place cells in, 233 nn. 23 and 24; research on, 130–31; sex differences in, 137–49; and spatial memory, 125, 129–37, 138–40, 139t., 140t.
historical explanation. *See* phylogeny
Hofmann, Hans, 77
homosexuality in males: as adaptive strategy, 151, 240n.7; cognitive profile of, 158, 240n.8
hormones. *See* sex steroid hormones
horns, natural selection and, 142
house sparrow, impact of humans on, 124
how-biology: agenda of, 7; of alternative reproductive tactics, 66–72, 73–74; of asexual reproduction, 24, 29, 33–35; on asexual vertebrates, 33–35; behavioral ecology and, 126; benefits of, 197–200; evolutionary neurobiology and, 129, 173; on female orgasm, 18; of mimicry in birds, 122–23; sensory exploitation, 110–12, 123; of sex change, 43–48, 50, 213 nn. 44 and 45; of sexual dimorphism, 140–48; of social suppression of reproduction, 84–88, 85f., 91–93, 100, 220n.46; of spotted hyena genitalia, 184–86, 190–91; value of, 8; *vs.* why-biology, 7–8
HPA. *See* hypothalamic-pituitary-adrenal axis
Hrdy, Sarah, 12
Hubbs, Carl, 25, 205n.28
human: alternative reproductive tactics in, 154; gender roles of, 152; impact on songbirds, 124; mate choice in, 152; SAT scoring and, 161–66; sexual development in, 185; spatial memory in, 152–54; special status of, 160, 244n.49. *See also* sexual dimorphism, human

human evolution: adaptationist view of, 150–54; alternative theories of, 239n.1

human symbionts, 124

Hume, David, 195–96

hunter-gatherer story, 150–51; credibility of, 171–72, 239 nn. 1 and 2; sexual dimorphism and, 152–54; tests of, 158

HVC (High Vocal Center): bird sex differences in, 118–19, 123; in birdsong production, 115, 117, 228 nn. 64–69, 229 nn. 75 and 76; birdsong versatility and, 117–18; experience-expectance in, 236n.45

Hyaena brunnea (brown hyena), 177

Hyaena hyaena (hyena, striped), 177

hybridogenesis, 27f.

hyena, brown (*Hyaena brunnea*), 177

hyena, spotted (*Crocuta crocuta*): adaptationists on, 176–82, 188–91; androgen levels in, 185–86, 190, 251n.19; birthing process in, 186–89, 188f.; female genitalia in, 176, 253n.33; female genitalia disadvantages in, 186–91, 188f; female genitalia function in, 178–80, 185–86, 188–91; greeting ritual of, 176, 178–81; how-biology of, 184–86, 190–91; hunting practices of, 180; reputation of, 175–76; social structure in, 99, 175, 181–82, 182–84; social suppression of reproduction in, 80–81, 83

hyena, striped (*Hyaena hyaena*), 177

hypogonadism, 69

hypothalamic-pituitary-adrenal (HPA) axis, 89, 90f.; interaction with HPG axis, 91–93, 92f., 221n.48

hypothalamic-pituitary-gonadal (HPG) axis, 67, 68f.; interaction with HPA axis, 91–93, 92f., 221n.48; in mammalian reproduction, 155; in sexual maturation, 67–72; in social suppression of reproduction, 84–88

hypothalamus: gonadotropin-releasing hormone and, 67–71; in mammalian reproduction, 155; neuron migration to, 71; sex change and, 48; sex steroid hormones and, 30; sexual dimorphism in mammals, 155–56; stress response and, 89, 91

ICHH. *See* immunocompetence handicap hypothesis

immune system, testosterone and, 109, 224 nn. 27 and 29

immunocompetence handicap hypothesis (ICHH), 109–10, 224 nn. 27 and 29

imperfections, design-in-nature argument and, 4–5

independent assortment, 203n.4

infanticide: in lions, 183; in primates, 12–13; in spotted hyenas, 184, 189, 251n.13 , 252 nn. 20 and 30

insects, social suppression of reproduction in, 81

intentional stance, 197–98

intergender hitchhiking, 141–43, 160–61, 237 nn. 59 and 62

intersexual selection, in birdsong, 106–7, 111–12

Jacklin, C., 157

Jacobsen, Ben, 85

jay, spatial memory in, 134, 134t.

jay, Mexican (*Aphelocoma ultramarina*), 134t., 234n.31

jay, pinyon (*Gymnorhinus cyanocephalus*), 134t, 234n.31

jay, scrub (*Aphelocoma coerulescens*), 134t., 234n.31

Jerison, H., 132–33

jigsaw model of social hierarchies, 98–99

Junco hyemalis (northern and slate-colored juncos), 128

junco, northern (*Junco hyemalis*), 114, 128

Kallmann's syndrome, 69, 71

Kama Sutra, 10

kangaroo rats (*Dipodomys*), 138–39, 140t.

Khajuro, India, 10

Kimura, Doreen, 246n.77

King, John, 115

Kinsey, Alfred, 16–18

Kruuk, Hans, 178–80

Labroides dimidiatus (cleaner wrasse), 39–40; mating practices of, 49; sex change in, 40

Lake Tanganyika, 75

lamprey (cyclostomes), 212n.42

language, history of research on, 167–68, 169

Laniarius funebris (East African bush shrike), 119, 229n.76

Leibniz, Gottfried Wilhelm, 194–95

Lepomis macrochirus (bluegill sunfish), mating practices of, 55–56

Leptodactylidae (neotropical frog), 111

Lewontin, R., 194

lion, African (*Panthera leo*), 177, 181, 182–83

Lisbon, earthquake of 1755, 195

little-striped whiptail (*Cnemidophorus inornatus*), 28; hormone activity in, 31, 32; sex act in, 19–20

lizard, side-blotched (*Uta stansburiana*), alternative reproductive tactics in, 59

lizard, whiptail. *See* whiptail lizard

Lloyd, Elizabeth, 11, 13, 17, 18

lordosis, in mammals, 155

Lorenz, Konrad, 178–79

lungless salamander (Plethodontidae), 234n.28

Lycaon pictus (African wild dog), 80–81, 83, 177

lyrebird (Menuridae), 231n.81

Maccoby, E., 157

Macropodus opercularis (paradisefish): sex determination in, 46, 47; social hierarchy in, 97–98

male: definition of, 154; nipples in, 17, 161, 202n.24; reproductive biology in, 154–55; testosterone in, 84–86, 85f, 91 (*see also* testosterone). *See also* homosexuality in males; sexual dimorphism

male human: gender roles of, 152; mate choice in, 152; reluctance of, to ask directions, 1, 153. *See also* sexual dimorphism, human

mamillary body, 132f.

mammals: adaptation in, 67; alternative reproductive tactics in, 60–61, 66–67, 71–72; biological reproductive requirements, 154–55; mating practices of, 155; neocortex mass in, 133; sex chromosomes in, 44; sexual development in, 44, 67–69, 71–72; social suppression of reproduction in, 78–82. *See also* sexual dimorphism, mammalian

Man and Woman: A Study of Human Secondary and Tertiary Sexual Characteristics (Ellis), 157

marsh tit (*Parus palustris*), 128, 134, 134t., 135–36

marsh warbler (*Acrocephalus palustris*), 231n.86

marsh wren (*Cistothoras palustris*), 117, 118, 229n.75

Mashona mole rat (*Cryptomys darlingi*), 220n.31

masturbation: and orgasm, 13; as sperm block, 14; in virgins, 15–16

materialism: anemic, 94, 199; misguided, 94–95; robust, 94, 166, 172, 174, 199

mathematical ability, sex difference in, 153, 161–66, 245n.58, 247 nn. 78 and 79, 248n.80

mating behavior: of baboons, 179; of bluegill sunfish, 55–56; of bluehead wrasse, 56–57; of Caribbean stoplight parrotfish, 56–57; of cleaner wrasse, 49; of elephant seals, 49–50, 55, 199; evolutionary role of, 10–15; of mammals, 155; of mouth-brooders, 75–76, 76f., 110, 179; in parthenogenic reproduction, 28–33; of platyfish, 70; of salmon, 56; of whiptail lizards, 19–20, 28–33

Maynard-Smith, John, 5, 20, 34, 62, 197–98

Mayr, Ernst, 7, 201n.11

meadow vole (*Microtus pennsylvanicus*), 138, 139t.

medaka (*Oryzias latipes*), sex chromosomes in, 44

meiosis, 21, 24

Melospiza georgiana (swamp sparrows), 113, 116, 122

memory, spatial. *See* spatial cognition

men. *See* male human

Menuridae (lyrebird), 231n.81

Merriam's kangaroo rat (*Dipodomys merriami*), 139, 140t.

Mexican jay (*Aphelocoma ultramarina*), 134t., 234n.31

Michigan School of Sociobiology, 248n.82

Microtus (vole), 137–38, 139t., 146–47

Midas cichlid (*Cichlasoma citrinellum*): care of young, 77; sex determination in, 46, 47

migration of birds, 127

mimicry: Batesian, 250n.5; in birds, 104–5, 112–13, 120–23; interspecies forms of, 179

mimic thrush (Mimidae), 103, 104, 113

Mimidae, 102–3, 104, 113, 120

Mimus polyglottus (mockingbird): adaptationists on, 104–5, 123; reasons for mimicry by, 104, 120–23; sex differences in song, 118; song learning in, 116–17; versatility of, 104, 112

Mirounga (elephant seal): mating behavior of, 49–50, 55, 199; species of, 213 nn. 48 and 49

Mirounga anguistirostris (northern elephant seal), 213 nn. 48 and 49

Mirounga leonina (southern elephant seal), 213 nn. 48 and 49

misguided materialism, 94–95

mixis: definition of, 203n.3; processes in, 21

mockingbird (*Mimus polyglottus*): adaptationists on, 104–5, 123; reasons for mimicry by, 104, 120–23; sex differences in song, 118; song learning in, 116–17; versatility of, 104, 112

modularity thesis, 169–71, 174

mole rat: social suppression of reproduction in, 81–82; species of, 219n.29

Moller, Anders, 223 nn. 21 and 22, 224n.27

molly (*Poecilia*): asexual reproduction in, 25–27; habitat of, 25; sex chromosomes in, 44

molly (*Poeciliopsis*), 26–27, 205n.32

molly, Amazon (*Poecilia formosa*), 25–27, 206n.33

molly, sailfin (*Poecilia latipinna*), 25–26, 205n.29

Molothrus ater (brown-headed cowbird): brain structure of, 125; human impact on, 124; nesting practices of, 124–25; sex differences in, 125; song repertoire of, 115

Molothrus badianus (bay-winged cowbird), 140

Molothrus bonariensis (shiny cowbird), 140

Molothrus rufoaxillaris (screaming cowbird), 140

monogamy, breech of. See extra-pair copulations

Moore, Michael, 68–69, 216n.45

Morgenstern, O., 62. See also evolutionary game theory

Morris, Desmond, 12, 13

mouth-brooder (*Haplochromis burtoni*): mating practices of, 75–76, 76f., 110, 179; sexual competition in, 76–77; social rank in, 84–88; social stress in, 91–93; social suppression of reproduction in, 77–78, 83, 100

Muller, Henry J., 22

Muller, Martin, 189–90

Muller's rachet, 22

musth, 78–79

mynah (Sturnidae), 113, 120

The Naked Ape (Morris), 12, 13

naked mole rat (*Heterocephalus glaber*), social suppression of reproduction in, 81–82

Natural History (magazine), 17

natural selection: burden of proof in, 6; genealogical limits on, 8–9; nonteleological view of, 3–4, 254n.16; as religion, 196–97; for sexual dimorphism, 141–43; *vs.* sexual selection, 105–6; speed of evolution and, 204n.11; teleological view of, 3, 5–6

Natural Theology, 4–5, 17, 195–96

nematocysts, 37, 208n.2

neocortex, mass of in mammals, 133

neotropical frog (Leptodactylidae), 111

nervous system (mammalian), sexual dimorphism in, 156

neuroethology, 116, 119, 125–26

neurons: migration of GnRH neurons, 70–72; origins of, 71; testosterone and, 144–45; use-it-or-lose-it phenomenon in, 142–43

New World sparrow (Emberizidae), 113

nipples, male, function of, 17, 161, 202n.24

nocturnal emissions, evolutionary role of, 15–16

northern elephant seal (*Mirounga anguistirostris*), 213 nn. 48 and 49

northern junco (*Junco hyemalis*), 114

note of bird song, 103

Nottebohm, Fernando, 117

Nucifraga columbiana (Clark's nutcracker), 128, 234n.31

nutcracker jay, 134t.

nutcracker, Clark's (*Nucifraga columbiana*), 128, 234n.31

olive baboon (*Papio anubis*): social suppression of reproduction in, 80–81, 83; social system of, 79–80

ontogenesis, 205n.26

Onuf's nucleus, 156

open-ended song learning, 116–17, 120, 122, 230n.81

optic tectum, 234n.28

optimal foraging theory, 126–27

optimal skew theory, 219n.17

orangutan (*Pongo pygmaeus*), alternative reproductive tactics in, 61, 72

Oreochromis (tilapia), sex manipulation in, 44

organizational/activational dichotomy, 216n.45

orgasm. *See* female orgasm

Oryzias latipes (medaka), sex chromosomes in, 44

outcrossing, 21

Paley, William (Bishop), 4–5, 7–8, 166, 194, 195–96

Pan panmiscus (bonobo), sexual behavior in, 11

Panthera leo (African lion), 177, 181, 182–83

Papio anubis (olive baboon): social suppression of reproduction in, 80–81, 83; social system of, 79–80

paradisefish (*Macropodus opercularis*): sex determination in, 46, 47; social hierarchy in, 97–98

parallel distributed processing (PDP), 169

paranoic explanatory style, 3; Darwinian, 3–7, 11–13, 83, 95, 129, 166, 174, 178, 197–98; Freudian, 166–71, 174

parasites: in mating tactics, 54; peacock plumage and, 109; sexual selection and, 25–27

Paridae, spatial memory in, 127–28, 134, 135–36

parrotfish, Caribbean stoplight (*Sparisoma viride*), mating behavior of, 56–57

parthenogenic reproduction: definition of, 206n.37; hormonal activity in, 30–33, 30f., 207n.51; mechanisms of, 28; sexual activity in, 28–33; in whiptails, 27–28

parthenogenic vertebrates, origins of, 33

Parus atricapillus (black-capped chickadee), 127–28

Parus caeruleus (blue tit), 128, 134, 134t., 135–36

Parus major (great tit), 128, 134, 134t.

Parus palustris (marsh tit), 128, 134, 134t., 135–36

PDP. *See* parallel distributed processing

peacock (*Pavo cristatus*): adaptationists view of, 107–8; sexual selection in, 105–11

penis, as homologous to clitoris, 17

pH, sex determination by, 45

phallus display: as aggression, 80; as greeting ritual, 176, 178–81

Philomachus pugnax (European ruff), alternative reproductive tactics in, 59–60, 64, 65–66

phoebe, eastern (*Sayornis phoebe*), 230n.81

phrenology, compared to Freudian explanatory style, 170; history of, 234n.29

phylogeny: genealogical limits on, 8–9; neo-Darwinians and, 196; and sex change, 49–50; sexual reproduction and, 24–25, 33–35, 205n.21

Physalaemus, mating practices of, 111

Physalaemus coloradorum, 111, 225n.35

Physalaemus pustulosus, 111

pine vole (*Microtus pinetorum*), 138, 139t.

pinyon jay (*Gymnorhinus cyanocephalus*), 134t., 234n.31

place cells, 233 nn. 23 and 24

placodes, 71

plainfish midshipman. *See* toadfish

plastic song stage of birdsong learning, 113–14, 114f., 121, 122

platyfish (*Xiphophorus maculatus*): alternative reproductive tactics in, 66; mating practices of, 70; sensory exploitation in, 111; sexual development in, 69–70

Plethodontidae (lungless salamander), 234n.28

POA. *See* preoptic area

Poecilia (molly). *See* molly (*Poecilia*); molly (*Poeciliopsis*); molly, Amazon (*Poecilia formosa*); molly, sailfin (*Poecilia latipinna*)

Poecilia mexicana, asexual reproduction in, 26

Poecilia reticulata (guppy), 110

Poeciliidae: alternative reproductive tactics in, 66; sensory exploitation in, 111; sexual development in, 69–70

Poeciliopsis latidens, 205n.32

Poeciliopsis lucida, 26, 205n.32

Poeciliopsis monacha lucida, 26–27

Poeciliopsis occidentalis, 205n.32

poleax hypothesis, 13

Pongo pygmaeus (orangutan), alternative reproductive tactics in, 61, 72

Porichthys notatus (toadfish): mating strategies in, 51–52, 53–54, 62–63, 65–66, 73; sexual development in, 70

PR. *See* progesterone receptors

preoptic area (POA), sex steroid hormones and, 30–31, 86–88

primates: extra-pair copulations in, 12; hormone levels in, 207n.51; orangutan reproductive tactics in, 61, 72; sexual behavior in, 11. *See also* baboon
principle of proper mass, 132–33
problem of evil, 4–5, 194–96
progesterone, in reproductive physiology of whiptail lizards, 30–33, 30f., 208n.57
progesterone receptors (PR), 31
protandry, 39, 40–41; in size-advantage model, 41–43, 48
Proteles cristatus (aardwolf), 177
protogyny, 39, 40–41, 47, 212n.42; alternative reproductive tactics and, 56–57; in size-advantage model, 41, 48
proximate explanations, 7
Pseudotropheus zebra, social hierarchy in, 96
psychological sex differences, 152–53
psychology: history of as discipline, 166–71; stancing in, 197–98
The Psychology of Sex Differences (Maccoby and Jacklin), 157
public, its interest in sex differences, 1
pygmy chimpanzee. *See* bonobo

Quiscalus quiscula (grackle), 112, 120

Rana catesbiana (bullfrog), alternative reproductive tactics in, 57–58
receiver bias, 225n.30
receptive sexual behavior. *See* lordosis
recombination: definition of, 203n.3; in meiosis, 21
Red Queen theory, 24, 204n.15
red-sided garter snake (*Thamnophis sirtalis parietalis*), alternative reproductive tactics in, 58–59
reductionism in biological sciences, 95
red-winged blackbird (*Agelaius phoeniceus*), 117, 120, 237n.62
relative plasticity hypothesis, 216n.45
religion: eroticism in, 10; paranoic explanatory style of, 3; teleological thinking in, 4–5, 17, 192–94
reproduction. *See* asexual reproduction; sexual reproduction; social suppression of reproduction
reproductive physiology, of whiptail lizards, 29–33, 30f., 206n.38, 207n.51, 208n.57
Rivulus marmoratus, 212n.38
robust materialism, 94, 166, 172, 174, 199

rodents, alternative reproductive tactics in, 60–61
rough and tumble play, 155
RU486, 208 nn. 57 and 58
ruff, European (*Philomachus pugnax*), alternative reproductive tactics in, 59–60, 64, 65–66
rufous and white wrens (*Thryothorus rufalbus*), 118
run-away selection, 107, 123
rusty rumped whiptail (*Cnemidophorus burti*), 28
Ryan, Michael, 110, 111

St. Paul, on eroticism, 10
salamander, lungless (Plethodontidae), 234n.28
salmon: mating practices in, 56; sex manipulation in, 44
sandpiper (Scolopacidae), alternative reproductive tactics in, 59–60
SAT. *See* Scholastic Abilities Test
satellite male, 58, 215n.15
Sayornis phoebe (eastern phoebe), 230n.81
Scholastic Abilities Test (SAT), 161–66, 244n.52, 245n.60
Scolopacidae (sandpiper), alternative reproductive tactics in, 59–60
screaming cowbird (*Molothrus rufoaxillaris*), 140
scrub jay (*Aphelocoma coerulescens*), 134t., 234n.31
SDN. *See* sexually dimorphic nucleus
sea bass (Serranidae), 212n.38
seal, elephant (*Mirounga*): mating behavior of, 49–50, 55, 199; species of, 213 nn. 48 and 49
seal, northern elephant (*Mirounga angustirostris*), 213 nn. 48 and 49
seal, southern elephant (*Mirounga leonina*), 213 nn. 48 and 49
sensitive period in birdsong learning, 113, 227n.51
sensorimotor learning phase of birdsong, 113–14, 114f., 115, 121
sensory biases in animals, 110–11, 225n.32
sensory drive, 225n.30
sensory exploitation hypothesis, 110–12, 123
sensory learning phase of birdsong, 113, 114f., 115, 121
sex: technical definition of, 21; vernacular definition of, 21

sex change: and alternative reproductive tactics, 56–57; in anemonefishes, 37–38, 39,
42–43, 47; evolutionary history and, 49–
50; in fishes, 37–38, 39–40, 42–48; how-
biology of, 43–48, 50, 213 nn. 44 and 45;
protandry *vs.* protogyny in, 39, 40–41, 47;
reproductive types in, 214n.13; size-
advantage model of, 38, 41–43, 42f, 48–
49; why-biology of, 50; in wrasses, 40, 48,
213n.45; in zebrafish, 45, 46–47. *See also*
sex manipulation
sex chromosomes: evidence of, 209n.13; in
fishes, 43–44, 46; heteromorphism of,
209n.12; in mammals, 44
sex determination, factors in, 45–47
sex manipulation in fishes, 44, 45–46
sex steroid hormones: cognition and, 163–
66, 246n.77; in parthenogenic reproduction, 30–33, 30f, 207n.51; receptor abundance and, 207 nn. 49 and 51; regulation
of, 70; in sexual differentiation, 143–48;
in social suppression of reproduction, 84–
88, 85f., 99–100; in songbirds, 230n.78; in
whiptail lizards, 29–33, 30f., 206n.38,
207n.51, 208n.57
sexual behavior. *See* mating behavior
sexual biology, components of, 9
sexual development: in asexual vertebrates,
24; of fishes, 43–44, 67, 69–72, 211n.38;
and freemartin effect, 210n.20; in humans,
185; in mammals, 44, 67–69, 71–72;
organizational/activational dichotomy in,
216n.45; in platyfish, 69–70; in toadfish,
70; in vertebrates, 24, 69–72
sexual differentiation: gene expression in,
145–46; hormones in, 143–48; process of,
143
sexual dimorphism: evolutionary neurobiology on, 173; evolution of, 159–66; in hippocampus size, 137–49; how-biology of,
140–48; natural selection and, 141–43;
sexual selection and, 106–7; in songbirds,
1, 118–19, 123, 142, 229n.76
sexual dimorphism, human, 1–2; in brain
structure and chemistry, 23, 24, 156–57,
241 nn. 22, and 25; evolution of, 152–54;
mathematical ability and, 153, 161–66,
245n.58, 247 nn. 78 and 79, 248n.80;
proof of, 240n.8; psychological, 152–53; in
spatial cognition, *see* spatial cognition in
humans

sexual dimorphism, mammalian: in behavior,
155; in brain structure and chemistry,
155–57; physical forms of, 154–55;
psychological forms of, 155–57; selection
for, 156
sexual double standard, 14–15, 152
sexual hang-ups, 34
sexually antagonistic traits, 141, 161
sexually dimorphic nucleus (SDN), 156
sexual parasitism, 25–27
sexual reproduction: adaptation and, 23; adaptationists on, 20–24, 33–35; advantages
of, 20, 22–23; definition of, 21; evolutionary origins of, 22–25; inertia of, 22, 24,
27; phylogenetic perspective on, 24–25,
33–35, 205n.21; prevalence of, 20; requirements for in mammals, 154–55; stress
effects in, 91
sexual selection: and birdsong versatility,
111–13, 118–20, 122–23; categories of,
106; Darwin on, 105–6, 222n.10; definition of, 106; good genes and, 107–10,
123, 223n.12; and hippocampal sex differences, 140–49; in human cognition,
160–61; inter- and intra-sexual, 106–7,
111–12; *vs.* natural selection, 105–6; nonteleological accounts of, 123; parasites,
25–27; in peacocks, 105–11; and sexual
dimorphism, 106–7; theories of,
107–11
Shakespeare, William, 18
she-male, in red-sided garter snakes, 58–59,
215 nn. 20 and 21
Sherman, Paul, 186
shiny cowbird (*Molothrus bonariensis*), 140
siblicide, in spotted hyenas, 182, 184, 189,
251n.13 , 252 nn. 20 and 30
side-blotched lizard (*Uta stansburiana*), alternative reproductive tactics in, 59
Silverman, Irwin, 158–59
size-advantage model of sex change, 38, 41–
43, 42f., 48–49
skew theory, optimal, 219n.17
Skinner, B. F., 167–68
slate-colored juncos (*Junco hyemalis*), 128
snake, garter (*Thamnophis sirtalis*), alternative
reproductive tactics in, 58–59
SNB. *See* spinal nucleus of the
bulbocavernosus
sneaking: in bluegill sunfish, 55–56; in bluehead wrasse, 56; in evolutionary game the-

ory, 62–63; in toadfish, 51–52, 53–54; in tree lizards, 59

social influences on biological phenomena: barriers to recognition of, 94–95, 99–101; in birdsong learning, 115; in cognitive sex differences, 165–66, 173; examples of, 235n.35; gene expression and, 99–100; in sex determination, 45–46. *See also* social suppression of reproduction

social living, costs and benefits of, 88–89

social stress, 88–100; in baboon society, 90–91; health effects of, 93; in mouthbrooders, 91–93; reproductive effects of, 91; stress axis for, 89; and stress response, 89

social structure: jigsaw model of, 98–99; in spotted hyenas, 99, 175, 181–82, 182–84; structuring mechanisms in, 95–99; transitivity of, 98

social suppression of reproduction: in baboons, 79–81, 83; in elephants, 78–79; how-biology of, 84–88, 85f., 91–93, 100, 220n.46; in insects, 81; in mammals, 78–82; in mouth-brooders, 77–78, 83, 100; in naked mole rats, 81–82; optimal skew theory, 219n.17; significance of, 83; in social carnivores, 80–81, 83; why-biology of, 93

sociobiology: atomistic reductionism in, 95, 100–101; on human evolution, 152

Sociobiology (Wilson), 126, 201n.11

song, of birds: regional dialects of, 115; sex differences in, 118–19, 123; sexual selection and, 111–13, 119–20, 122–23. *See also* versatility of birdsong

song acquisition circuit, 115

songbirds: categories of, 116; as dawn chorus, 103–4; definition of, 102; duetting in, 117, 226n.37, 229n.75; impact of humans on, 124; mimicry in, 104, 120–23; rearing of, 124–25; sexual dimorphism in, 1, 118–19, 123, 142, 229n.76; song analysis for, 102–3; taxonomy of, 102; versatility of, *see* versatility of birdsong

song learning in birds: age-limited *vs.* open-ended, 116–17, 120, 228n.64, 230n.81; brain structure and, 115–16, 117, 120; dialect adjustments to, 115; heterochrony in, 230n.80; and socially conditioned adjustments, 115; stages of, 113–15, 114f.; tutoring model of, 114–15

song production circuit, 115

song sparrow (*Melospiza melodia*), 114, 228n.64

sonograms, 102–3, 103f.

southern elephant seal (*Mirounga leonina*), 213 nn. 48 and 49

"The Spandrels of San Marco and the Panglossian Paradigm" (Gould and Lewontin), 194

Sparisoma viride (Caribbean stoplight parrotfish), mating practices of, 56–57

sparrow: song learning in, 113–14, 114f., 116, 122, 230n.80

sparrow, house, impact of humans on, 124

sparrow, New World (Emberizidae), 113

sparrow, song (*Melospiza melodia*), 114

sparrow, swamp (*Melospiza georgiana*), 113, 116, 122

sparrow, white-crowned (*Zonotrichia leucophrys*), 103f., 114

spatial cognition: in birds, 127–29; hippocampus and, 125, 129–37, 138–40, 139t., 140t.; selection for, 141–48

spatial cognition in humans: sexual dimorphism in, 152–54, 157–66, 243 nn. 36 and 44, 246n.77, 247 nn. 78 and 79; sexual selection in, 160–61; social factors in, 165–66, 173; testosterone and, 163–66, 246n.77

sperm block, 14

sperm limitation, 209n.9

sperm retention maximization, 13–14

spinal nucleus of the bulbocavernosus (SNB), 156

spotted hyena (*Crocuta crocuta*): adaptationists on, 176–82, 188–91; androgen levels in, 185–86, 190, 251n.19; birthing process in, 186–89, 188f.; female genitalia of, 176, 253n.33; female genitalia disadvantages in, 186–91, 188f.; female genitalia function in, 178–80, 185–86, 188–91; greeting ritual of, 176, 178–81; how-biology of, 184–86, 190–91; hunting practices of, 180; reputation of, 175–76; social structure in, 99, 175, 181–82, 182–84; social suppression of reproduction in, 80–81, 83

SSSM. *See* standard social science model

stancing, 197–98

standard social science model (SSSM), 248n.81

starling (Sturnidae), 113, 120–22, 231n.82

stem cells, 145

steroid hormone receptors: androgen receptors, 31, 144–45, 147; estrogen receptors, 31, 144–45, 147, 207n.49; progesterone receptors, 31; role of, 31; steroid levels and, 207 nn. 49 and 51
stimulus, supernormal, 110
stress: health effects of, 93; reproductive effects of, 91. See also social stress
stress axis, 89
stress response, 89
striped hyena (Hyaena hyaena), 177
structuralists, on sexual reproduction, 34
Sturnidae (starlings and mynahs), 113, 120–22, 231n.82
subsong stage of birdsong learning, 113, 114f.
Sulidae (booby), mating rituals of, 28
sunfish, bluegill (Lepomis macrochirus), mating practices of, 55–56
supernormal stimuli, 110
swamp sparrow (Melospiza georgiana), 113, 116, 122
swordtail (family Poeciliidae), 66; sensory exploitation in, 111
syllable, of bird song, 103
Sylviidae (European warbler), 228n.69
symbiosis, in anemonefishes and anemones, 36–37
Symons, Donald, 17, 18

Taeniopygia guttata (zebra finch), 103f., 114f., 116–17, 118–19
tangled bank theory, 23–24, 204n.13
teleology: adaptationists and, 2–6; burden of proof in, 6–7; critique of, 2–3; in evolutionary psychology, 172–73; heuristic value of, 197–200; history of, 4–6; parthenogenic reproduction and, 29; in religion, 4–5, 17, 192–94; scientists' view of, 3–4; stancing in, 197–98, and ultimate explanations, 7, 201n.11. See also why-biology
teleost fishes: alternative reproductive tactics in, 51–54, 62–67, 71–73; gonad structure in, 212n.42; sex change in, 37–38, 39–40, 42–48; sex chromosomes in, 43–44, 46; sex manipulation in, 44, 45–46; sexual development of, 43–44, 67, 69–72, 211n.38
temperature, sex determination by, 45
termite, social suppression of reproduction in, 81

testosterone: brain chemistry and, 144–48; in brain ecology, 146–48; in dominant males, 84–86, 85f., 91; and human sexual development, 185; and immune system, 109, 224 nn. 27 and 29; and peacock plumage, 109; in sexual differentiation, 143–48; and social hierarchy, 96; spatial cognition and, 163–66, 246n.77; stress response and, 91; in whiptail parthenogenesis, 30–31
Thalassoma bifasciatum (bluehead wrasse): mating practices of, 56–57; sex change in, 48, 213n.45
Thamnophis sirtalis (garter snake), alternative reproductive tactics in, 58–59
Thornhill, Randy, 15, 18, 223 nn. 21 and 22, 249n.98
thrasher (Mimidae), 102–3, 120
throwing accuracy tests, 158
thrush (Turdidae), 113, 117
thrush, mimic (Mimidae), 103, 104, 113
Thryothorus leucotis (buff-breasted wren), 118–19
Thryothorus ludovicianus (Carolina wren), 118–19
Thryothorus nigricapillus (bay wren), 118–19
Thryothorus rufalbus (rufous and white wrens), 118
tilapia (Oreochromis), sex manipulation in, 44
Tinbergen, Niko, 110, 125–26
tit, 128, 134–35, 134t.
tit, blue (Parus caeruleus), 128, 134, 134t., 135–36
tit, great (Parus major), 128, 134, 134t.
tit, marsh (Parus palustris), 128, 134, 134t., 135–36
tit, willow, 134, 134t.
toadfish (Porichthys notatus): mating strategies in, 51–52, 53–54, 62–63, 65–66, 73; sexual development in, 70
Tooby, John, 169, 170, 171
transfer, of mathematical knowledge, 162
translocation, 203n.5
transvestitism, in red-sided garter snakes, 58–59
treefrog, alternative reproductive tactics in, 54–55, 65
tree lizard (Urosaurus ornatus), 216n.45; alternative reproductive tactics in, 59
tree of life. See phylogeny

triploid genomes, 28, 31–32, 206n.33,
    207n.51
tutoring model of birdsong learning, 114–15

upsuck hypothesis, 13–14
use-it-or-lose-it phenomenon, 142–43
Uta stansburiana (lizard, side-blotched), alter-
    native reproductive tactics in, 59

ventromedial hypothalamus (VHM), 30–31,
    207n.42
Verbal Behavior (Skinner), 168
versatility of birdsong: brain structure and,
    117–18, 120; evaluation of, 102–3; sexual
    selection and, 112–13, 118–20, 122–23
vertebrates: asexual, *see* asexual vertebrates;
    biological reproductive requirements in,
    154–55; sexual development of, 24,
    69–72
VHM. *See* ventromedial hypothalamus
Victorian sexual mores, 10–11
virgins: male orgasms in, 15–16; masturba-
    tion in, 15
vole (*Microtus*), 137–38, 139t., 146–47
Voltaire, François M. A. de, 194–95
von Neumann, J., 62
Vrba, E., 185–86, 191

warbler, European (Sylviidae), 228n.69
warbler, marsh (*Acrocephalus palustris*),
    231n.86
Warner, Robert, 41
waylayers, 54–55, 58, 60, 61, 64–65
Weisman, August, 23
Weiss, Kenneth, 6
West, Meridith, 115
wet dreams, evolutionary role of, 15–16
whip-poor-will (*Caprimulgus vociferus*), 104
whiptail lizard (*Cnemidophorus*): asexual re-
    production in, 20, 27–33; evolution of
    asexual reproduction in, 28, 29; male role
    adoption in, 29, 32; reproductive physiol-
    ogy of, 29–33, 30f., 206n.38, 207n.51,
    208n.57; sex act in, 19–20
white-crowned sparrow (*Zonatrichia leuco-
    phrys*), 103f., 114, 228n.64
why-biology: agenda of, 7; of alternative re-

productive tactics, 66, 73–74; on asexual
    reproduction, 29, 33–35; in evolutionary
    psychology, 172–73; on female orgasm, 18;
    vs. how-biology, 7–8; ideological roots of,
    192–97; of mimicry in birds, 122–23; self-
    containment of, 201n.11; of sex change,
    50; of sexual dimorphism, 143, 159–66; of
    social suppression of reproduction, 93. *See
    also* adaptationists; teleology
Wickler, Wolfgang, 178–80
Williams, George, 6, 23, 24, 179–80
Wilson, Edward O., 126, 201n.11, 248n.82
wolf, social suppression of reproduction in,
    80–81
Wrangham, Richard, 189–90
wrasse, sexual dimorphism in, 1
wrasse, bluehead (*Thalassoma bifasciatum*):
    mating practices of, 56–57; sex change in,
    48, 213n.45
wrasse, cleaner (*Labroides dimidiatus*), 39–40;
    mating practices of, 49; sex change in, 40
wren (Troglodytidae), 113, 117, 118–19
wren, bay (*Thryothorus nigricapillus*), 118–19
wren, buff-breasted (*Thryothorus leucotis*),
    118–19
wren, Carolina (*Thryothorus ludovicianus*),
    118–19
wren, marsh (*Cistothoras palustris*), 117, 118,
    229n.75
wrens, rufous and white (*Thryothorus
    rufalbus*), 118
Wundt, Wilhelm, 167

Xiphophorus maculatus (platyfish): alternative
    reproductive tactics in, 66; mating prac-
    tices of, 70; sensory exploitation in, 111;
    sexual development in, 69–70

Zahavi, A., 223n.23
Zebra danio (zebrafish): cloning of, 44–45,
    210n.21; sex determination in, 45, 46–47
zebra finch (*Taeniopygia guttata*), 103f., 114f.,
    116–17, 118–19
zebrafish (*Zebra danio*): cloning of, 44–45,
    210n.21; sex determination in, 45, 46–47
Zonotrichia leucophrys (white-crowned spar-
    row), 103f., 114, 228n.64